About Island Press

Island Press is the only nonprofit organization in the United States whose principal purpose is the publication of books on environmental issues and natural resource management. We provide solutions-oriented information to professionals, public officials, business and community leaders, and concerned citizens who are shaping responses to environmental problems.

In 2000, Island Press celebrates its sixteenth anniversary as the leading provider of timely and practical books that take a multidisciplinary approach to critical environmental concerns. Our growing list of titles reflects our commitment to bringing the best of an expanding body of literature to the environmental community throughout North America and the world.

Support for Island Press is provided by The Jenifer Altman Foundation, The Bullitt Foundation, The Mary Flagler Cary Charitable Trust, The Nathan Cummings Foundation, The Geraldine R. Dodge Foundation, The Charles Engelhard Foundation, The Ford Foundation, The Vira I. Heinz Endowment, The W. Alton Jones Foundation, The John D. and Catherine T. MacArthur Foundation, The Andrew W. Mellon Foundation, The Charles Stewart Mott Foundation, The Curtis and Edith Munson Foundation, The National Fish and Wildlife Foundation, The National Science Foundation, The New-Land Foundation, The David and Lucile Packard Foundation, The Pew Charitable Trusts, The Surdna Foundation, The Winslow Foundation, and individual donors.

About Save-the-Rewoods League

The Save-the-Redwoods League is a national nonprofit conservation organization founded in 1918 to protect the magnificent redwoods of California, to foster a better understanding of the value of America's primeval forests, and to support conservation of our forests.

The League purchases Coast Redwood and Giant Sequoia forest and watershed land for public parks. Its program has produced impressive concrete results: more than $100 million in donations have purchased 130,000 acres of redwood land for the California Redwood State Parks, Redwood and Sequoia National Parks, and many local parks and reserves.

Save-the-Redwoods League gratefully acknowledges its more than 100,000 current and past donors who have worked together for over eighty years to save the redwoods.

THE REDWOOD FOREST

To the many creatures, known
and unknown, of the redwood forest.

THE REDWOOD FOREST

History, Ecology, and Conservation of the Coast Redwoods

———

Edited by

Reed F. Noss

———

Save-the-Redwoods League

ISLAND PRESS

Washington, D.C. • Covelo, California

Library of Congress Cataloging-in-Publication Data
The redwood forest : history, ecology, and conservation of the coast
  redwoods / edited by Reed F. Noss.
     p.   cm.
   "A project of Save-the-Redwoods League."
   Includes bibliographical references and index.
   ISBN 1–55963–725–0 (cloth). — ISBN 1–55963–726–9 (pbk.)
   1. Redwood. 2. Redwood—Ecology. 3. Forest ecology. 4. Forest
conservation. I. Noss, Reed F. II. Save-the-Redwoods League.
SD397.R3R455   2000
634.9'758—dc21   99-16799
                                                              CIP

Printed on recycled, acid-free paper

Manufactured in the United States of America
10 9 8 7 6 5 4 3

CONTENTS

FIGURES

TABLES

BOXES

Appendices

FOREWORD

Once when I was conducting a tour of an ancient north coast redwood forest, a woman turned to me and asked why the trees weren't red. It was her first visit to the coast redwoods, and I guess she wasn't expecting moist, dark-colored tree trunks with green moss and gray lichen clinging to them. At the time, I was startled by her question, but over the years, I've learned that there are as many subjective expectations and impressions of the redwoods as there are people. One expectation most people share, however, whether or not they have ever visited the trees: They want these ancient giants to live beyond the span of human life.

There is a world of difference between a planted or second-growth forest of healthy, young, redwoods and a genuine old-growth forest of ancient redwoods that have survived fire and wind and drought and flood through the centuries. Today, when we speak of only 4 percent of the redwood forest remaining, we are talking about ancient or old-growth redwood forest. That kind of primeval forest is home to an extraordinary array of biological processes that accumulate only over long periods of time. This biological richness and ecological complexity does not occur and cannot be replicated in a young forest—a fact that becomes abundantly clear when we look at the work of Stephen Sillett, of Humboldt State University, who is carrying out a truly fascinating program of research high in the canopy of old-growth redwoods.

Because people love the redwoods, Save-the-Redwoods League for eighty-one years has been buying redwood forest lands and creating parks. Though the League and other groups have protected many fine redwood areas, the overall acreage acquired so far amounts to only a small portion of the entire two-million-acre modern-day range of the redwoods. Everyone recognizes that it is impossible to preserve the entire range, but we can make an earnest effort to assess their current status, to consider all their known physical aspects, and, on the basis of our accumulated knowledge, to develop a reliable long-term protection program.

The League recognizes the need for protected areas. We will therefore continue to acquire lands in fee title for park use. We also recognize the need to maintain privately owned forestland for wood production. The issue is not whether society should or should not use redwood for construction but whether this society can set responsible standards for harvesting and whether it has the ability to curtail excessive consumption of resources and inadvisable destruction of natural habitat. If the sale of redwood products were halted, consumers would turn to other wood—probably forcing heavier cutting of other forests. Given what we have learned so far, it is clear that the key to perpetuating the redwood forest involves understanding how it functions. Once we share that knowledge, we can hope to work cooperatively toward maintaining a healthy, viable ecosystem through a multilayered system of protections.

This book is an important element of just such an analysis and set of proposals. It describes the scientific basis for the League's current Master Plan for the Redwoods. The master plan integrates scientific findings with a stakeholders' report, GIS mapping, and an analysis of current conditions and issues as the basis for enlightened long-range redwood forest preservation policies.

Knowledge about the redwoods continues to grow. Over the years, several reports have been written about the state of the redwoods. In October of 1920, for example, the League sponsored a U.S. Forest Service study of the need for a national redwood park. In 1963, a major compendium of scientific data about the redwoods was assembled at Humboldt State University for use by the National Park Service. The League worked with the California Department of Parks and Recreation on a Master Plan for the Redwoods in 1965. More recently, a wide range of dedicated individuals continue to conduct in-depth research in a variety of specialized areas.

This large and growing body of knowledge will be more valuable to land planners now that it has been gathered into the one convenient source of this book. In addition, through advances in computer technology and need assessment techniques, it has become possible to look at the entire range of the redwoods in a new light. Dr. Reed Noss and his diverse team of experts have done an outstanding job of compiling the most current information about the redwoods. His knowledge of conservation biology as applied to the redwoods gives us the insight we need to make rational management decisions.

I am confident that the information in this book, together with the recommendations of the Master Plan for the Redwoods, will be used by people for years to come as they make decisions about the redwoods. We hope this information will be updated as new information about the redwoods becomes available. Meanwhile, I hope you enjoy this book and enjoy learning about the redwoods as much as I have.

When we first behold the redwoods, it is with a sense of amazement that beckons us to know and respect the natural world around us. Let me leave you with the words of three wise and well-informed observers of the redwoods:

> An understanding of the relationship between the stages of forest growth and the abundance and variety of wildlife will help suggest some compromises between maximum wood yield and optimum wildlife habitat. . . . The welfare of a sharp-shinned hawk is inseparable from the welfare of the small vertebrates on which it feeds, and it is impossible to consider the one without the other.
>
> —A. S. Leopold

> Man must recognize the necessity of co-operating with nature. He must temper his demands and use and conserve the natural living resources of this Earth in a manner that alone can provide for the continuation of his civilization.
>
> —Fairfield Osborn

> There are certain values in our landscape that ought to be sustained against destruction or impairment, though their worth cannot be expressed in money terms. They are essential to our life, liberty and pursuit of happiness; this nation of ours is not so rich it can afford to lose them; it is still rich enough to afford to preserve them.
>
> —Newton B. Drury

Mary A. Angle
Former Secretary and Executive Director
Save-the-Redwoods League

PREFACE

I was surprised when Save-the-Redwoods League contacted me in mid-1997 and asked me to organize and lead a team of scientists in compiling everything there is to know, biologically and ecologically, about the coast redwood forest. Although I had conducted studies, mainly of birds, in other forest ecosystems, and had published articles on sustainable forestry, I had never studied redwoods and could make no claim to expertise in any aspect of this ecosystem. I had visited several redwood parks and was fascinated and awed by them, but I did not even live in the redwood region and felt like an outsider.

Apparently, this lack of specific expertise was viewed as a blessing—I had no preconceived notions about redwood ecology nor any biases about whose research was most credible. I would have to ask around and form fresh opinions. The main purpose of this effort was, and is, to assist the League in preparing its master plan. The information our team assembled on the ecology, conservation, and natural and cultural history of the redwoods was to be used as a general reference for anyone interested in this great forest. It was also to be used as a guide for making decisions about which lands should be acquired and about the appropriate management of redwood forests—both of reserves and of lands devoted to timber production and other human uses. The task was becoming somewhat less daunting. Regional conservation planning is something I have been involved with for years. Furthermore, my experience as an editor—for instance, of the journal *Conservation Biology*—would come in handy in such a broad-ranging literary endeavor.

My first task, then, was to identify the experts on as many aspects of the redwood forest as possible. I talked with many people, including staff of the League and scientists at several institutions (mainly in California and Oregon) to come up with a list of potential contributors. Although, sadly, some areas of expertise (bryophytes, for example) are not represented in this book because I could find no one willing and able to write about them authoritatively, the more than thirty authors represented in these pages are leading authorities on a vast number of topics in redwood ecology and natural history.

The original "science document" for Save-the-Redwoods League's master plan was to be written within six months. At the time, I suspected that deadline was unreasonable because most of the potential authors are busy people with countless commitments. Furthermore, I know that getting the most out of authors requires some promise that what they write actually will be read, rather than gathering dust on a shelf. Unpublished reports have a knack for dust-gathering. For this reason, I suggested to the League that we might as well try to make a book out of this project. If we truly intended to produce the authoritative text on redwood ecology and conservation, it should be something that many people—including a publisher—would find of interest. To this end, I contacted my friend Barbara Dean at Island Press, who had handled two of my previous books. Barbara encouraged me to submit a formal book proposal. I did, the reviews were positive, and Island Press accepted our proposal.

Meanwhile, the writing of chapters was proceeding more slowly than I had hoped. Mary Angle, then secretary and executive director of Save-the-Redwoods League, was remarkably patient and agreed that quality was more important than rapidity. We still had Island Press deadlines to meet, though, and there were many times (up to the last few days, in fact) when I feared it would be impossible. I thank the authors for generally coming close to my editorial deadlines and allowing me to survive this project without developing ulcers.

The introductory chapter 1 provides some information on the purpose and scope of this book. With a general science and conservation audience in mind, we have tried to avoid being overly technical. At the same time, however, we strove to provide sufficient technical explanation for the book to be useful to professionals and students in the biological and natural resources sciences. Although we avoided unnecessary jargon, scientific terminology is often the most precise and accurate way to express a concept; therefore, we use some terms that may be unfamiliar to the general reader. Most of these terms are defined when first introduced; the uninitiated may find the glossary helpful. A species list with common and scientific names is also provided. Common names are generally used in the text, except for species that have no well-accepted common names, or when referring to genera or families (for example, salamanders of the genus *Plethodon* in the Plethodontidae).

Any project this comprehensive and sometimes specialized requires abundant help from reviewers. Aside from the authors, most of whom reviewed considerable material from the book that they did not write themselves, I thank the following people for providing reviews of one or more chapters: M. Angle, D. Axelrod (deceased), M. Barbour, B. Bingham, R. Hartley, B. Harvey, T. Lewis, A. Lind, T. Lisle, J. Moles, P. Morrison, D. Perry, L. Reid, and K. Rodrigues. Furthermore, the authors of chapter 6 (aquatics) wish to thank N. Karraker, A. Lind, L. Ollivier, E. Bell, and G. Hodgson for their help and insights during the development of the chapter. I am indebted to Lisa Loegering, who assisted me throughout this project and handled much of the correspondence with authors, besides the unenviable task of assembling the literature cited and list of species names, often working into the wee hours of the night. The excellent work of copyeditor Jane Taylor and production editorial supervisor Christine McGowan is also greatly appreciated. Finally, as with my previous Island Press books, I am enormously grateful to Barbara Dean and Barbara Youngblood for seeing this manuscript through the process of becoming a book.

Chapter 1

MORE THAN BIG TREES

Reed F. Noss

Humans delight in superlatives. Big things, in particular, impress and inspire us: Mount Everest, the Grand Canyon, dinosaurs, whales . . . redwoods. Redwoods (i.e., the coast redwood, *Sequoia sempervirens*) deserve all the lavish terms used to describe them. No one with an open mind could walk through an old-growth redwood forest without being humbled. No thoughtful person could stand beneath one of these immense trees, gaze up into its canopy, and help but think that here is a remarkable organism—so much more than all the board-feet of lumber that men might cleave from it. Not only are the coast redwoods among the largest living trees, they are among the largest living organisms ever to inhabit the earth. Their close ancestors have been here since other giants—including the dinosaurs—came and went. An entire forest of these trees is one of the most remarkable expressions of nature's productive capacity. And it is beautiful, truly beautiful.

A redwood forest is more than big trees. From the bewildering variety of life and past life (e.g., woody debris) on the forest floor to the intricate community of fungi, lichens, liverworts, vascular plants (including trees several meters tall), earthworms, millipedes, mollusks, insects, and salamanders tens of meters up in redwood canopies, the redwood forest is a complex ecosystem. And like virtually every ecosystem that contains something of commercial value to humans, the redwood forest has declined markedly in quality and extent over recent decades. Although the coast redwood, as a species, is in no immediate danger of extinction, the old-growth redwood forest has declined in area by more than 95 percent since European settlement (U.S. Fish and Wildlife Service 1997) and is in danger of being lost as an intact, functioning system.

Evidence is mounting that redwood forests, like many other ecosystems, cannot perpetuate themselves as small, isolated fragments in human-altered landscapes. Such fragments lose their diversity over time and may even risk losing the ability to grow new, giant trees. For example, large patches of old-growth redwoods, with their complex crown structure, capture significant quantities of water from fog. This water may be crucial to their ability to grow large under conditions of moisture stress (see chap. 2). Will redwoods of the stature we see today ever return to these impoverished landscapes? Perhaps not. Furthermore, we are learning that young stands of redwoods—even stands approaching two hundred years old—lack many of the ecological attributes and habitat values of older forests. This should come as no surprise when we are dealing with a tree that can grow more than 100 m tall and 5 m thick, and whose potential life span is more than two thousand years. Redwood forests will be conserved only when we recognize them for all their values, including those values we are only beginning to understand and some that we will never fully comprehend.

The Value of Redwoods

This book, written in support of Save-the-Redwood League's master plan, is meant to provide scientific guidance for saving the redwoods by assembling available information pertinent to their conservation and management. This mission assumes that redwood forests have value. These values, already touched on above, range from the directly economic to the aesthetic, recreational, and spiritual. An increasing number of people believe that the redwood forest has intrinsic value independent of humans and that people have no right to diminish this value. We will not address subjective values to any great depth in this book, not because we think such values are unimportant—indeed they are what motivate us to pursue work in conservation biology—but because it is a large enough task merely to pull together the available information on more tangible, scientific topics. More important, we believe that a profound respect for redwood forests emerges from an understanding and appreciation of their fascinating natural history, and that this respect provides the most solid foundation for a conservation ethic.

The value of redwood forests can be perceived from several frames of reference. Viewing the redwood forest within a broad context is essential. One element of that context is the historic. Knowing that today's redwoods are relics of an ancient lineage extending back to the age of the dinosaurs and once covering large portions of North America and other continents gives us pause to reflect on where redwoods have been and where they are going. Could this venerable lineage, which has persisted over vast spans of time while the continents drifted about and while innumerable co-occurring species arose, prospered, and then declined to extinction, be terminated before our eyes? The extinction of the coast redwood species may be unlikely any time soon, but the ecological extinction of the old-growth redwood forest is no fantasy.

As detailed in chapter 2, the brief interval over which modern humans have wrought extreme changes to the redwood forest contrasts with a much longer history, during which changes in distribution and abundance of redwood and its relatives were no less dramatic but which took place much more slowly. Although modern redwood associations, often with codominant species, such as Douglas-fir and tanoak, may or may not have had close analogs in the past, the millennial redwood forests surviving today reflect a set of environmental and biogeographic conditions from a particular time in the past. Environmental conditions have never been static over long periods, and the conditions under which today's old growth developed may not be replicated in the foreseeable future—thus, the "climax" redwood forests we know today may be impossible to recreate. Some ecologists might call them historical artifacts. Nevertheless, a lineage that has persisted for so long and in the face of so much global turmoil demands our admiration—and our careful stewardship—if we want it to keep evolving into the future.

The redwood forest also can be placed within a broader context by considering its present qualities in comparison with other forest types and regions, across the continent and worldwide. Such considerations are vitally important for making conservation decisions on large scales. World Wildlife Fund (WWF) recently completed a conservation assessment of the terrestrial ecoregions of the United States and Canada (Ricketts et al. 1999), the results of which are being compared with similar assessments conducted for other continents. The redwood region stood out as a globally significant ecoregion in this assessment.

The WWF study had several goals, which included identifying ecoregions that support globally outstanding biological and ecological qualities, assessing the types and immediacy of threats to ecoregions, and identifying appropriate conservation activities for each of 116 ecoregions in the United States and Canada. Biological distinctiveness was determined through an analysis of species richness, endemism, distinctiveness of higher taxa (families, orders, etc.), unusual ecological or evolutionary phenomena, and global rarity. Conservation status was based on an assessment of habitat loss, habitat fragmentation, degree of protection, and current and potential threats. Each ecoregion was evaluated for biological distinctiveness and threat only in comparison with other ecoregions of the same major habitat (e.g., temperate coniferous forest). The Northern

California Coastal Forests ecoregion (which generally corresponds to the redwood region discussed in this book; see fig. 1.1) placed in the highest class in all three assessments—biological distinctiveness, conservation status and threat, and overall conservation priority—when compared with other ecoregions of the same major habitat type globally.

We should bear in mind that the WWF assessment evaluated all habitats in the redwood region, and as we will see (e.g., chap. 3, 4, and 5), the redwood forest is not particularly high in species richness and endemism. The redwood

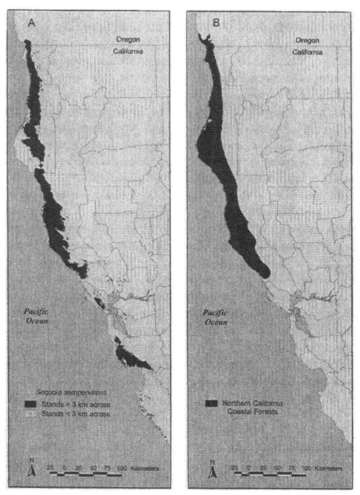

Figure 1.1. The Redwood Region as discussed in this book *(A)* and as compared to the Northern California Coastal Forests Ecoregion of World Wildlife Fund *(B)* (Ricketts et al. 1999).

region ranked high, in large part, because of the impressive (indeed, unsurpassed) biomass, structural complexity, and other unique ecological qualities of redwood forests. Associated habitats in the region, including coastal grasslands, chaparral, oak woodlands and savannas, wetlands, and diverse aquatic communities, add to the richness and distinctiveness of the region. Moreover, the redwood forests face a high level of threat to their persistence in a natural state (Ricketts et al. 1999). The high ranking by WWF suggests that land management in this globally significant and highly imperiled ecoregion should be subject to more intense scrutiny than in regions with more modest biological and ecological values.

Simply put, the redwoods region has much to lose if managed unsustainably. There is ample reason to believe that management of redwood forests over the past century and a half has not been sustainable—especially when sustainability is interpreted in the properly broad sense of sustaining all species, structures, and processes of the forest ecosystem. This dire situation necessitates three kinds of action: (1) protection of the most biologically significant remaining stands of redwoods, both old growth and second growth, representing the natural range of variation of redwood forest types and within a configuration of reserves adequate to maintain ecological integrity over time; (2) restoration of many areas of degraded redwood forest to something resembling natural conditions; and (3) truly sustainable management of appropriate redwood stands for timber and other values.

Purpose and Scope of This Book

This book is meant to provide the scientific basis for the three kinds of action just listed: protection, restoration, and sustainable management of the redwood forest. We focus largely on the redwood forest but recognize that the biological richness of the redwood region springs from the rich mosaic of terrestrial, riparian, and aquatic habitats on the landscape. We seek to give conservation organizations the information, technical tools, and broad perspective they need to evaluate redwood sites and landscapes for conservation, while providing public and private foresters and land managers with relevant information for managing redwoods and associated biological communities wisely. Through our review of the natural history and ecology of the redwood forest, we hope to acquaint readers—professional and amateur alike—with this unique ecosystem and stimulate a deep appreciation of its many values. We hope that many of you will be inspired to venture into the forest, with proper respect, to acquaint yourselves more intimately with its many microhabitats and creatures.

The chapters are organized to provide a reasonably comprehensive account of the redwood forest, including its geologic and cultural history, natural history, ecology, management, and conservation. We focus on the coast redwood species but discuss its close relatives when appropriate. Chapter 2 reviews the history of the redwood lineage, from the probable origin of the family in the Triassic period, through its diversification and subsequent decline in the Cenozoic era, to its drastic diminishing by logging in recent decades. We also review the history of redwoods conservation, addressing efforts to preserve both the Sierran redwood (giant sequoia) and the coast redwood.

Chapters 3 and 4 address the subjects of redwood trees, communities, and ecosystems. Chapter 3 focuses on the redwood vegetation and the variation in assemblages over the range of the species, as well as the vascular plants, fungi, and lichens associated with redwoods; this chapter also discusses the remarkable communities of organisms associated with redwood canopies. Chapter 4 takes a closer look at redwoods—their life history, architecture, genetics, environmental relations, and disturbance regimes. In chapter 5, we address the terrestrial fauna of redwood forests, with particular attention to some of the species threatened by logging. Chapter 6, a review of the aquatic ecosystems of the redwood region, draws attention to the effects of logging and other landscape modification on aquatic fauna, especially fish and amphibians. Chapter 7 discusses conservation planning, as applied to redwood forests on a landscape scale; it includes a demonstration of a method for selecting "focal areas" for conservation action. Chapter 8 reviews management alternatives for redwood forests, with emphasis on silviculture but also including restoration and management of redwood parks. Chapter 9, the conclusion, summarizes the lessons of previous chapters.

We hope this book, as a whole, will serve as a model for similar projects involving other ecosystems. There are many fascinating and imperiled ecosystems worldwide—forests, savannas, grasslands, shrublands, aquatic and marine ecosystems, and others (Noss et al. 1995; Noss and Peters 1995)—deserving book-length treatments of their own. Only when a significant number of people develop a deep understanding and appreciation of these ecosystems will we have a decent chance of saving them. There is little time to waste.

Chapter 2

History of Redwood and Redwood Forests

John O. Sawyer, Jane Gray, G. James West,
Dale A. Thornburgh, Reed F. Noss, Joseph H. Engbeck Jr.,
Bruce G. Marcot, and Roland Raymond

The history of any taxon, community, or ecosystem provides a context for interpreting its current distribution and status. Narrowly distributed species have always interested naturalists. One pattern that begs explanation is the restriction of redwood to a relatively narrow coastal strip from central California to extreme southwestern Oregon. Sometimes a limited geographic distribution reflects a short history—the species has evolved only recently and, perhaps, is in the process of expanding its range. Or a restricted distribution may reflect a limitation in available habitat (e.g., species on islands or islandlike physical habitats, such as serpentine outcrops). In still other instances, the present range of a species may be but a tiny remnant of a much broader distribution in the past. This is the case for redwood and its close relatives.

This chapter reviews the history of redwood and closely related species, from their antecedents more than 100 million years ago to their restricted status

Author contributions: Sawyer, general organization; Gray, pre-Holocene; West, Pleistocene-Holocene transition and Holocene; Thornburgh, logging history; Noss, section introductions, transitions, and general writing and editing; Engbeck, conservation history; Marcot, box on paleoecology; Raymond, box on Yuroks.

today. We summarize the early development of redwood and its close relatives, through their expansion to become among the most widely distributed conifers during the Cretaceous and Tertiary periods, and through their drastic diminishment over the past few million years as mountain ranges and rain shadows formed and climates became more severe. We also review the history of human occupation and use of the redwood forest, from the first recorded settlements in the early Holocene, through the subsistence uses of the Yurok people, to the generally unsustainable logging practices that characterize the twentieth century. This long history of redwood and its associated species, both plants and animals, provides an important perspective for forest managers today (box 2.1). The chapter concludes with a ray of hope—a summary of the redwood preservation movement and its accomplishments to date. We hope these accomplishments continue, so the long history of redwood does not come to a premature end.

Box 2.1. Why Should Managers Care About Paleoecology?

The longevity of an individual redwood tree reminds us that history has shaped today's redwood forests. We might perceive these forests as unchanging during the brief visit afforded by a hike, a scientific study, a management plan, or even a human lifetime, but that would be illusory. Redwood forests are dynamic, changing with shifting climates, coastlines, ecological influences, and the evolution and dispersal of organisms.

Noss (1992; echoed by Grumbine 1992, 1994) described the conservation of biodiversity in terms of maintaining the ecological integrity of ecosystems. This requires maintaining viable populations of native plant and animal species and ensuring their long-term evolutionary potential. Little quantitative work has been done to analyze population viability of even the rarer species found in redwood forests. This is particularly true for those unseen life-forms so critical to forest productivity: the belowground soil bacteria, earthworms, mesoarthropods—such as springtails and soil mites—microfungi, nematodes, and protozoa; and the bryophytes, lichens, vascular plants, insects and other life-forms-inhabiting tree canopies (see chap. 3). The long-term viability, evolutionary development, and response to human activities of these creatures are largely unknown.

Why should forest managers care about paleohistory? Because it helps to interpret the *past*: how species populations evolved, shifted, and mixed with other species over time. Because it helps to explain the *present*: why the once extensive redwood with its ancient lineage now occupies only coastal environments. And because it helps to portend the *future*: how species currently found in the redwood forests might respond to human-influenced deforestation, other disturbances, and changes in regional climate; how the mix of species associated with redwood may change over time; and which species might not make it through these changes.

Of particular interest are organisms—species and subspecies of plants and animals—found only in the redwood region. Such endemics can be of two kinds: paleoendemics, relicts left over from ancient climates and conditions; and neoendemics, organisms relatively newly evolved in the redwood region. Endemics of both kinds are not necessarily restricted to redwood forests. Endemism is relatively low in the redwood forest and moderate in the redwood region (Ricketts et al. 1999, and see chap. 5). Within the redwood region are many landscapes and plant communities, dominated, for example, by redwood, Douglas-fir and tanoak, beach pine, bishop pine, and Sitka spruce, as well as chaparral, northern coastal scrub, and grasslands. Organisms may inhabit one or more of these vegetation types, with some restricted to one or a few types. The fog shrew, for instance, is a mammal near endemic to the redwood region, found in redwood and other coniferous forests but more commonly in alder/salmonberry, riparian alder, and skunk cabbage marsh habitats (Maser et al. 1981). On the other hand, the red-bellied newt has been found mainly in redwood forests (Petranka 1998) but also occurs in other habitats (H. Welsh, pers. obs.).

This chapter explores the past but helps us interpret the present, such as understanding the current distributions of relict or paleoendemic organisms. For example, during the Pliocene many plants and animals of the redwood region migrated from the northern Sierra Nevada to their present location, leaving behind some populations in the moister sites of the Sierra Nevada (Axelrod 1973). These likely Sierran paleo-remnants include some amphibians, which are among the terrestrial vertebrates most sensitive to changes in humidity and precipitation regimes. In the redwood region, the red-bellied newt and Del Norte salamander are examples of paleo-remnants. The latter may be a paleoendemic (Welsh 1990), a remnant species that was in the coastal mountains before the redwoods arrived, while the former may have come (speculatively) from the Sierra.

Among reptiles, at least four subspecies of garter snake are endemic or near-endemic to the redwood region; most occur in a variety of vegetation types or aquatic habitats (chap. 5; table 5.2). One of these subspecies—the San Francisco garter snake—is currently listed as federally endangered. The differentiation of subspecies of garter snakes in the redwood region suggests that evolutionary divergence of forms is an ongoing and active process—but a process that is potentially imperiled, given the status of some taxa. Evolutionary differentiation may reflect the diversity of vegetation types along California's central and northern coasts, both in the past and today.

The redwood forest's species (and subspecies) richness can be attributed to the collective histories of the species currently living there. Some have similar histories, but most have different histories originating in locales and environments other than redwoods (e.g., among plants, madrone's close relatives are in Mexico, tanoak's in Asia.). Some species living in the redwood region, such as Roosevelt elk, are likely invaders, which did not evolve within the redwood forest but have come to find suitable habitat in the region as their ranges shifted with time. Roosevelt elk occur only in the northern part of redwood's range; most of the redwood region lacks elk. Other species, like the Del Norte salamander, may be true paleoendemics or relicts.

(continues)

Box 2.1. *(continued)*

A few species, perhaps including the red-bellied newt and California tree vole, may be neoendemics.

As this chapter discusses, the current plant species composition of redwood forests may be only 4,000 years old or less. This short time may explain why so few species or even subspecies are neoendemics—they have not had time to evolve. Hence, although Del Norte salamanders, red-bellied newts, California tree voles, and Roosevelt elk coexist in some of the same coastal forests, their origins and reasons for appearance are species-specific. Determining such histories can help forest managers interpret the current distributions and habitat-selection behaviors of species, which may aid their conservation. Species of limited distribution that occur predominantly in the redwood region deserve special conservation attention, as it is mainly here where they will be saved or lost.

Before the Holocene

Redwood *(Sequoia sempervirens)* is the sole surviving species of the genus *Sequoia,* a member of the conifer family Taxodiaceae, popularly known as the baldcypress or redwood family. *Sequoia* lingering in the foggy, coastal belt from central California to southern Oregon, and relict populations of Sierran redwood or giant sequoia *(Sequoiadendron giganteum)* on the humid slopes of the western Sierra Nevada in California are true "living fossils." Like ginkgo, native now only to China, and lungfishes, found now only in Australia, Africa, and South America, redwood and Sierran redwood have existed largely unchanged morphologically for millennia—isolated remnants of a once robust lineage that circled middle and high latitudes in the Northern Hemisphere in the geological past.

Ancestry of Redwood

The present relictual distribution of redwoods, like that of other taxodiaceous conifers, is principally the result of post-Eocene climatic changes that culminated in massive late Pliocene-Pleistocene glaciations over the past 2 million years. During the late Mesozoic and early Cenozoic (Paleocene, Eocene, and Oligocene epochs, grouped by geologists into the Paleogene; table 2.1), when Coniferophytes attained greatest abundance and diversity and widest geographic distribution (Miller 1977, 1988), taxodiaceous conifers were common and diverse in the Northern Hemisphere and, to a lesser extent, in the Southern Hemisphere. The possible extinction of redwood and other members of the family in the next few centuries would terminate a phylogenetic history that began in the early Mesozoic for the family Taxodiaceae, and via putative antecedents, extends to the late Paleozoic some 250 million years ago (Miller 1977, 1988).

Table 2.1. The Geologic Time Scale. (Data from the Geologic Society of America; plant history from Tidwell [1998] and from text)

Era	Period		Epoch	Millions of Years Ago Began	Events in Redwood History and Plant Life
Cenozoic	Quaternary		Holocene	0.01	Climatic instability; humans occupy entire coast of California in early Holocene; *Sequoia* most abundant in mid-Holocene (ca. 5,500 years ago), with range contracting thereafter; heavy logging begins in nineteenth century
			Pleistocene	1.6	Glacial cycles; *Sequoia* as far south as Santa Barbara in late Pleistocene
	Tertiary	NEOGENE	Pliocene	5.3	Principal uplift of Sierras, Cascades, and coast ranges; *Sequoia* still as far north and east as Columbia River gorge in early Pliocene; floras with *Sequoia* in California much like modern floras, but included summer-wet genera (e.g., elm, holly)
			Miocene	23.7	Climate becomes dry over much of western North America; *Sequoia affinis* forests in western and eastern Oregon and Idaho; *Sequoiadendron* forests in western Nevada much like modern Sierran forest
		PALEOGENE	Oligocene	36.6	*Sequoia* as far south as Colorado at beginning of epoch, then eliminated a few million years later; *Sequoia* in Oregon and Montana in Middle and Late Oligocene; *Sequoia* and *Sequoiadendron* segregated
			Eocene	57.8	Climate in western United States subtropical to temperate; *Sequoia affinis* in Nevada and Idaho
			Paleocene	66.4	*Sequoia affinis* in Wyoming
Mesozoic	Cretaceous			144	First angiosperms; peak of diversification and distribution of Taxodiaceae, including *Austrosequoia, Taxodium, Sequoia,* and *Sequoiadendron*
	Jurassic			208	Seed ferns, cycadeoids, and conifers dominant; possible *Sequoia*
	Triassic			245	Conifers increase, seed ferns and true ferns decrease; oldest Taxodiaceae
Paleozoic	Permian			286	Treelike lycopods and *Calamites* decline and go extinct; early conifer ancestors of Taxodiacene and Cupressaceae present
	CARBONIFEROUS	Pennsylvanian		320	Coal-forming swamps present in eastern United States
		Mississippian		360	Progymnosperms, treelike lycopods, *Calamites,* seed ferns, and true ferns present
	Devonian			408	Lycopods, horsetail relatives, seed ferns, and early progymnosperms present
	Silurian			438	Land plants first appear
	Ordovician			505	Marine algae
	Cambrian			570	Marine "algae"
	Precambrian			4,600	Simple "algae" and fungi

The early history and phylogenetic relationships of the Taxodiaceae with other conifers are matters of conjecture. Links are suggested with Cupressaceae (cypress family) and Pinaceae (pine family), whose evolutionary histories closely parallel in time the fossil record of the Taxodiaceae (Miller 1988). Cladistic analysis (a system of classification based on a statistical analysis of characters) of fossil and modern seed cones provides broadest support for an alliance between Taxodiaceae and Cupressaceae (Miller 1988), whose modern members are largely distinguished by differences in leaf characters and leaf arrangement. The mixing of taxodiaceous and cupressaceous features in a possible ancestral complex of Permian and Triassic conifers (Miller 1977, 1988) strongly supports shared ancestry between these families. Because of many similarities, there has been continuing debate whether Taxodiaceae and Cupressaceae should be merged into a single family (Miller 1988; Brunsfeld et al. 1994).

As with other conifers, the history of the redwood family traces loss of generic and specific diversity over time. Living species are largely confined to humid, warm-temperate to subtropical, generally frost-free habitats in Asia and North America. Apart from *Sequoia*, contemporary genera in the family include *Athrotaxis* of Tasmania, *Cryptomeria* (Japanese cedar), *Sciadopitys* (umbrella pine) and *Taiwania* of Japan, *Glyptostrobus* (water-pine), *Metasequoia* (dawn redwood) and *Cunninghamia* (China fir) of China (*Cunninghamia* also occurs in Taiwan), and *Sequoiadendron* and *Taxodium* (baldcypress) of North America. Most are monotypic genera, consisting of a single species with restricted geographic distribution. None of the contemporary genera has species on more than one continent, although fossil floras often include several genera of the family. *Metasequoia*, *Sequoiadendron*, and *Sequoia* are all often called "redwoods." As the fossil record indicates, *Sequoia* and *Sequoiadendron* (once placed in genus *Sequoia*, as *Sequoia gigantea*) are closely related (Chaney 1951a; Miller 1977, 1988).

Fossils attributed to the redwood family include leafy shoots, wood, male (pollen) and female (seed) cones, seeds, and pollen grains. Cones have proved particularly valuable in circumscribing the Mesozoic and early Cenozoic evolutionary history of the family (Miller 1988). The windborne pollen is widespread in Mesozoic and Cenozoic rocks, but seldom attributable to specific genera and not consistently distinguishable from pollen of the closely related Cupressaceae or from Taxaceae (the yew family). Megafossils are thus more reliable for circumscribing the Mesozoic and Cenozoic evolutionary history of the family.

The Earliest Redwoods

Pre-Mesozoic ancestry for the Taxodiaceae may trace to late Carboniferous and Permian conifers (ca. 320–245 million years ago) commonly included in the order Volziales, a group that lived from the late Carboniferous to Jurassic. Some Volziales are sometimes called "transitional conifers" because their ovulate (seed-bearing) cones document modifications that could have led from cones of the most primitive Paleozoic conifers (Cordaitales) to cones characteristic of modern conifers. The exact role of the Volziales and transitional conifers in the evolution of modern conifer families remains elusive (Miller 1988:481).

Taxodiaceae as a recognizable family can be traced into the early middle Triassic of Antarctica based on seed cones from a silicified peat (Yao et al. 1997). These cones combine features of several taxodiaceous genera: They may represent an ancestor of younger members of the family, such as *Sciadopitys* or *Cryptomeria*. Jurassic and possibly older fossil woods with such names as *Taxodioxylon, Sequoioxylon,* and *Cupressinoxylon* could belong either to the redwood or cypress families. These woods provide evidence for the antiquity of both families and their shared ancestry (Vaudois and Prive 1971; Stewart 1983).

Leafy imprints (Manchuria) and a seed cone (France) from the Jurassic superficially resemble genus *Sequoia* (Chaney 1951a; Miller 1977). This Jurassic *"Sequoia"* antedates all other records for contemporary genera within the family. Many of the oldest fossils are known principally as isolated, nondiagnostic leafy impressions without evidence of reproductive structures; thus, their affinities remain uncertain despite names that may imply a relationship to modern genera. Nevertheless, the fossil record indicates that the redwood family existed by the middle Triassic (ca. 240 million years ago) and that *Sequoia* may have existed by the Jurassic (ca. 208–144 million years ago).

Cretaceous Redwoods

The acme of diversification and distribution of the Taxodiaceae occurred in the Cretaceous period (ca. 144–66 million years ago) but mainly involved lineages that are extinct today (Miller 1977, 1988). Taxodiaceous conifers of the Cretaceous, as well as those of the Jurassic and early Cenozoic, often combine features of different present-day genera or share some features with modern genera, but nevertheless are sufficiently distinct to warrant separate classification. Several examples of this pattern are found in the evolutionary history of *Sequoia* and *Sequoiadendron*.

One remarkable example is the late Cretaceous *Austrosequoia* from Australia. In most features of seed cone morphology, *Austrosequoia* closely resembles modern *Sequoia,* but the attached leaves resemble Tasmanian *Athrotaxis,* the only other member of the family with fossil records in the Southern Hemisphere (Peters and Christophel 1978). Close taxonomic relationship between *Athrotaxis* and *Sequoia* suggested by morphological evidence thus has fossil support (Peters and Christophel 1978). Besides its apparent relationship to *Sequoia,* the presence of *Austrosequoia* in Australia and of *Athrotaxis*-like remains in the Northern Hemisphere adds to the puzzle of *Sequoia's* biogeographic history (Miller and LaPasha 1983).

Cretaceous "scaly foliage" reminiscent of living *Sequoiadendron* was long attributed to *Sequoia* before separation of *Sequoiadendron* into a distinct genus (Chaney 1951a). If such foliage indeed represents *Sequoiadendron,* as Chaney (1951a) and Axelrod (1986) suggested, it probably was not from a species ancestral to contemporary *Sequoiadendron.* On the other hand, such scaly foliage may represent a "mixed" ancestral genus, such as the Swedish late Cretaceous *Quasisequoia. Quasisequoia* has foliage shoots that resemble extant *Sequoiadendron,* but seed cones that resemble cones of both *Sequoiadendron* and *Sequoia* in some features but differ from both in other features (Srinivasan and Friis 1989). Extinct fossil genera with "mixed" characters that cannot be assigned to living genera, such as *Athrotaxites, Austrosequoia, Cunninghamiostrobus, Parataxodium,* and *Quasisequoia,* may or may not be ancestral to modern genera (Miller and LaPasha 1983; Miller 1988; Miller and Crabtree 1989).

In addition to these extinct genera, all modern genera in the family have been identified primarily from nondiagnostic sterile foliage imprints in the late Cretaceous (Miller 1977). Recently, however, *Taxodium* has been unequivocally identified on the basis of leaves, pollen and seed cones, seeds, and pollen from late Cretaceous (ca. 70 million years ago) rocks of Canada (Aulenback and Le Page 1998). Cretaceous foliage attributed to *Sequoia* and *Sequoiadendron* from western North America, and foliage and cones (*Sequoiadendron* and *Sequoia:* Smiley 1966, 1969) from Arctic northwestern Alaska (ca. 69–70 degrees N. Lat.), are also among the oldest remains attributed to extant taxodiaceous genera (Chaney 1951a; Axelrod 1986).

In summary, all modern genera of taxodiaceous conifers, including redwood, have been identified in the Cretaceous, mainly from leafy shoots without evidence of confirming reproductive structures. Genera with characters similar to contemporary genera or that share characters with several contemporary genera may represent an extinct group of Taxodiaceae or lineages ancestral to living genera.

Redwoods in the Tertiary

The extensive Northern Hemisphere distribution of taxodiaceous conifers in the late Cretaceous continued into the early Cenozoic and even into the late Cenozoic (Miocene and Pliocene epochs grouped by geologists into the Neogene; table 2.1) in some mid-latitude regions. In western North America, distributional contraction and loss of species and genera appear to have begun in the middle to late Oligocene, about 30 million years ago.

At one time, *Sequoia* was believed to have been the most widely distributed taxodiaceous conifer in the Cenozoic, with a nearly circumpolar arctic distribution, as well as an extensive mid-latitude distribution. The discovery of living *Metasequoia* in central China led to reevaluation of these fossils (Chaney 1951a); the vast majority of putative *Sequoia* fossils at middle latitudes, as well as *all* fossils originally assigned to redwood at high northern latitudes, were judged to be

Metasequoia. Chaney (1951a) suggested this evidence implied that redwood had a middle-latitude origin in accord with the climatic requirements of the living coast redwood.

Mid-latitude Tertiary records for redwood that are still accepted include Turkestan-Kazakhstan (Axelrod 1979); western Europe, where redwood was more common than dawn redwood (Chaney 1948, 1951a) and persisted into the Pliocene (Schloemer-Jager 1958); and Japan, where redwood persisted into the early Pliocene (Tanai 1967; Chaney 1967). Nearly all taxodiaceous conifers had been eliminated from Japan by the late Pliocene (Tanai 1967), approximately 2 to 3 million years ago.

Since Chaney's (1951a) prediction of a mid-latitude origin for redwood, a few, high–northern latitude records for *Sequoia* have been identified. Besides the Cretaceous foliage and cones from northwestern Alaska, noted above, these include warm-temperate Paleogene floras in Greenland, Alaska (Schloemer-Jager 1958), and Spitzbergen (Schloemer-Jager 1958, Schweitzer 1974, 1980), Miocene(?) floras of Iceland with *Magnolia, Sassafras, Acer* (maple), and other temperate taxa (Friedrich 1966; Chaney 1967), and a possible late early and/or middle Miocene warm-temperate, summer-wet flora from southern Alaska (Wolfe et al. 1966; their table 2). Although *Sequoia* existed widely in the Cenozoic, including a few exceptional records in high northern latitudes that may be the same species as *S. sempervirens,* most mid-latitude and high–northern latitude records are dawn redwood rather than redwood.

Redwood in the Tertiary of Western North America

Of the six or more species of *Sequoia* once recognized in western North American fossil assemblages, Chaney (1951a) accepted only two: *S. dakotensis,* confined to the late Cretaceous of Canada, Montana, Wyoming, and North Dakota, and *S. affinis.* Chaney considered both to be of "*sempervirens*-type." Because the more widely distributed *S. affinis* bears close morphological similarity to *S. sempervirens,* Chaney considered it either a modern analogue or conspecific (the same species) with *S. sempervirens.*

The first western North American identification of *S. affinis* is in Paleocene (ca. 66–57 million years ago) floras of Wyoming; it persisted there into the middle Eocene, or approximately 50–42 million years ago (Chaney 1951a; Axelrod 1966). *Sequoia affinis* is also identified in several late Eocene floras of interior upland regions in northeastern Nevada and Idaho (Chaney 1951a; Chaney and Axelrod 1959; Axelrod 1966, 1976). All are mixed deciduous hardwood-conifer forests with a more temperate "upland" aspect than contemporary "lowland" floras of coastal Oregon and California.

In the Florissant flora of transitional Eo-Oligocene age, *Sequoia* was found as far south as the central Rocky Mountains of Colorado (MacGinitie 1953), where it occupied riparian and lake shoreline habitats. Large-rooted stumps attest to groves where redwood grew with summer-wet genera found today in the southern Appalachians and central and southern China. The regional vegetation included plants whose closest living relatives are now found in drier (subhumid), sclerophyllous woodland and chaparral in the southern Rockies, central Texas, and northeastern Mexico (MacGinitie 1953). Redwood apparently was eliminated from Colorado a few million years later (Axelrod 1976, 1977), another indication of early vegetational modernization in the Cordilleran region (Axelrod and Raven 1985).

Middle and late Oligocene floras (ca. 32–23 million years ago) with *Sequoia* are found in western Oregon (Lakhanpal 1958; Wolfe and Brown 1964; Axelrod 1966) and late Oligocene to early Miocene floras (ca. 26–20 million years ago) with redwood are known in Montana (Chaney 1951a; Becker 1969, 1973). These Paleogene and early Neogene floras indicate mixed conifer–hardwood forests whose nearest living analogs now occur in humid coastal forests of California and Oregon, in summer-wet, temperate (or mild) deciduous forests of eastern Asia and eastern North America, and in the contemporary local vegetation (Lakhanpal 1958; Wolfe and Brown 1964; Axelrod 1966; Becker 1969, 1973; Raven and Axelrod 1978). Becker (1969), for example, reported subhumid taxa in Oligocene floras of Montana associated with mesic and mild-temperature deciduous and gymnospermous taxa now restricted to eastern North America and eastern Asia. Some genera in Oligocene redwood forests of western Oregon also had a subtropical and even tropical character (Wolfe and Brown 1964).

By the late Oligocene and Miocene (ca. 26–5 million years ago), if not earlier, *Sequoia* and *Sequoiadendron* had segregated distributions. *Sequoia* forests that appear largely ancestral to modern coastal forests were restricted to coastal slope regions of milder, more mesic climate in Oregon and Idaho, both to the east and west of the present Cascade crest. Miocene floras with *Sequoiadendron*, in contrast, had strong Sierran affinities. These floras are found in more extreme climates to the south and interior in Idaho, Nevada, and west-central California and represented drier, warmer conditions (Axelrod 1976, 1986; Raven and Axelrod 1978). By the late Miocene, few "exotic" taxa (i.e., taxa no longer native to western North America) were associated with the *Sequoiadendron* forests, although a few summer-wet taxa persisted as late as the middle Pliocene (ca. 4–3 million years ago) in western Nevada in a "near modern" *Sequoiadendron* forest like that living today on the western slopes of the Sierra Nevada (Axelrod 1962, 1986). Disruption of the *Sequoiadendron* forest into the present discontinuous groves in the central and southern Sierra seems likely to have occurred within the past 10,000 years in response to drier and warmer climate (Axelrod 1973, 1976, 1986).

Miocene-age floras with *S. affinis* are found in eastern and western Oregon (Chaney 1951a; Chaney and Axelrod 1959; Wolfe and Brown 1964) and Idaho

(Chaney 1951a; Smiley and Rember 1985). These are mesophytic, warm-temperate, mixed evergreen, conifer-hardwood forests with common summer-wet genera (Chaney 1959; Smiley and Rember 1985; Gray 1985), as well as winter-wet genera that are now members of the coastal forest. A few subtropical and very warm temperate genera, holdovers from earlier Oligocene floras in western Oregon, were extinct in that area by the end of the early Miocene (Wolfe and Brown 1964).

Few post-Miocene to pre-Holocene floras document the restriction of redwood to its present distribution in coastal Oregon and California (Chaney 1951a). In a late Miocene or Mio-Pliocene flora from the western slope of the Sierra Nevada, *Sequoia* occurred with a few mild-winter and remnant eastern American summer-wet genera, while sclerophyllous plants dominated on drier slopes (Axelrod 1962, 1977). Also well outside its contemporary range, redwood was found in an early Pliocene flora in the Columbia River Gorge to the west of the present Cascade axis. Genera in this flora are now found on the Pacific Coast and in temperate, summer-wet hardwood forests (Chaney 1944; Gray 1964). This was perhaps the northernmost occurrence of *Sequoia* in the Pliocene. No Pliocene floras to the east of the Cascade crest in central Washington or central Oregon include either *Sequoia* or *Sequoiadendron* (Chaney 1951a; Smiley 1963).

In Neogene (Miocene, Pliocene) forests of western North America, redwood mixed with plant species with which it no longer grows, as well as with genera it presently occurs with in Pacific Coast forests. These late Cenozoic redwood forests contracted in distribution and changed in composition over time with a decrease in rainfall, a shift in seasonal precipitation from summer to winter, and more extreme summer and winter temperatures, particularly east of the Cascade and Sierra crests. Some geologists associate these climatic changes with global tectonic uplift and mountain-building of the post-Eocene, particularly in the Tibetan Plateau, which culminated in the last major ice age (Raymo and Ruddiman 1992; Ruddiman et al. 1997) and with local uplift and mountain-building in western North America. Although elevation of the western Basin and Range in western North America may have been as much as 1–1.5 km higher than its present altitude 15 to 16 million years ago (Wolfe et al. 1997), principal uplift of the Sierra Nevada, Cascades, and Coast Ranges that effected major changes in rainfall to their lee, is commonly assumed to have occurred in the past 3 to 5 million years (Axelrod 1977, 1986; Raven and Axelrod 1978; Axelrod and Raven 1985).

Sequoia Floras of Coastal California

The last significant representation of redwood on the West Coast associated with summer-wet, broad-leaved hardwoods now extinct in coastal vegetation, occurred in late Pliocene floras (Axelrod 1976, 1977). Among these is the Upper

Wildcat or Garberville flora of northern California. This flora appears to provide evidence of predominantly floodplain vegetation, with redwood represented only by logs transported from nearby upland forests. A few summer-wet relics from Miocene forests still persist in this flora, indicating summer rains on the West Coast until about 2 to 3 million years ago (Chaney 1951a; Axelrod and Demere 1984).

The late Pliocene Sonoma flora north of San Francisco Bay near Santa Rosa resembled modern coastal conifer forests of northwest California with *Sequoia*, *Abies* (fir), *Pseudotsuga* (Douglas-fir), *Picea* (spruce), and *Tsuga* (hemlock), but also included *Ilex* (holly), *Persea* (avocado), and *Ulmus* (elm), whose closest relatives are now restricted to summer-wet climates (Axelrod 1944a, 1950, 1976, 1977; Axelrod and Demere, 1984; Chaney 1951b). The somewhat younger flora near Napa, about 55 km southeast of Santa Rosa, with uncommon redwood, numerous sclerophyllous plants, and one or two summer-wet genera, appears to represent a drier, cooler flora similar to vegetation at the inner limits of the north coast forest (Axelrod 1950, 1976, 1977; Axelrod and Demere 1984). The older middle Pliocene Mulholland flora from the Oakland-Berkeley Hills east of San Francisco indicates drier, warmer conditions than in the late Pliocene and included no redwood (Axelrod 1944b, 1973; Chaney 1951b).

Redwood also occurred in the late Pleistocene Carpinteria flora, older than 40,000 years, near Santa Barbara in southern California (Chaney and Mason 1933; Chaney 1951b; Axelrod 1967, 1981; Raven and Axelrod 1978). Logs of redwood at Carpinteria, originally interpreted as probable sea-drift from the north (Chaney and Mason 1933), are now believed to have been transported by streams from interior forests. This would indicate a considerable southern shift of redwood forest (Axelrod 1967, 1976, 1977, 1981; Raven and Axelrod 1978) because the Carpinteria flora most resembles the Monterey pine forest about 320 km to the north, near the present southern limit for redwood. Temporary extension of redwood and Monterey pine well to the south of their present limit indicates an interval of increased precipitation and lower temperatures in southern California (Chaney and Mason 1933) in the late Pleistocene. This occurrence at Carpinteria is the southernmost extension known for *Sequoia* in the Pleistocene (Chaney 1951b).

During the Cenozoic, therefore, redwood forests changed over time as they were affected by climatic changes and concomitant tectonic uplift and mountain-building. Two major consequences were segregation of the comparatively depauperate contemporary redwood community from the diverse, mesic, mixed conifer-deciduous hardwood forests of the Paleogene and Neogene and ultimate restriction of *Sequoia* to its present coastal distribution, probably toward the end of the Pleistocene.

Transition to the Holocene

The Holocene follows the last major phase of continental glaciation; it extends from about 10,000 to 12,000 years ago up to the present day (Wright 1983). As a formal chronostratigraphic unit (Hedberg 1976), the Holocene marks the transition from glacial to nonglacial conditions 10,000 years ago at a location in northwestern Europe. In reality, the transition from the late Pleistocene glacial climate to the Holocene varied significantly from one region to another depending on the region's climatic history (Watson and Wright 1980).

Today, California has many regional climates, ranging from those associated with temperate rain forests of the northwest to hot arid deserts of the southeast portion of the state. The Pleistocene and Holocene climatic histories of each region are distinct and, at various levels of resolution, not necessarily congruent with one another. For example, the large Pleistocene mountain glaciers in the Sierra Nevada had different effects on the regional climates of the east and west sides of the range (West 1998). The ocean also has affected the climate by moderating and dampening temperature fluctuations along the coastal margin (Johnson 1977); this is the result of a wind-driven coastal upwelling system, which provides today's cool, foggy summer conditions typical of the coasts of central and northern California (Bakun 1990). The eastward increase in continentality (i.e., more seasonality in weather), enhanced by increases in elevation, has had a significant effect on the distribution of California's vegetation (Major 1988). As observed in other areas (Davis 1984), the response of plants to climate change has been individualistic depending on the climatic variable controlling each taxa on different spatial and temporal scales (Davis 1981).

The late Pleistocene and Holocene have been dynamic times, with varying precipitation and periods of cold alternating with periods of warmth. The transition from the late Glacial Maximum (22,000 years ago) of the Pleistocene to the Holocene appears to have occurred in a steplike manner with several reversals, the most significant being the period called the Younger Dryas. The Younger Dryas, named for the genus *Dryas* in the rose family, was a cold period that lasted from about 12,800–11,600 years B.P. (before present). Expressions of the Younger Dryas have been observed in the fossil records of marine sediments from basins off the California coast (Kennett and Ingram 1995) and in the fossil pollen record of Clear Lake (Adam 1988; G. J. West unpub. pollen counts of core CL-73-5).

The Younger Dryas ended as abruptly as it began, and once again temperatures increased significantly, continental and mountain glacial ice melted, and sea levels around the world rose rapidly (Broecker and Denton 1990; Porter 1983; Bloom 1983a, 1983b). At the same time, major changes took place in the earth's atmosphere. Between the late Glacial Maximum and the pre-industrial period, CO_2 levels increased by more than 30 percent, methane increased, and water vapor levels rose (Jouzel et al. 1993; Sowers and Bender 1995; Yung et al. 1996). With the exception of the Younger Dryas (Watson and Wright 1980), the transition from the late Glacial Maximum to the Holocene may have been caused by changes in solar radiation as a result of changes in the orientation of the earth's axis (COHMAP 1988; Hays et al. 1976).

The Holocene has not been static; significant changes in temperature and precipitation occurred, though less extreme than those during the transition from the late Glacial Maximum (Wright 1983). Temperatures apparently reached their highest values before 5000–7000 B.P., in a period termed the *Hypsithermal* by Deevey and Flint (1957). Glaciers disappeared and did not return to the Sierra Nevada until later in the Holocene. Briefer climatic episodes, such as the Medieval Warm period (Stine 1994) and Little Ice Age (Burke and Birkeland 1983), are reflected in a number of records in California. To varying degrees, these changes since the late Glacial Maximum were instrumental in shaping the current West Coast's flora and fauna, including redwood's range and abundance.

Cold temperatures during the late Glacial Maximum probably restricted the distribution of redwood to more southerly coastal regions and small, protected coastal areas within its current range. The presence of spruce at Bolinas Lagoon, fir and Douglas-fir at Point Reyes Peninsula (Rypins et al. 1989), and fir at Laguna de las Trancas during the late Pleistocene indicated significantly lower temperatures, more effective moisture, and a longer rainy season in central coastal California. For Laguna de las Trancas, the difference would be analogous to shifting climatic conditions southward 260 km (Adam et al. 1981). In the interior and uplands, values for pine pollen reach their highest levels in the late Pleistocene (Adam 1988; West 1993).

The Holocene

Both macrofossil and pollen data have been used for tracing the Holocene record of redwood (fig. 2.1). Also, charcoal is often present in Holocene sediments, providing information on fire history. Although the record has significant geographic gaps and few temporal constraints, the data are sufficient to outline redwood's Holocene history.

Fortunately, redwood pollen grains from Holocene sediments are distinct and can be identified readily with a light microscope. Further, the pollen usually does not have a large dispersal range (Heusser 1983, chap. 4), although it can be transported by water, as evidenced by its presence in marine sediments off the coast of California (Gardner et al. 1988; Heusser and Balsam 1977), and to some extent by wind. Heusser (1983) found that redwood pollen numerically dominated the pollen rain (ranging from 40–90 percent of the total pollen sum) in contemporary redwood forests.

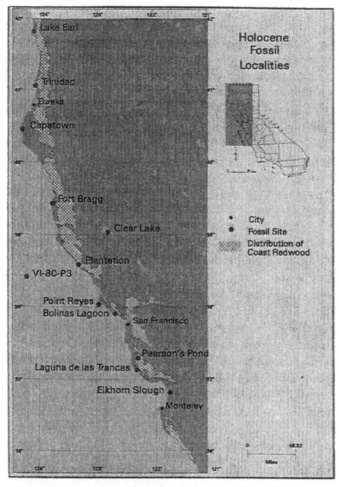

Figure 2.1. Holocene fossil localities (from West 1986).

Early Holocene

The early Holocene in coastal California was characterized by rapid increases in temperature and sea level and changes in precipitation. Temperature estimates suggest a warming of 2° to 8°C (Adam et al. 1981; Adam and West 1983; Heusser et al. 1980). The boundary conditions of a general circulation model (COHMAP 1988) showed an orbitally induced increase in solar radiation in summer and a decrease in winter from 12,000–6000 B.P. Average solar radiation for mid-latitude regions, peaking between 10,000 and 9000 B.P., was 8 percent higher in summer and 10 percent lower in winter than it is today (Kutzbach and Guetter 1986). Consequently, seasonality in climate (continental conditions) was greater between 10,000 and 9000 B.P. than any other time within the Holocene. From 9000 to 6000 B.P., the subtropical high was stronger in July than today. At 9000 B.P., the westerly jet stream arched farther northward, away from northwestern California; as a consequence, moisture-bearing storms would have had a higher-latitude track than during the late Pleistocene.

Because they are subject to changing CO_2 levels in the atmosphere, precipitation values determined from isotopic ratios in fossil vegetation records are not straightforward. Water-use efficiency by an individual plant is almost directly proportional to the level of CO_2 for a given regime of temperature and humidity (Farquhar 1997). Further, varying amounts of atmospheric CO_2 can favor or inhibit the growth of various taxa, depending on their photosythentic pathway. Precipitation estimates derived from plant fossils suggest less moisture since the late Pleistocene in the southern North Coast ranges of California (Adam and West 1983) and greater moisture in the Pacific Northwest (Heusser et al. 1980). This apparent contradiction is consistent with atmospheric global circulation models, upon which scenarios of climatic change were based.

During the late Glacial Maximum, sea level was some 120 ± 60 m below present day level (Bloom 1983a) and about 20,000 km[2] of land in California was exposed (Bickel 1978). Between 10,000 and 8000 B.P., sea level increased at a rate of about 2 cm/yr (Atwater et al.1977). The rise in the sea level began at San Francisco Bay about 11,000 B.P.; for Elkhorn Slough, Monterey County, approximately 8000 B.P. (Schwartz et al. 1986); and for of Bolinas Lagoon, Marin County, 7770 ± 66 B.P. (Bergquist 1977/1978).

The first evidence of human occupation within the range of redwood occurred during the early Holocene. Along the southernmost coast of Oregon, mussel shells associated with chipped-stone tools and other cultural debris yielded radiocarbon dates averaging 8200 B.P.; charcoal and shell from a deep midden deposit on the Sonoma County coast indicated a date of 8600 B.P. (Moss and Erlandson 1995).

MARINE RECORD (Core VI-80-P3). Redwood pollen was present in late Pleistocene and Holocene marine sediments recovered about 100 km west of the present-day mouth of the Russian River, Sonoma County (Gardner et al. 1988). From about the late Glacial Maximum to 12,000 B.P., redwood pollen values were low, then showed a sharp increase in the early Holocene, peaking about 5500 B.P. After this time, redwood pollen values showed a slight decline. These trends in redwood pollen values suggest that redwood responded rapidly to climate changes at the end of the Pleistocene and was more abundant in mid-Holocene times than before or after 5500 B.P.

LAGUNA DE LAS TRANCAS. A 2.1-m core from Laguna de las Trancas, a marsh atop a landslide in northern Santa Cruz County, yielded a pollen record for the period from about 30,000 to 5000 B.P. (Adam et al. 1981). Three pollen zones were recognized: pine, pine-fir, and redwood. Redwood pollen was present, but not abundant, in the pine and pine-fir zones. In the redwood zone, redwood pollen values exceeded 50 percent of the total, whereas pine values dropped sharply and fir was no longer represented. Adam et al. (1981) interpreted the redwood zone as spanning the latest Pleistocene to middle Holocene; however, no radiocarbon ages were obtained. From the same marsh, Heusser (1982) stated that the base of the redwood zone is 8,000 years old. Sedimentary charcoal particles increase at 8000 B.P. and remain abundant until about 2000 B.P.

Middle Holocene

At times before 5000–7000 B.P., the interior regions of California and the Great Basin were generally warmer, with longer summers than today. As a result of greater oceanic-continental temperature contrasts, a stronger sea-to-land atmospheric gradient developed, which generated upwelling winds. Under these conditions of greater upwelling (van Geen et al. 1992), the extent of maritime summer fog was probably greater in middle Holocene times than today. The mean jet stream was probably north of its current track by a few degrees latitude.

The rate of sea level rise slowed between 7000 and 4000 B.P. (Bloom 1983b) and since about 6000 B.P. has been about 1–2 mm/yr (Atwater et al.1977). By about 3000 B.P., the majority of the coastal shelf exposed during the late Glacial Maximum had become inundated.

The numerous archaeological sites of this period suggest that the entire coast of California was occupied by humans (Jones 1992). Whereas the interpretation of the archeology of this period is proceeding at a rapid rate, much is not known, specifically population levels and the interaction of native peoples with their environment. High densities of archaeological sites in the North Coast ranges just inland from the redwood forests suggest intensive use of that region's natural resources (Hildebrandt and Hayes 1993).

ELKHORN SLOUGH. Elkhorn Slough is a drowned river valley filled with estuary sediments of Holocene age. In an exploratory study, redwood pollen, most likely washed in from upland forests, was found throughout a sedimentary section that extends back more than 5500 B.P. (West 1988). Redwood pollen values peaked about 4000 B.P., then declined markedly in the upper, youngest part of the section. This latter shift may have been caused by intensive logging that occurred in the region over the past century. Changes in the drainage related to fault movement also may have affected the pollen spectra.

BOLINAS LAGOON. Holocene-age sediments here may contain redwood pollen values in varying amounts, but their specific relative and absolute fluctuations are masked because they are grouped with related taxa within the TCT (Taxodiaceae, Cupressaceae, Taxaceae) category (Bergquist 1977/1978). The preponderance of TCT pollen at this locality very likely came from redwood. Spruce pollen was present in sediments older than 8400 B.P., but the data are inadequate to indicate redwood presence. Bergquist (1977/1978) attributes an increase in TCT pollen above the spruce levels to three possibilities: (1) the landward movement of redwood in response to rising sea level, (2) the geographic spread of redwood, or (3) possible decomposition of redwood pollen at depth. The abundance of pollen of other taxa known to be sensitive to degradation in the lower levels does not support the latter possibility.

The higher redwood pollen values during mid-Holocene times observed at three fossil localities indicate that redwood may have been more widely distributed then than today. The mid-Holocene expansion of redwood may indicate a response to increased upwelling as the result of the Hypsithermal warming of interior areas creating an increased pressure gradient. At the same time, in the interior and uplands, oak pollen values increase significantly, suggesting a warmer, drier, more Mediterranean-type climate for the period (Adam 1988; West 1993).

Late Holocene

Beginning about 4000 B.P., upwelling of California coastal waters has apparently decreased (van Geen et al. 1992) because of a diminished seasonality of climate in interior California and the Great Basin. The decrease in winds was the result of a reduction in the sea-to-land atmospheric gradient. The rate of sea-level rise slowed to less than 1–2 mm/yr, but some tectonic subsidence dating back to mid-Holocene times is evident along parts of the northwest coast of California where redwood and other forest types were drowned (Clarke and Carver 1992; Page et al. 1982). Nevertheless, the effect of sea-level changes during the late Holocene on the overall distribution of redwood was insignificant.

Human populations within the redwood region were at high levels during the late Holocene. The peoples and cultures of northwestern California were less "Californian" than they were an extension of the Northwest Culture Area from coastal Alaska southward. These people were primarily fishermen, rather than hunter-gatherers, and their populations were almost exclusively along major salmon streams. To the south, in the North Coast ranges, populations were some of the densest in California (Baumhoff 1978), subsisting more from gathering and hunting than from fishing.

PLANTATION. Plantation is located on the San Andreas fault within a redwood forest just inland from Salt Point. Sag ponds along the fault contained sediments that yielded a 2,000-year redwood pollen record (West, unpub.). Three distinct kinds of redwood forests were indicated: (1) an open-canopy redwood forest with a shrub understory of wax-myrtle, Ericaceae species, and tanoak; (2) a closed-canopy redwood forest with little understory; and (3) a historic postlogging redwood–Douglas-fir–pine forest. Pine pollen grains present in the record were likely from coastal pine species. Charcoal present in the pond sediments attested to fires, and at least one local fire was recorded.

PEARSON'S POND. In a 3,000-year pollen record from a landslide pond in southern San Mateo County, redwood was the major arboreal pollen present (Adam 1975). From about 3000 to 2000 B.P., redwood pollen values were around 20 percent. Redwood pollen was most abundant from about 2000 B.P. to the historic period, where it averaged about 30 percent. During the historic period, redwood pollen values show a marked decrease, whereas Douglas-fir pollen values increased. Adam (1975) attributed the decrease of redwood to logging.

POINT REYES. Redwood was the most abundant tree pollen type recorded in a 275-cm sedimentary core from Wildcat Lake, located near the southwest tip of Point Reyes National Seashore, Marin County (Russell 1983). An A.D. 1275 age for the base of the record was determined by extrapolating from the historic sedimentation rate, as determined by the presence of introduced weed pollen (sheep sorrel and plantain), to the prehistoric section of the core. Only eighteen samples were examined from the section, so it is difficult to evaluate the significance of the fluctuations in redwood pollen values; however, no marked trends were evident. Because redwood trees are not now present in the immediate area of Wildcat Lake, it was surmised that redwood pollen was blown in from a distance. Fluctuations in the abundance of charcoal fragments suggested four fires for the period of record.

LAKE EARL, CAPETOWN, AND FORT BRAGG. Heusser (1960) reported relatively low redwood pollen values in three undated northern California sedimentary sections—Lake Earl, Del Norte County; Capetown, Humboldt County; and Fort Bragg, Mendocino County. Increased pollen densities were found in the upper parts of the sections; however, because of the possibility of differential destruction of the pollen grains, Heusser concluded that no climatic significance could be ascribed to the change.

During the late Holocene, the range of redwood may have shrunk in response to lower temperatures and reduced coastal upwelling. Taxa such as Douglas-fir and tanoak in the North Coast ranges (West 1990, 1993) were able to expand their distributions during this period. Disturbances associated with historic logging apparently altered the composition of some redwood forests and allowed for the invasion of Douglas-fir (Adam 1975; West, unpub.). Sedimentary charcoal fragments associated with redwood pollen attest to fire in the redwood forests during the Holocene. The correspondence of Holocene human populations and the sedimentary charcoal evidence of fire is probably more than coincidence. The Yurok people, the major aboriginal group inhabiting the redwood region before and during European settlement, are known to have burned the forests regularly (box 2.2).

Box 2.2. The Yurok Period

The Yurok people believe they have resided in the lower Klamath River Basin since time immemorial. Before and immediately after the first major European invasion into the area (1820–1850), at least 3,000 Yurok people occupied approximately 780,000 ha of land. The ancestral boundary ranged from Little River in northern Humboldt County up the coast to Wilson Creek in Del Norte County. The boundaries extended eastward following a series of creeks and met on the Trinity River approximately 2 km from its confluence with the Klamath River.

Historical documents, archaeological sites, and current forest conditions indicate that Yurok people were present for more than fifty generations. They lived in seventy or more villages throughout their ancestral territory, with from one family to fifty or more individuals in each village. Their houses were made predominantly of redwood planking, with at least twenty-five houses in some of the largest villages. Although village sites were chosen for numerous reasons, accessibility to local food sources was a paramount consideration. Village size was subject to change based on natural disasters, war, famine, disease, marriage, and personal problems among the inhabitants.

For the most part, the diet of the Yurok people revolved around salmon and acorns. Forests and grasslands also provided berries, bulbs, game birds and mammals, grain, nuts, and a multitude of herbaceous plants. Through gathering and trade, salmon was supplemented by eel, sea lion, seal, seaweed and shellfish from both fresh and salt waters, smelt, steelhead and other kinds of trout, sturgeon, surf fish, and an occasional whale. In addition, the forest and waterways provided products used for the construction of boats, containers, homes, nets, tools, utensils, weapons, and products for ceremonial, medicinal, and spiritual purposes. Acorns were used almost daily. The most sought acorn was that of tanoak, but in times of low supply, black oak and canyon live oak were used. Large tanoak crops are cyclic, so acorns were stored for extended periods.

To enhance tanoak crops in the inland Douglas-fir–tanoak forest, the Yurok people learned to apply a regular, low-intensity fire regime (i.e., burning every two years at a consistent time of the year). This burning regime provided many advantages. It established large tracts dominated with tanoak, enhanced the quantity of future acorn crops, created firebreaks around villages, reduced encroaching conifers, made for easier access and gathering, inhibited the spread of disease and pests, created an understory rich in succulent herbaceous plants (especially bear-grass), and helped reduce fuel loads and prevent the spread of fire. Bear-grass was burned annually, with the next season's new, flexible leaves highly coveted by basket weavers.

The Yurok people also used fire to enhance tanoak populations in the northern redwood forest. Here, the tended tracts were not as large as in inland areas. Some areas, however, were converted into prairies, favoring grass and bear-grass. Villages were usually located outside the deepest woods, and many forests were managed using annual fires. Fire was used to create, enlarge, and maintain openings in the forest and to reduce plants to an ash rich in nutrients for spreading on cultivated plots. Fire also enhanced populations of plant species used in basketmaking, berry picking, and medicine. The largest redwood trees were not harmed by this regime, although the species composition and structure of the forest were undoubtedly affected through the influence of fire on plants in the ground and shrub layers.

Since European Settlement

The generally slow pace of vegetation change over geologic time and the relatively light and benign impacts of Native Americans through most of the Holocene came to an abrupt end soon after the first Europeans arrived in the redwood region. Increasingly, intense logging left an indelible imprint on the forest, threatening to eliminate many habitat components and species associated with the old-growth stage (see later chapters). The logging continues, but conservationists too have left their legacy.

Oxen and Horse Logging

Logging of redwoods began in earnest early in the nineteenth century. The early logging of large redwoods was slow and cumbersome because of the large, heavy logs and the lack of suitable equipment. The large trees were felled with hand axes and split into small chunks with the use of hardwood wedges and iron-bound wooden mallets. These sections were then hauled to a sawmill.

After the Civil War, adequate saws were available for felling and bucking redwood trees into logs. To reduce breakage of boles, early loggers often felled the trees on hand-prepared beds of limbs and understory shrubs. Because redwood bark is very stringy, thick, and tough, it was necessary to hand-peel logs in the woods before they were taken to the mill. The accumulation of limbs, shrubs, and peeled bark made moving logs out of the forest difficult. After the bark was peeled from logs, the accumulated slash was burned. Large logs were rolled downhill by hand to the creek bottoms or to constructed skid trails, then dragged by teams of oxen or horses to mills or to a watercourse, where the logs could be floated off during winter rains. This style of logging was very slow, allowing only small areas to be cut each year. Photographs taken in the 1800s show that even after burning of the slash, much debris remained on the logging

Table 2.2. Percentage of Sprouts from Old-Growth Trees and Snags in the Arcata City Forest, Arcata, California.

	No Sprouts	1 Sprout	2–7 Sprouts
Trees	27	18	53
Snags	37	19	44

Note: The forest was logged with oxen in the 1880s.

area (Carranco and Labbe 1975). Incomplete burns required that most areas be burned repeatedly until the slash was reduced sufficiently. Repeated burning favored growth of redwood and hardwoods, which could sprout after each burn, and selected against Douglas-fir, grand fir, Sitka spruce, and western hemlock.

The early logging took place primarily in low-elevation coastal forests, where redwood stands had been managed to some degree by Native Americans (box 2.2). Logging photographs from the early 1800s show stands dominated by large redwood trees (Carranco and Labbe 1975). Some photographs, however, show stands with a mixture of Douglas-fir, grand fir, redwood, Sitka spruce, and western hemlock. Species other than redwood were not harvested from these sites, but they frequently were damaged by the felling, burning, and yarding of redwoods. Probably in some areas, high numbers of seedlings other than redwood were present, but were destroyed by repeated burns.

The Arcata City Forest in Arcata, Humboldt County, was logged in the 1880s as described above, including yarding by oxen and repeated burning. The current (1998) canopy is dominated by an even-aged stand of 113-year-old redwood trees with a tree density higher than that of old-growth stands. A few younger spruce, grand fir, and western hemlock trees coexist in the stand. The majority of redwood trees are single stems, not clumps of stems, indicating that redwood seedlings, when burned, produce a single sprout (see chap. 4). Most sprouts came up in a ring around the base of a stump; only 1.5 percent were on top of a stump. Large redwood snags show a higher incidence of sprouting (table 2.2).

Steam-Railroad Logging

Following the development of the steam engine, logging methods changed sufficiently to create a different kind of second-growth forest. Large trees were felled by crosscut handsaws and bucked into logs; in most operations, bark was peeled off the logs. Long skid trails were used to drag logs from slopes and ridges to railroads located along the creeks. Steam engines were skidded on sleds through the forests, or used to skid cabled logs to the main skid roads. Larger steam engines were used to ground-cable skid logs to the railroads. These skid trails were up to 5–6 km long. Accumulated slash and peeled bark powered the engines, which were called "steam donkeys." Accidental slash burns occurred regularly and were not suppressed because fire had little effect either on standing or felled trees.

This harvesting system, which produced extensive areas of disturbed soil, favored establishment by tree species other than redwood. For example, stands in the Jacoby Creek Forest, Arcata, Humboldt County, were logged in 1913 using steam engines. In 1998, they consist of a few large, residual redwoods, some redwood sprouts, and mainly 80–85-year-old Douglas-fir, grand fir, and western hemlock. These species, which were discouraged by the intensive burning of the early logging period, increased later in areas where burning was less frequent. From the 1920s to 1940s, some companies developed tree seedling nurseries, and land managers planted redwood seedlings following logging.

Tractor-Truck Logging

After World War II, and the development of larger trucks and tractors, redwood logging gradually shifted from railroad logging to tractor-truck logging. Numerous roads were constructed though the forest. This type of harvest, combined with improved technology in the mills, enabled loggers to cut a higher proportion of the trees, including trees that would have been considered culls in the past. This method reduced the seed source for redwood and Douglas-fir and left more of the less-valuable western hemlock trees in the northern region (see chap. 3). Tractor logging also resulted in more mineral soil exposed at the end of operations than did previous logging systems. These effects favored Douglas-fir, grand fir, and especially western hemlock regeneration. Some of these forests were planted with redwood seedlings.

Attempts at Conversion to Pasture

During the steam engine–railroad era, logging camps were established in the woods so workers could dwell close to the logging operations. To provide fresh meat for the camps, attempts were made to raise cattle in formerly forested areas. These recently logged sites were repeatedly burned and seeded with grass. This practice favored redwood over the other tree species because of the ability of redwood seedlings, saplings, and stumps to sprout after fires. Because logging camps were continually relocated to follow shifting logging operations, new pasturelands were established regularly near new logging camps. This practice favored the establishment of pure second-growth redwood stands.

Selection Cuts

After the end of World War II, some landowners had loggers select individual trees to harvest rather than clear-cut entire hillsides. This new method, usually a "take the best and leave the rest" harvesting operation, became known as "selection harvest" or "selection logging" (O'Dell 1996). For redwood, the method might best be considered a "single-tree selection system."

In northern redwood forests (see chap. 3), the selection method was problematic because of blowdown of trees left after harvesting. Selection harvest has continued into the present time in central and southern redwood forests, both in old-growth forests and older second-growth stands. In some ways, selective harvest was superior economically to clear-cutting because it allowed a landowner to leave lower-value or presently unmerchantable trees for possible future harvest.

In the northern redwood forests, selection harvesting tended to convert the redwood-dominated forests to mixed forests with grand fir, redwood, Sitka spruce, and western hemlock. Central and southern redwood forests tended to convert to forests with sprouting California bay, madrone, redwood, and tanoak (Helms 1995; Adams et al. 1996). Libby (1996) hypothesized that continued use of the single-tree selection system, which gives trees other than redwood a competitive advantage, will result in redwood being largely or completely eliminated from the southern redwood forests.

Tax Cuts

Tax laws probably have had as much influence on how redwood forests are harvested than have any regulations dealing directly with logging. Before the timber yield tax became effective, in 1977, standing timber was taxed every year on its assessed value under the *ad valorem* tax system. If a landowner cut at least 70 percent of the volume from an area, however, the remaining timber was taken off the tax rolls until the stand regenerated back to a merchantable condition. This rule created an incentive for small landowners to harvest 70 percent or more of the volume from as much land as possible, thus removing this land from the tax rolls. It also created a standing inventory of untaxed residual timber.

The estate inheritance tax of 55 percent of the current land and timber value forced many small landowners to harvest their forests when they inherited the land. Most of these forests were harvested using a "high-grading-diameter-limit-cut method"; that is, cut all merchantable trees above a certain size. This system tended to favor regeneration of grand fir, Sitka spruce, tanoak, and western hemlock.

Large, Catastrophic Fires

Large, catastrophic fires are rare in redwood forests, but some have occurred (see chap. 4). A fire in 1945, for example, burned 124,000 ha in northern Humboldt County. Most of the land affected contained old-growth forests that recently had been selectively logged. Here, the fire killed most of the remaining Douglas-fir, grand fir, and western hemlock and converted most stands to shrubs and red alder among dense clumps of redwood sprouts. The landowners reforested the burned stands with Douglas-fir and redwood seedlings.

A portion of the same fire burned through unlogged old-growth forests of Redwood Creek and Bridge Creek. Fire did little damage to the large redwood trees but killed most of the smaller grand fir and western hemlock. Forty years later, these stands have a cohort of western hemlock with a few redwood basal sprouts under the crowns of the old trees.

Timber Harvest Under Early Forest Practice Regulations

California's first Forest Practices Act was passed in 1945 with the requirement that ten to twenty seed trees/ha be left following timber harvest. The intention of this regulation was to assure that all cutover areas would regenerate with seeds from local trees. Some landowners would leave the required 10–20/ha as redwood seed trees and aerially spread Douglas-fir seed at 4–5 kg/ha. The usual source of Douglas-fir seed was trees in Oregon. After twenty years, the result in many areas was a very dense stand of Douglas-fir (5,000–10,000 trees/ha). In most of these sites today, few or no redwood seedlings exist, even though an average of 43 percent of the redwood stumps sprouted and formed clumps. Redwood density is typically 90 trees/ha, less than in the original stands.

Conversion to Forest Plantations

During the time when state regulations required leaving seed trees, some timber owners applied to the state for alternative plans. The most common practice was to clear-cut, burn the logging slash, and plant genetically improved, nursery-grown, two-year-old redwood or Douglas-fir seedlings at 1,600 trees/ha. If potentially competing trees (red alder) or shrubs (blue blossom) began to dominate a site, herbicides were used to kill them or reduce their growth. The redwood or Douglas-fir plantations were usually precommercially and commercially thinned several times before another cycle began. Because of their even age and spacing, these stands typically had a dense, uniform canopy with little understory. Nevertheless, sprouting trees (California bay, madrone, redwood, and tanoak) appeared, irregularly spaced, in these stands and, along with the faster height growth of the redwood sprouts, created a relatively complex vertical and horizontal structure. Some of these forests are inhabited today by flying squirrels, red tree voles, fishers, spotted owls, and other species often associated with old-growth conditions. The presence of residual old-growth components from the former, natural forest increases habitat suitability for these species (Noon and Murphy 1997; see also chaps. 5 and 8). Habitat quality can be expected to decline as rotations shorten and the structural legacies decay over time.

Effect of Change in Timber Values

After World War II, with the improvement of the economy in California and the nation, lumber and other forest products increased in economic value. Mills specializing in products other than lumber—fiberboard, particle-board, pulp, and studs—became established in the redwood region. This development increased the merchantability of logs from tree species that were considered culls during earlier logging periods. Consequently, all live trees were removed, along with downed logs, snags, and even the high stumps left after the first harvest. These intensive methods, which are still used, reduce the biological and structural diversity that would carry over to future stands; they also reduce habitat for cavity nesters and other animal species that require snags or downed logs. Old-growth species therefore are less likely to recolonize these stands.

Conversion to Nonforest Uses

FARM-RANGELAND. Conversion of forest to farms or rangeland was usually not successful in this region because of the large size of the stumps and the vigorous sprouting ability of redwood. Early attempts at clear-cutting, burning, seeding grass, and continued hacking of the redwood sprouts to produce pastures or crop fields were rarely successful. Most of the areas where such conversion was attempted are now pure redwood stands. Some small landholdings remain with large, burnt redwood stumps scattered throughout pastures. Sometimes the clearing was more successful, and former redwood stands are now vegetable fields (M. Angle, pers. comm.).

RURAL HOME SITES. Conversion of forests to rural home sites is increasing throughout the redwood region. Increased forest practices regulations, which include restricted use of herbicides, more controls on stream siltation and road building, and requirements for riparian leave-tree zones, along with more public concern for biodiversity and aesthetics, have pressured landowners in some areas near urban centers or major highways to sell forestland for rural subdivisions. Most of these sites are 5–20 ha, with one or two houses plus other buildings, some lawn or pasture, and numerous roads and driveways. Most rural homes have bright night lighting. The forest, which may have been logged in the recent past, is further modified: trees are left in small clumps or patches and stands are kept free of shrubs, ground-layer plants, tree litter, and slash to create city park–like conditions and reduce fuel levels.

URBAN CONVERSION. The creation of housing densities of twelve houses/ha essentially eliminates the forest ecosystem. A few scattered trees may be left in and among the houses. A few birds, insects, slugs, snails, and other species characteristic of redwood forests may persist indefinitely in these urban areas. As pointed out by the California Native Plant Society (1997), "As we enter the new millennium, nearly 30 percent of California's 6,300 native plants are rare, threatened or limited in distribution, with the main threat being residential development." This statement applies to some degree to the redwood region, although logging and associated impacts remain the largest threats (see chap. 5, 6, and 8).

Today's Forest

Redwood's long history in the geological record extends to its present-day restriction to a strip of coastal California and southwestern Oregon. The terms *relict, paleoendemic* (Stebbins and Major 1965), and *living fossil* afford special status to this species, and rightly so. Yet, extinction of the redwood within the foreseeable future is unlikely. Even within the past 10,000 years, the species has shown both expansions and contractions of its range. Today, redwood appears to be retreating in the southern part of its range and expanding to the north (chap. 3). The species generally thrives within and beyond its natural range, even along freeways in southern California.

Despite the long history of redwood and related species, the history of the present-day redwood forest is relatively brief. The current species composition and structure of old-growth redwood forests are recent, considering the longevity of redwood and its companion trees, though similar combinations of species have occurred in the past. Two of the important tree species in contemporary forests, Douglas-fir and tanoak, became so over the past 4,000 years; however, Douglas-fir and redwood occurred together in Pliocene and older floras, and tanoak is known from the Sonoma flora of the Pliocene (J. Gray, pers. comm.; also see earlier sections, this chapter). Nevertheless, today's old-growth redwood forests mostly reflect the history of environmental change over the past 2,000 to 4,000 years (chap. 4). Many of today's ancient trees were established as seedlings during the Medieval Warm period (900–600 years ago), when Yurok people managed many northern redwood stands (see box 2.2).

Younger redwood forests reflect a shorter history. Many of the trees established after logging by oxen or steam engines began their lives under a climate closer to that of the Little Ice Age (150–350 years ago) than to current conditions. Seedlings of redwood and co-occurring species, which established after contemporary logging methods, began life under different conditions of recruitment. The subsequent forests differ in species composition from the forests encountered by the first European settlers and from those that regenerated after early logging. We can surmise that redwood forests will continue to change in their species composition and structure as the individual species that compose these forests respond in their own ways to continued environmental change. The prospect of rapid global warming and other abrupt anthropogenic changes over the next century raise many questions about what the forests of the future will look like. If they receive careful stewardship, however, we have confidence that redwood forests of some kind or another will endure.

The Redwood Preservation Movement

The most recent phase in the long history of the redwoods is the ongoing effort to save a portion of these forests for their inherent values and for the inspiration, education, and pleasure of future generations. This new phase, the redwood forest preservation movement, is in step with the broader conservation movement in North America that developed during the late nineteenth and twentieth centuries. Below we summarize the efforts to protect both the Sierran redwood (giant sequoia) and the coast redwood forests.

In the Beginning

The first great milestone was reached in June 1864 when President Abraham Lincoln signed congressional legislation that established the world's first redwood forest preserve for giant sequoia. The idea for such a preserve had been put forward by a group of prominent Californians—"gentlemen of fortune, of taste, and refinement," as Senator John Conness described them in his introductory remarks supporting the bill. They argued that the Mariposa Grove and nearby Yosemite Valley should remain forever in public ownership so the people of the nation could freely visit those places for recreation, refreshment, renewal, and inspiration (Conness 1864).

"It is a scientific fact," said Frederick Law Olmsted in his report on the newly established reserve, "that the occasional contemplation of natural scenes of an impressive character . . . is favorable to the health and vigor of men and especially to the health and vigor of their intellect beyond any other conditions which can be offered them." And yet, as Olmsted went on to point out, "without means are taken by government to withhold them from the grasp of individuals, all places favorable in scenery to the recreation of the mind and body will be closed against the great body of the people" (Olmsted 1865).

California's matchless redwoods had already captured the imagination of people across the nation and around the world during the 1850s—when two magnificent 100-m-high giant sequoias in the Calaveras Grove of Big Trees were cut down to create traveling exhibits. The bark from the second tree—the "Mother of the Forest"—was stripped off, shipped east, and then reassembled to form an imposing 10-m-diameter, 37-m-high display in the Crystal Palace exhibition hall in New York City. The exhibit was later moved to the Crystal Palace in London, England, where it remained on display until 1866. In both cities the display was a smash hit, though these tree-cutting exploits also inspired criticism. One American magazine lamented the loss of the living tree: "In its natural condition, rearing its majestic head towards heaven, and waving in all its natural vigor, strength and verdure, it was a sight worth a pilgrimage to see; but now alas, it is only a monument to the cupidity of those who have destroyed all there was of interest connected with it" (Gleason's Pictorial Drawing-Room Companion 1853).

Despite such sentiments, the Mariposa Grove remained the only protected grove until 1890, when years of effort by John Muir, Robert Underwood Johnson of Century Magazine, and others—including George Stewart, editor of the newspaper in Visalia, California—finally led to the creation of Sequoia National Park. President Benjamin Harrison signed legislation in 1890 that set aside what is now the heart of Sequoia and Kings Canyon National Parks; later that same year he signed legislation calling for the creation of Yosemite National Park surrounding the state reserve that protected the valley itself. Then, in 1893, President Harrison created the Sierra Forest Reserve, an area of some 4 million hectares that included most of the giant sequoia groves in the central and southern Sierra. Those groves are managed today by the U.S. Forest Service (Dilsaver and Tweed 1990).

In 1900, the two Calaveras Big Tree Groves were again the center of widespread attention. After failing to persuade the federal government to take over the groves as a national park, their owner had given up and sold them to "a lumberman from Duluth." Newspapers throughout California featured the story and public outrage was immediate. As Muir pointed out, the Calaveras groves "were the first discovered and are the best known. Thousands of travelers from every country have come to pay them tribute of admiration and praise. Their reputation is world-wide." And so it was only natural that people would want them to be protected. Meetings were held, speeches were given, letters were written, and a distinguished delegation of Californians was soon on its way to Washington, D.C., to lobby for preservation of the two groves (Engbeck 1973).

Federal action was not forthcoming. Weeks and months and years went by while the stalemate continued. Private land had never been acquired by the federal government to create a national park, and many members of Congress feared that federal acquisition of the Calaveras groves would set a dangerous precedent. In fact, Congress managed to avoid setting that precedent for several decades—until the 1960s (Wirth 1980).

Saving the Coast Redwoods

The entire redwood preservation effort up to 1900 had been focused on saving the giant sequoia of California's Sierra Nevada. Now, as a new century began, it was time to save some portion of California's coastal redwood forest.

The campaign got under way in April 1900 when Andrew P. Hill, a well-known painter and photographer, led a group of educators, writers, and women's club members on a three-day exploratory expedition to the Santa Cruz Mountains—to the area we know today as Big Basin. While there they decided to form an organization, the Sempervirens Club, to lobby for preservation of the area as a public park. Because the campaign to create a national park at Calaveras Big Trees was still in progress, the movement to save Big Basin was aimed at state government. That campaign was crowned with success in 1902 when the state's first coastal redwood park was created (Taylor 1912; Hill and Hill 1927; Engbeck 1980).

One year later, William Kent acquired a canyon full of redwoods in Marin County to keep the canyon from being logged and then being submerged beneath a man-made reservoir. In 1908, the Kent family gave that land to the people of the nation on condition that it be named in honor of John Muir and preserved in its natural state for public enjoyment. President Theodore Roosevelt issued the decree that made Muir Woods a national monument (Engbeck 1995).

After World War I, the demand for lumber, together with renewed highway construction, began to threaten the very heart of California's central and northern redwood forest. Previously protected by its relatively remote location, the matchless redwood forest along the south fork of the Eel River had long been admired by a small handful of naturalists and by residents of Eureka and other communities in Humboldt County. For years those people had been pleading, without success, for government action to protect selected areas of the forest they knew and loved. Finally, in 1917, Stephen T. Mather and Horace Albright of the newly created National Park Service persuaded three distinguished naturalists to investigate the situation firsthand (Engbeck 1973; Schrepfer 1983).

Madison Grant was a founder of the New York Zoological Society. John C. Merriam was a professor of paleontology at the University of California at Berkeley and was about to be named president of the Carnegie Institution of Washington, D.C. Henry Fairfield Osborn was curator of the American Museum of Natural History in New York City. They were alarmed by what they discovered (Drury 1972). The new state highway along the Eel River was about to bring large-scale, intensive logging to the area. Already some roadside logging activity was under way. Trees that were one thousand or even two thousand years old were being cut down to make railroad ties, shingles, fence rails, and other split products. As Madison Grant later described it, that kind of logging was like "breaking up one's grandfather clock for kindling to save the trouble of splitting logs at the woodpile." Grant, Merriam, and Osborn immediately wrote to the governor of California asking for state intervention (Grant 1919).

Save-the-Redwoods League

Grant, Merriam, and Osborn agreed to form a new organization that would conduct an intelligent, thoroughly informed, and ongoing campaign to save the redwoods. That organization, the Save-the-Redwoods League, was formed in March 1918. Grant, Merriam, and Osborn were its founders and Secretary of Interior Franklin K. Lane its first president. Along with Osborn, Stephen T. Mather, William Kent, and E. C. Bradley provided major gifts that set the organization in motion.

Beginning in 1919 and 1920, under the leadership of Merriam (who served as president of the League from 1921 to 1944) and Newton B. Drury (who served the League in various capacities from 1919 to 1978), the League clearly defined its objectives, quickly built up a large membership, conducted field surveys, and set out to obtain government support for its long-range program. It soon became apparent, however, that the federal government was not willing to act on behalf of the redwoods. Undeterred, the League set out to work with California state government (Schrepfer 1983).

The League's long-range plan called for the creation of four state parks, each featuring a different aspect of the old-growth central and northern redwood forests. Those parks—Humboldt, Prairie Creek, Del Norte, and Jedediah Smith Redwoods State Parks, created in the 1920s and 1930s and rounded out little by little in later years—soon came to symbolize and otherwise serve as the cornerstone of the California State Park System, a statewide system of parks launched and shaped largely by the League and its allies during the 1920s.

Major floods in 1955 and 1964 caused extensive damage to several parks, including the world-famous Rockefeller Forest in Humboldt Redwoods State Park. As a result, both the League and the California State Park System increased their emphasis on preserving whole watersheds—and nearly whole ecosystems—to protect more adequately what was left of California's primeval redwood forest.

Today, after eighty years of continued work, including the provision of major financial assistance toward the creation of Emerald Bay State Park, Point Lobos State Reserve, and some twenty-seven state redwood parks, the League is still pursuing its idealistic mission, still quietly adding to its extraordinary record of accomplishment. The current value of the 48,000 ha of redwood forest that the League has turned over to the California State Park System and other public agencies is more than $5 billion (Save-the-Redwoods League 1996).

In 1968, after decades of Congressional delay and public frustration, the federal government finally agreed to create a national park for the redwoods. Land acquisition for the new park occurred in 1968 and 1978 and involved about 39,000 ha at a cost of about $195 million plus an exchange of land—then the most expensive land transaction in national park history. (For perspective, the acquisition and construction cost of a single mile of six-lane freeway can exceed $200 million.)

The Ongoing Movement

In 1998, the federal government and the State of California agreed to acquire an area in Humboldt County known as the Headwaters Forest. This area includes the largest remaining privately owned old-growth redwood grove, several smaller areas of old growth, and a buffer zone. This $530 million transaction, just completed at the time of this writing, demonstrates the continued strength of the redwood forest preservation movement (Harris 1995).

The story of the movement to save the redwoods is largely a success story—a story of grand accomplishment. But it is not over. The job is not yet done.

There are still forests of giant sequoia and redwood, both on the coast and in the Sierra Nevada, that deserve to be preserved as publicly owned natural areas. As will be explained in chapter 7, some kinds and subregions of redwood forest are better represented in protected areas than others. And there are state and national parks and other redwood forest areas—already "saved"—that still need to be rounded out so their boundaries match watershed boundaries or other environmental realities. Moreover, the regulations that govern human activity on privately owned redwood forest areas need to be designed to adequately respect wildlife and other public trust values.

To accomplish all of these objectives in a balanced and reasonable way, and to foster unity of action by an increasingly wide range of groups and individuals who care about the redwoods, an up-to-date master plan for the redwoods is currently being developed that will guide the movement well into the twenty-first century. This book is part of that master planning effort—part of the ongoing movement to "save the redwoods."

Chapter 3

CHARACTERISTICS OF
REDWOOD FORESTS

*John O. Sawyer, Stephen C. Sillett, James H. Popenoe,
Anthony LaBanca, Teresa Sholars, David L. Largent, Fred Euphrat,
Reed F. Noss, and Robert Van Pelt*

Redwood trees grow in an interrupted 724-km belt along the Pacific Coast from the southwestern tip of Oregon (42°09' N. latitude) to southern Monterey County in California (35°41' N. latitude), once covering some 647,500–770,000 ha (Olson et al. 1990; U.S. Fish and Wildlife Service 1997). Redwoods, now largely second or third growth, are still found throughout this range in a variety of ecological settings. They can occur in pure stands or with a variable number of associated tree species. Individual trees can grow among pygmy cypresses or sand dune herbs on the Fort Bragg marine terraces, along coastal bluffs with low shrubs, with *Sphagnum* mosses in coastal fens, or with evergreen hardwoods, chaparral shrubs, or even coastal prairie grasses. Common conifer associates are Douglas-fir and western hemlock, but grand fir, Coulter pine, Monterey pine, Port-Orford-cedar, Sitka spruce, sugar pine, and western red-

Author contributions: Sawyer, general organization and lead author; Sillett, section on canopies and editing throughout; Popenoe, section on soils; LaBanca, information on Redwoods National Park; Sholars, sections on flora, lichens, and exotic plants; Largent, section on fungi; Euphrat, box on classification; Noss, writing and editing throughout; Van Pelt, information on tree mass and volume.

cedar also mix with redwoods. Common hardwood associates are bigleaf maple, California bay, canyon live oak, coast live oak, golden chinquapin, madrone, and tanoak. The assemblages and habitats in which redwood grows are diverse, but the species is found most commonly in tree-dominated stands where it achieves sufficient size, longevity of stems (both alive and dead), and shade production to influence ecosystem processes significantly. We recognize such a stand as a redwood forest (Barbour et al. 1992; Johnston 1994; Sawyer 1996; Ornduff 1998).

Redwood forests are renowned for their extremely high volume of standing biomass. In some stands, the accumulations of wood are truly astounding. The largest volume of wood per unit area that we have seen published was for a relatively young stand (Hallin 1934). This 260-year-old stand had an estimated 178 trees/ha, a volume of 10,856 m³/ha, and a standing biomass of 3,568 metric tons/ha. More recently, a value of 9,785 m³/ha of volume was published by Westman and Whittaker (1975). This estimate corresponds to a standing biomass of 3,300 metric tons/ha; however, the estimate was made on the basis of a single small (less than 0.3 ha) sample. A 1.44-ha plot measured by Fujimori (1977) on the lower Bull Creek in Humboldt Redwoods State Park, Humboldt County, had 100 trees. The estimated volume was 10,817 m³/ha and 3,461 metric tons/ha. Mulder and de Waart (1984) presented a stand profile of this forest. During the summer of 1997, a 3-ha plot was established overlapping Fujimori's plot (Van Pelt and Franklin, in prep.). The wood volume averaged 8,072 m³/ha, and estimated aboveground stem biomass was 2,583 metric tons/ha. Young-growth redwood stands are nearly as impressive. On good second-growth sites, trees approach 70 m in height and stand volumes of 3,600 m³/ha within 100 years (Olson et al. 1990).

Redwood forests are more than just vast volumes of high-quality wood. They include—along with the living trees—standing snags, downed woody debris (decaying logs and branches), and a myriad of plant, fungal, and animal species that make use of these habitat components. In addition, redwood trees may last as long as logs as they do as living stems, so both are important components of ecosystem structure and function (Franklin et al. 1981).

There has been little research specifically on the amounts of woody debris and rates of decay in redwood forests (Agee 1993), but the few studies suggest some patterns (see chap. 4). Studies at Prairie Creek and Humboldt Redwoods State Parks by Pillars and Stuart (1993) showed that leaves and twigs have a much shorter half-life than branches and trunks. Bingham and Sawyer (1988) reported a volume of 957 m³/ha and a log mass of 200 metric tons/ha for woody debris in coastal slope forests at Prairie Creek Redwoods State Park, Humboldt County. Redwood logs made up 84 percent of the coastal slope forest's log mass. Finney (1991) estimated very different values, 10 metric tons/ha and 280 metric tons/ha, for two frequently burned stands near the inland edge of redwood's range. An unpublished estimate of woody debris in the old-growth slope stands dominated by redwood at Angelo Coast Range Preserve, Mendocino County, is 25–29 metric tons/ha (Bingham 1992). Greenlee (1983) reported 186 metric tons/ha of woody debris in redwood stands in the Santa Cruz Mountains. Strikingly, the lowest estimates of woody debris mass in redwood forests are two to five times greater than those reported for other temperate or tropical forests (Franklin and Waring 1980).

Stand history appears to be crucial in determining the amount of woody debris present today in many areas. On many private lands, downed logs are removed soon after they have fallen from accessible parts of old-growth stands or cut on the site for shakes and other products. A slope forest near an old logging camp had only 12 metric tons/ha of woody debris (Bingham 1992). Manicured stands are expected in parks that have intense human use, such as city parks or heavily used campground or day-use areas of state and national parks. For example, a mature Douglas-fir–tanoak stand near Willits in Mendocino County is unusual for its lack of large tanoak trees, as they were removed to make the area more "parklike."

Variation in Redwood Forests

Stands with an important component of redwood trees vary substantially in their structure, species composition, environmental conditions, and ecosystem processes. Considering this variation, conservation and preservation of redwoods is best considered in local or subregional settings. As an organizing approximation we identify three major subregions (sections): northern redwood forests from southwestern Oregon to east of Humboldt Bay in Humboldt County, California, central redwood forests from southern Humboldt County to northern San Francisco Bay, and southern redwood forests from Alameda to Monterey County (fig. 3.1). Within these sections, a variety of redwood forest types can be classified (box 3.1)

Northern Redwood Forests

Northern redwood forests are regionally extensive, blanketing most of the landscape of this section. The landscape is a set of parallel (northwest to southeast) ranges of folded and faulted strata with rounded crests of uniform height. Variously metamorphosed rocks of the Franciscan Formation (Mesozoic age) are the most extensive parent materials underlying northern redwood forests. Local areas of shelf and slope sedimentary rocks are found along the coastline. Slow or relatively slow streams and rivers in alluvial or weak bedrock channels flow directly to the Pacific Ocean. Elevations range from sea level to 950 m.

The climate of the northern redwood forests may be characterized by 1961–1990 normals from Klamath and Orick weather stations (table 3.1; NOAA 1997). The winter and most of the spring and fall are characteristically

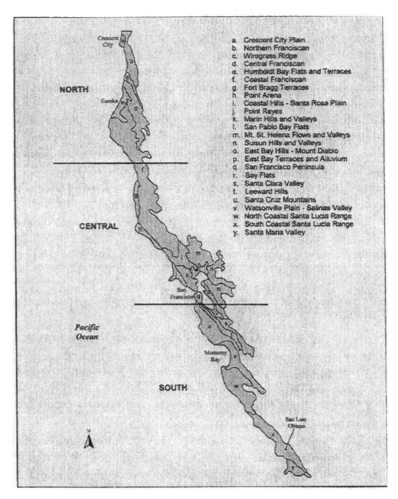

Figure 3.1. The natural distribution of coast redwood, showing three major sections and twenty-five subsections.

cool and wet. Annual precipitation may exceed 2,500 mm in many areas of Del Norte and northern Humboldt Counties. Snow is commonplace, especially in late winter and early spring, although it usually melts quickly. Summers are characterized by cool, sunny days interspersed with periods of high fog, which sometimes persists throughout the day.

In many ways, northern redwood forests are ecologically more similar to the temperate rain forests of the Oregon and Washington Coast Ranges and the Olympic Peninsula than to redwood forests to the south. Northern redwood forests are mainly conifer dominated, with redwood mixing with Douglas-fir, grand fir, Port-Orford-cedar, Sitka spruce, western hemlock, western redcedar

Box 3.1. Classification of Redwoods

A vegetation classification can be seen as a language created to organize and communicate how plants vary across the landscape (Whittaker 1978; Pickett and Kolasa 1989). Classifications can be casual or formal; each one reflects a particular scale of description and the philosophies and purposes of the people who created it. The redwood forest has long been recognized as a special kind of vegetation (Whitney Survey 1865; Jepson 1910; Shantz and Zon 1924). The Save-the-Redwoods League was created to preserve redwood forests—but what exactly is to be preserved?

An ideal classification of redwood forests would have spatial and temporal levels that thoroughly describe the variation within the forests. Every stand is unique in its exact combination of plants and animals, as well as in its physical environment and history. Recurring patterns of similarities and differences seen in a sample of stands provide a basis for classifying them (Braun-Blanquet 1932/1951; Gauch 1982).

If descriptions were taken in a precise and consistent way, we would notice that no two stands are identical in all ways. Species composition (also referred to as *floristics*) has two components: (1) species presence (a list of species); and (2) some measure of each species' relative importance in the stand. A measure commonly used for describing importance is crown cover, as it can be evaluated both on the ground and aerially (Mueller-Dombois and Ellenberg 1974). Structure, also referred to as *physiognomy*, refers to plant density, height, and growth form, that is, the way plants are distributed horizontally and vertically within the stand (Dansereau 1957). Structure can be evaluated for the stand as a whole (e.g., basal area of all trees) or by species (e.g., redwood seedling cover) (Bonham 1989).

The most useful classifications of vegetation are based on field data and explicit, consistent rules (Sokal 1974). For example, what is meant when someone decides that a given stand fits the class "redwood forest?" Often this determination is based on personal judgment using unspecified criteria. How many redwoods are necessary for a stand to be a redwood forest and not some other kind of forest? If all the trees in the stand are redwood, then there would be little question, especially if the trees are sufficiently close to each other that their canopies touch. But what if the stand is a mixture of Douglas-fir, redwood, and western hemlock; a mixture of madrone, redwood, and tanoak; or a mixture of California bay, Douglas-fir, and redwood trees? Are all these redwood forests?

Two basic ways are available for obtaining vegetation data and classifying them: remote sensing and ground-based methods (Kent and Coker 1992). Aerial methods are associated with vegetation mapping, whereas ground-based sampling methods are often associated with the creation of classifications. However, vegetation mapping units (the individual patches depicted on a map) and vegetation classification units (the vegetation type described at a given hierarchical level of a classification) are not necessarily one and the same (Küchler 1967, 1975). A vegetation map is a symbolic representation of visually distinct groupings of plants. A classification can be more descriptive, because it can involve more compositional and

(continues)

Box 3.1. *(continued)*

structural details, but is not designed to show the extent and range of a vegetation unit. Nevertheless, the best vegetation maps are presented in the context of a classification.

In this book, we consider all stands in which redwood is a sole, dominant, or important component in the tree canopy. An existing classification (Sawyer and Keeler-Wolf 1995) based on cover of current vegetation and UNESCO rules (Mueller-Dombois and Ellenberg 1974) is consistent with national efforts, such as those of The Nature Conservancy (Grossman et al. 1994), the Ecological Society of America, and the USDA Forest Service (Bailey 1995, 1996):

Formation class: Closed forests
    Formation subclass: Mainly evergreen forests
        Formation group: Temperate and subpolar coniferous forests
            Formation: Evergreen giant coniferous forests
                Series: Redwood forests
                    Association: Redwood/oxalis forest

Here, the series—representing all stands in which species in the most extensive layer (ground, shrub, or tree) have similar cover—is the basic unit of classification. Subseries may be used to account for variation in canopy composition. In other cases, tree canopies may have a consistent makeup among stands, but the shrub or ground layer compositions vary. This variation can be handled by considering a series to be composed of subtypes called "associations" (Daubenmire 1968; Allen 1987), one example of which (Redwood/oxalis forest) is shown above.

In this chapter, redwood forests are considered geographically according to the U.S. Forest Service's ECOMAP classification (Miles and Goudey 1997) (see fig. 3.1), within which classifications based on floristics can be embedded and redwood stands can be mapped. Remote sensing from satellites (U.S. Landsat, French SPOT, Indian Remote Sensing—IRS) and aircraft (photography, video, digital imagery), coupled with geographic information systems (GIS), provides a convenient means to map the distribution of vegetation across landscapes.

Vegetation mapping at the series level requires identification of the dominant species with the highest cover. Some series characterized by leaf shape (needleleaf or broadleaf) may be mapped from satellite remote sensing, but many of these mapping units actually may be species mixtures. Most authorities agree that aerial mapping at the series level is possible only from aircraft, and not all overstory species differences are distinguishable even then. Mapping at the series level, therefore, requires a combination of remote sensing and ground based mapping techniques. Ground information may come from site visits and from previously developed relationships between species distributions and such ecological parameters as geographic region, elevation, aspect, and geology. Ground-level studies are essential for mapping and describing vegetation at the association level. Furthermore, although forest age is often distinguishable by remote sensing, at least in terms of canopy closure and tree size, adequate characterization of old-growth stands also requires ground surveys (see chap. 4).

Because redwood forests are the focus of conservation efforts not just for trees, but for wildlife and other ecological components, it is important not only to understand the bases of classifications, but also their uses. Characterization of vegetation is often used as a proxy for ecological condition, such as with the Wildlife-Habitat Relationship approach developed for California (Mayer and Laudenslayer 1988). In this system, for example, the presence of large redwoods and other trees, along with a dense canopy, is considered an indicator of the presence of animals such as fisher, marbled murrelet, marten, and spotted owl (see chap. 5).

Extensions of this approach include designation of stream habitat quality based on canopy characteristics, soil type, or watershed characteristics. On a local scale, the quality of the stream and soil environments may be significantly different from what is predicted by general habitat models. Stream channels may be altered for flood prevention; old-growth stands may lack downed wood because they were managed to be "parklike"; or fire exclusion may have changed habitat structure and species composition in the ground and understory layers. Such site characteristics are critical for understanding the habitat quality of local areas for wildlife and can be incorporated into classifications when appropriate.

Table 3.1. Thirty-Year Normal Annual Temperature and Annual Precipitation for Eight California Stations Associated with Redwood Forests Arranged in a North-South Gradient, 1961–1990

Station	Location	Elev. (m)	Annual Temperature (°C)			Annual Precipitation (mm)		
			High	Mean	Low	High	Mean	Low
Klamath Del Norte Co.	lat. 41°31'N long. 124°31'W	6	18.3	15.6	12.6	3,159	2,003	1,124
Orick Humboldt Co.	lat. 41°22'N long. 124°01'W	50	18.5	15.0	11.8	2,798	1,679	868
Scotia Humboldt Co.	lat. 40°29'N long. 124°06'W	43	18.9	15.9	13.0	2,244	1,222	617
Richardson SP Humboldt Co.	lat. 40°02'N long. 123°47'W	150	20.6	16.5	12.1	3,247	1,742	756
Willits Mendocino Co.	lat. 39°25'N long. 123°20'W	412	20.7	15.7	11.1	2,326	1,261	547
Anguin Napa Co.	lat. 38°35'N long. 122°29'W	555	20.9	16.9	12.5	2,257	1,041	324
Kentfield Marin Co.	lat. 38°17'N long. 122°02'W	6	21.5	17.3	13.2	2,397	1,231	523
San Gregorio San Mateo Co.	lat. 38°17'N long. 122°22'W	82	19.2	16.1	12.7	1,467	717	341

Note: High temperature = highest mean annual maximum temperature. Low temperature = lowest mean annual minimum temperature. High precipitation = highest precipitation total for a single year. Low precipitation = lowest precipitation total for a single year during the period.

and/or tanoak (Zinke 1988). Western hemlock's continuous range defines the extent of northern redwood forests. The shrub layer may or may not be dense with black huckleberry, rhododendron, salal, or salmonberry. The ground layer may be composed of a few herbaceous plants or be largely cloaked with sword fern.

In a west-to-east transect from the Pacific Ocean, the coastal bluffs are generally covered with coastal scrub and salt-pruned individual redwood or Sitka spruce trees (Olive et al. 1982). The coastal, fringing stands of trees are also commonly dominated by Sitka spruce. More inland and to the east, stands are first mixes of redwood, western hemlock, and other conifers and then of redwood, Douglas-fir, and tanoak. In this transect, the eastern boundary of redwood forests is rather abrupt where they meet the forests, chaparral, and grassland on ultramafic parent materials (e.g., serpentine) of the Klamath region. Elsewhere, northern redwood forests merge into Douglas-fir, Douglas-fir–tanoak, or western hemlock forests, and in a few instances quickly into Oregon white oak woodlands or grassland.

Subsections of the Northern Redwood Forests

The Crescent City plain and Humboldt Bay flats and terraces are landscapes of Quaternary marine and nonmarine terraces with substantial deposits of sand and alluvium (Miles and Goudey 1997). Forests of beach pine, bishop pine, and Sitka spruce are more common than redwood stands. Northern redwood forests occur only locally on the well-aerated terrace and alluvial flats. Forests in this subsection mix with grassland, coastal scrub, and sand dune vegetation.

The northern Franciscan, Wiregrass Ridge, and central Franciscan landscapes are north-trending terrains that differ geologically from each other, but are similar ecologically (Miles and Goudey 1997). They are stretches of low mountains with rounded ridges, steep to moderately steep slopes, and narrow canyons. Mass wasting, predominately as debris flow on steep terrain, and slide-and-flow processes on moderately steep slopes lead to a high frequency of landslides. Variously metamorphosed rocks of the Franciscan Formation are most extensive. Most soils in these three landscapes support northern redwood forests, with forests of varying ages covering most of the landscape. East of redwood's range, soils support Douglas-fir–tanoak forests, Oregon white oak woodlands, and grassland. Western hemlock reaches its contiguous range limit in the central Franciscan landscape.

Examples of Northern Redwood Forests

S. Sillett and his students established a 0.5-ha plot near Cal Barrel Road and five 0.1-ha stands along Godwood Creek in Prairie Creek Redwoods State Park. Tree location and size (i.e., height, trunk diameter, and crown radius) were recorded for trees more than 10 cm diameter at breast height (dbh). The Cal Barrel Road stand has a diverse canopy; yet, redwood accounts for 72 percent of the trees (107/ha) and 90 percent of the basal area (279 m²/ha). Other tree species include Douglas-fir (6/ ha, 8 m²/ha), western hemlock (10/ha, 8m²/ha), California bay (8/ha, 11 m²/ha), cascara (2 /ha, 0.3 m²/ha), and hazel (4/ha, 0.06 m²/ha). The canopy exhibits multilayered stratification (fig. 3.2). Broad-leaved trees and young conifers, especially western hemlock, occupy the lower layer. Redwood, Douglas-fir, and western hemlock occupy the middle to upper layers; a few tall (more than 91 m) redwood trees are emergent. Dense thickets of sword fern or salmonberry occupy the ground layer with little recruitment of redwood.

The Godwood Creek plot was selected in an area with tall Sitka spruce as well as redwood trees, to compare the crown architecture and epiphyte assemblages of these two species. The 0.1-ha plots are too small to permit a reliable evaluation of vertical stratification, but several patterns in stand-level canopy structure are evident along the inland-coastal gradient. The more sheltered, inland alluvial terraces support the tallest trees (110 m), and several other redwoods approach 105 m in height. Redwood height steadily diminishes toward the ocean. Sitka spruce trees of uniform height (more than 90 m tall) are present throughout the gradient. The world's tallest known western hemlock tree (78 m) and several tall Douglas-fir (more than 90 m) also occur here. Seedlings and saplings of redwood, Sitka spruce, and western hemlock are abundant, especially on nurse logs as well as within tall redwood tree crowns. All size classes of Port-Orford-cedar are prevalent, but the distribution of this restricted and imperiled conifer is quite patchy.

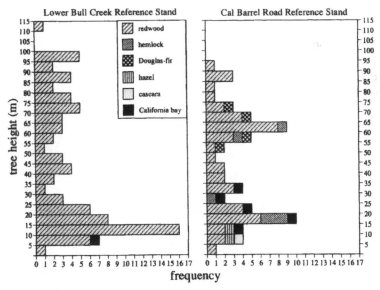

Figure 3.2. Frequency distribution of tree heights in two 0.5-ha reference stands containing old-growth redwood forest (see text for details).

Olive et al. (1982) mapped three forest communities in Del Norte, Jedediah Smith, and Prairie Creek Redwoods State Parks. The redwood type occupies alluvial flats and experiences repeated flooding and deposition of sediments. The Redwood–Douglas-fir–western hemlock type is found in sheltered drainages close to the coast. The Redwood–Douglas-fir type inhabits upland slopes and ridgetops. Madrone increases in importance inland and at higher elevations.

Lenihan (1986) sampled eighty stands in the Little Lost Man Creek Research Natural Area of Redwood National Park, Humboldt County. He described three forest associations that follow the same topographic gradient as those of Olive et al. (1982). A Redwood/deer fern association occupies the lower slopes and a Redwood/littleleaf Oregon-grape association occupies mid-slopes; on both slope positions Douglas-fir is nearly codominant with redwood. The Redwood/madrone association covers convex slopes and ridges; redwood and Douglas-fir tower over a lower tier of tanoak and madrone.

The studies by Olive et al. (1982) and Lenihan (1986), despite differences in methodology, agree in their descriptions of vegetation pattern. The terrace stands (Olive's redwood type) differed from those of uplands; Little Lost Man Creek has few terraces. The low and mid-slopes (Olive's Redwood–Douglas-fir–western hemlock type) were further divided based on differing understory composition (Lenihan's Redwood/deer fern and Redwood/littleleaf Oregon-grape associations). Both studies recognized the change in overstory tree composition along a gradient to the convex slopes and ridges, where Olive's Redwood–Douglas-fir type (= Lenihan's Redwood/madrone association) was found.

Allgood (1996) compared species composition of old-growth and harvested stands in the Yurok Experimental Forest, Del Norte County. Cluster analysis of thirty-six plots divided the formerly clear-cut stands from the old-growth stands, with the old growth having lower tree canopy cover and higher shrub cover. Selection-cut plots were similar to old-growth plots in overall shrub cover. Clear-cut plots resembled old-growth plots in the sapling layer.

Muldavin et al. (1981) and Lenihan et al. (1983) classified second-growth forests on logged lands acquired in the 1978 expansion of Redwood National Park. Both studies found a relationship between vegetation and the extent of disturbance. The general trend after timber harvest was a shift from short-lived annual or biennial herbs to shrub dominance within three years. The ten shrub-dominated types identified in these studies fell into two broad groups: remnant plants from the former stand covered the less-disturbed areas, and invasive shrubs dominated the more disturbed areas. As expected, plant associations within the low-disturbance, remnant group were arranged by site in a similar way to Lenihan's (1986) old-growth types. Plant associations within the invasive shrub group were assorted by moisture (dry, mesic, hydric sites) and by the character of the disturbance (burned, unburned, aerially seeded areas).

Muldavin et al. (1981) attempted to predict long-term, second-growth stand dynamics and suggested rehabilitation methods for accelerating development of old-growth structure and composition. The density ratio of canopy redwood to canopy Douglas-fir trees ranges from 10:1 to 3:1 in current old growth. The situation is reversed in current second-growth forests in Redwood National Park, an artifact of aerially seeding Douglas-fir. A linear regression analysis suggested a 1:1 ratio after stands reached 100 years of age, and a 2:1 ratio of redwood to Douglas-fir after 250 to 300 years. This analysis suggests that early thinning of Douglas-fir will produce stands at 60 years of age that have density ratios similar to those of old growth.

Central Redwood Forests

The central redwood forests are generally mixes of conifers and hardwoods with Douglas-fir, tanoak, and other hardwoods as common associates (Zinke 1988). In many ways, these forests are more similar compositionally and ecologically to the surrounding Douglas-fir–tanoak forests than to the northern redwood forests. The shrub layer is often dense with black huckleberry or salal. The ground layer usually has few herbs. River terraces differ greatly from the uplands in that the terrace stands are dominated by redwood, not mixes of conifers and hardwood trees. Shrubs are uncommon on terraces, and the ground is carpeted with redwood oxalis. Central redwood forests are regionally extensive.

In a west-to-east transect, the coastal bluffs are covered with coastal scrub and prairie or with stands of Sitka spruce, grand fir, or western hemlock. Inland stands are mixes of Douglas-fir, redwood, and western hemlock (Westman and Whittaker 1975). Farther to the east, stands are mixes of Douglas-fir, redwood, tanoak, and other hardwoods. The eastern boundary of redwood forest is broad, as redwood-dominated forests gradually change into Douglas-fir–tanoak forests, or— in a few instances—quickly into Oregon white oak woodlands or grassland.

The climate of the central redwood forests can be characterized by 1961–1990 normals from Scotia, Richardson Grove State Park, Willits, Anguin, and Kentfield stations (table 3.1; NOAA 1997). The winter, early spring, and late fall are usually cool and wet. Annual precipitation may exceed 2,500 mm in a few areas east of Cape Mendocino. Snow is uncommon, falling mainly in late winter and early spring and quickly melting. Summers are characterized by warm or hot sunny days. Foggy summer days are common along the coast, but inland fog is local and occurs mainly in the early morning.

Subsections of the Central Redwood Forests

Redwood forests extend over much of the central Franciscan subsection. The terrain is one of steep, low-to-high mountains, with King Peak, just south of Cape Mendocino, exceeding 1,230 m. The mountains of this subsection have rounded ridges, steep to moderately steep slopes, and narrow canyons. Mass wasting, predominately as debris flows and slides on the steep terrain and flow on moderately steep areas, makes for frequent landslides. The Van Duzen, Eel, Navarro, and Russian Rivers have substantial alluvial plains.

Most of the soils in the Franciscan subsection are derived from Mesozoic Franciscan Formation rocks, although those in the north coastal portion are derived from Cretaceous marine sedimentary rocks. Redwood is lacking on soils derived from Cretaceous rocks despite high precipitation and summer fog. Instead, Douglas-fir–tanoak, grand fir, or Sitka spruce forests, coastal scrub, and grassland are found. Zinke (1988) hypothesized that climate is responsible for this pattern. Steep slopes and thin soils with low water-holding abilities, however, may be a better explanation for redwood's absence here. To the east and south, central redwood forests are extensive, but on increasingly drier sites they change into Douglas-fir–tanoak forests, Oregon white oak woodlands, grassland, and chaparral. Redwood stands of the Eel River terraces support particularly massive trees.

The Fort Bragg terraces and Point Arena subsections are composed of Pleistocene and Cretaceous marine sedimentary rocks of the coastal plain that have been uplifted five times to form a set of terraces. Soils of the higher terraces are unsuitable for redwood and instead support stunted beach pine, bishop pine, or pygmy cypress stands. Soils are nutrient-deficient and are alternately saturated or desiccated in the rooting zone of plants (Jenney et al. 1969; Zinke 1988). Redwoods in these subsections grow in ravines, where mass wasting is common (Westman and Whittaker 1975). They also grow on the dunes and bluffs between each terrace and usually on the second terrace.

The coast hills–Santa Rosa Plain subsection is one of broad valleys and associated hills of Pliocene and Quaternary marine and nonmarine sediments and recent alluvium of the Cotati-Russian River valley. Stands of redwoods are localized. More commonly, oak-dominated forests and woodlands mix with grassland. As in southern Humboldt County, younger geologic strata generally lack redwoods.

Two additional subsections have a complex mosaic of redwood forests, Douglas-fir–tanoak forests, oak forests and woodlands, or chaparral and grassland. The Marin hill and valley subsection is an area of mountains with rounded ridges, steep to moderately steep slopes, and narrow canyons. Mount Tamalpais at 800 m dominates the area. Although most of the soils are derived from Franciscan Formation rocks, local areas of ultramafic rocks exist. Stands of redwoods in southern Marin County are most extensive on Mount Tamalpais.

The Mount St. Helena lava flows and valley subsection is one of Pliocene age volcanic flows or Franciscan and Cretaceous sedimentary rocks often associated with ultramafic materials. These rocks underlie three parallel ranges of low mountains separated by valleys. The most extensive stands of redwoods in this area grow in the mountains between Sonoma and Napa valleys.

Examples of Central Redwood Forests

Waring and Major's (1964) gradient analysis of redwood forest species in relation to water, nutrients, light, and temperature established a basis for contemporary research on central redwood forests. Their findings promote understanding of central redwood forests at a finer scale. Matthews (1986), after analyzing 124 plots taken over 4,000 ha, recognized five kinds of forests. The most distinctive type was the Redwood/redwood oxalis association on alluvial terraces. For years, ecologists and managers have informally differentiated these "park quality" alluvial redwood forests from upland stands.

S. Sillett and his students established a permanently marked 0.5-ha plot in old growth along lower Bull Creek in Humboldt Redwoods State Park. Tree location and size (i.e., height, trunk diameter, and crown radius) were recorded for trees more than 10 cm dbh. This stand is overwhelmingly dominated by redwood, which accounts for 99 percent of the trees (143/ha) and 99.98 percent of the basal area (271 m²/ha). California bay makes up the remainder. Redwood trees are evenly distributed continuously through the stand's vertical profile except for a large number of shorter trees 10–20 m tall (fig. 3.2). One clearly emergent tree is 10 m taller than any other tree in the stand. Extensive recruitment of redwood, following the 1955 and 1964 floods, has resulted in patches of 30–40-year-old juveniles up to 3 m tall (Sugihara 1992).

From the gradient analysis of Waring and Major (1964), it is difficult to appreciate the prompt topographic break between the flat terraces and the surrounding steep slopes. Just above the flat terraces, stands immediately change to a mix of Douglas-fir, redwood, and tanoak. Various names, such as "upland redwood forest," "mixed forest," or "redwood–Douglas-fir forest," have been used to distinguish these stands. The upland environment, however, is not as uniform as the terraces. Matthews (1986) differentiated four associations. The Redwood–Douglas-fir/salal association occurred on the lower slopes and raised benches with deep soils. The Redwood–Douglas-fir/black huckleberry association was found on open slopes with shallow soils. The Redwood–Douglas-fir/madrone association included stands on ridges and upper slopes with low cover in the shrub and ground layers.

The Heath and Marjorie Angelo Coast Range Preserve is a 3,200-ha old-growth reserve near Branscomb in Mendocino County, managed by the University of California and the USDI Bureau of Land Management. Much of the old growth is Douglas-fir–tanoak forest, but some stands have an important redwood component. Many ecological parallels exist between these and the upland redwood forests at Humboldt Redwoods State Park.

Young (40–100 yr), mature (100–200 yr), and old-growth (200–560 yr) stands at the Angelo Preserve differ in live and dead tree components (Bingham and Sawyer 1991). When young, forests are short (15–40 m) with a single-tiered canopy of California bay, black oak, canyon live oak, Douglas-fir, madrone, and tanoak; 30–75 percent of stand basal area is hardwood. Mature forests are up to

55 m tall with an indistinct two-tiered canopy; 5–45 percent of stand basal area is hardwood. Old-growth trees are more than 55 m tall with a distinct two-tiered canopy; 15–40 percent of stand basal area is hardwood. The Douglas-fir trees are twice the height of the tanoak trees and other hardwoods in old-growth stands. When redwood is present, the trees of the upper tier are taller and have more canopy cover. Fourteen years of change in these stands show little modification of tree species composition, with Douglas-fir and tanoak remaining codominants (Hunter 1997).

Southern Redwood Forests

Southern redwood forests occur along the central California coast. The stands are small to extensive, with redwood mixing with Douglas-fir and such hardwoods as California bay, coast live oak, and tanoak (Zinke 1988). Compositionally, ecologically, and genetically (see chap. 4) these forests are distinct from central redwood forests and especially from northern redwood forests. The shrub layer is sparse or lacking. The ground layer is generally sparse, except on wetter sites.

Redwood stands are one of several vegetation types that characterize the landscape in this section. Redwoods commonly are associated with raised stream terraces and lower slopes, especially those facing north (Henson and Usner 1993). Here, redwood trees should not be considered "riparian" because they cannot grow in soils that experience long periods of water saturation; instead, they are associated with normally dry and well-drained ravines. Boundaries between redwood forest and conifer, hardwood, and mixed forests, chaparral, or coastal scrub are abrupt. Clones of shrubby redwood stems grow on exposed, frequently burned sites.

The climate of the southern redwood forests may be characterized by 1961–1990 normals of the San Grogoria weather station (table 3.1; NOAA 1997). Winter, early spring, and late fall are generally cool and wet. Annual precipitation may exceed 1,000 mm on the western slopes of the coastal mountains. Winter snow is rare, and summers are cool and foggy.

Subsections of Southern Redwood Forests

The southern redwood forest section is characterized by parallel (northwest to southeast) ranges and valleys of folded and faulted strata with rounded crests of uniform heights. Cenozoic marine and nonmarine sedimentary rocks and alluvial deposits are extensive, although local areas of granitic and ultramafic rocks are also found. A few perennial streams and rivers in alluvial or weak bedrock channels flow directly to the Pacific Ocean. Dissected marine terraces are of limited extent along the coast, with recent alluvium in narrow flood plains. Elevations range from sea level to 740 m.

This section is composed of twelve subsections, but extensive tracts of southern redwood forests occur in only two: the Santa Cruz Mountains and northern Santa Lucia Mountains west of the San Andreas fault (Miles and Goudey 1997). Redwood forests in these subsections mix with Douglas-fir–tanoak forests, oak forests or woodlands, chaparral, coastal scrub, and grassland in a complex pattern. Redwood is found on western, ocean-facing sides of low-to-high mountains with rounded ridges, on steep to moderately steep slopes, and in narrow canyons. Redwood stands are confined to the slopes of the larger drainages and to the bottoms of many of the smaller canyons.

Examples of Southern Redwood Forests

Borchert et al. (1988) studied redwood stands in the larger drainages and many of the smaller drainages on the western slopes of the Santa Lucia Mountains in Monterey County. The sample involved 160 plots in deep, well-shaped canyons, ravines, gulches, and inner gorges near springs, seeps, and sheltered coves. Redwoods had been selectively harvested from these areas in the 1880s. Six plant associations (called "ecological types") were recognized. Four were along terraces or streamsides, or on mesic alluvial-colluvial fans; two associations were on uplands.

The redwood/bracken–chain fern, redwood/sword fern–trillium, and redwood associations are low-elevation types that differ in species diversity and cover of shrubs and herbs (note: association names have been shortened by removing accompanying soil names). The first two associations have a moderate cover of redwood oxalis, compared to the redwood association, which has few or no understory plants. Near-stream terraces at higher elevations support a redwood–bigleaf maple/California polypody association. The redwood/manroot–vetch association consists of nearly pure redwood stands of stunted, salt-spray trimmed redwood trees with canopies as short as 1 m. The middle and lower convex slopes of the canyons and draws support a redwood–tanoak/round-fruited sedge–Douglas iris association. Tree canopies are similar in structure to those of upland associations in the central redwood forests, but lack the Douglas-fir component of the more northern types.

Outlier Stands

Disjunct stands south and east of the major range of redwood have long interested ecologists and geneticists (see fig. 3.1). The southernmost patch of redwood is 3 km south of Salmon Creek in the Santa Lucia Mountains, southern Monterey County (stands in northern San Luis Obispo Co. are thought to have been planted). Individuals and small groups of redwoods in the Santa Lucia Mountains occur with sycamore and California bay in stringers of vegetation in canyons up to 1,077 m in elevation (Veirs 1996). Redwood occurrences to the east of the main distribution are found in Alameda, Napa, and Del Norte Counties. Four stands, one of which is extensive, grow in the Oakland hills. Five stands grow in hills east of Napa, some in ravines on dry, east-facing slopes surrounded by chaparral. In Del Norte County, a disjunct stand grows in the Klamath Mountains some 20 km east of the main redwood distribution limit (Griffin and Critchfield 1972). In Oregon, two isolated stands occur along the Chetco River and one along the Winchuck River; they are the northernmost stands of redwood (Franklin and Dyrness 1988).

Redwood Flora

A *flora* by definition is a list of the plant species within a geographic area. The flora of any area of redwood forest will differ depending on its location and physical environment; it will include exotic species (box 3.2) as well as natives. The floristic variation of the understory of the redwood forest is reflected in the different habitats that the forests occupy, from coastal alluvial flood plains, to canyons, to young or narrow terraces.

Box 3.2. Exotic Plants in Redwood Forests

Exotic plants, often called "weeds," can have a profound effect on the flora of a redwood forest. Although redwood is naturally the dominant tree along much of the coastal fog belt of northern California, most of the old-growth redwoods have been logged; hence, most sites today are in younger stages of stand development. Disturbances, such as clear-cutting or selective logging, give weedy species ample opportunity to become established.

Logging reduces canopy closure, changing the understory from a shady habitat with small patches of light to open sunlight. This change, coupled with soil disturbance from yarding and road building, significantly affects plant success. Depending on the type of silviculture and how timber operators manage their operations, the understory can be affected to a small or very large degree. In central redwood forests native species, such as hairy manzanita and blue blossom, commonly dominate sites after logging. In coastal Mendocino County these species persist for approximately twenty years, often starting out shrub-like and becoming small trees. Douglas-fir and redwood commonly grow under these shrubs and overtake them after approximately twenty years.

Both Australian fireweeds are common immediately after logging on sunny disturbed sites. These species persist as occasional individuals as the canopy closes and native species dominate. Pampas grass is a problem in Jackson State forest on logged sites and can persist for years on clear-cuts, as well as roadsides. Once the forest canopy closes, this grass is killed, but it competes with seedlings of native trees early on. Blue gum, the eucalyptus often planted in the region, sometimes spreads from sites, encroaching into redwood forests.

Scotch and French broom are persistent problems along highways in the redwood region. Roadsides that are kept open for motorist visibility are ideal habitat for these nitrogen-fixing legumes. They also compete with seedlings and small sprouts of redwoods after clear-cutting, until the canopy closes. Controlling these species is difficult, as they are prolific seeders and sprout back from cut crowns. Less noxious species such as forget-me-not have naturalized along small, shady streams and other moist sites.

In Redwood National Park Cape ivy and Scotch broom, which often invade coastal grasslands, have been designated for control by park personnel. Tansy ragwort populations on coastal grasslands currently are being reduced by a bio-control beetle release program. Pampas grass poses the greatest problem in areas clear-cut prior to 1978. English and Cape ivy can invade old-growth stands (Fritzke and Moore 1998).

The following are weed species that occur in redwoods stands and are on the California Exotic Pest Council's list of Exotic Pest Plants of Greatest Ecological Concern in California (adapted from Randall et. al 1998).

Centaurea solstitalis	yellow star-thistle
Cotoneaster pannosus	cotoneaster
Cotoneaster lacteus	cotoneaster
Cortaderia jubata	pampas grass
Cytissus scoparius	Scotch broom
Eucalyptus globulus	blue gum eucalyptus
Genista monspessulanus	French broom
Hedera helix	English ivy
Rubus discolor	Himalayan blackberry
Senecio mikanioides (Delaria odorata)	Cape ivy (German ivy)
Ulex europaeus	gorse

The following are exotic species in redwood forests that occur on the California Exotic Pest Council's List B: Wildland Pest Plants of Lesser Invasiveness (adapted from Randall et. al 1998).

Centaurea calcitrapa	purple star-thistle
Cirsium arvense	Canada thistle
Cirsium vulgare	bull thistle
Erechtites minima	Australian fireweed
Erechtites arguta	Australian fireweed
Mentha pulgium	pennyroyal
Phalaris aquatica	Harding grass
Senecio jacobaea	tansy ragwort
Spartium junceum	Spanish broom
Vinca major	periwinkle

An Example Flora

An example of a redwood flora is provided in appendix 3.1, a checklist of 180 species of vascular plants, both herbaceous and woody, in Van Damme State Park, located in central coastal Mendocino County. The park was logged in the early 1900s. The redwood forest there covers the watershed of a small creek, Little River. Redwood trees dominate the slopes, ridges, and edges of the creek. Grand fir, hemlock, Douglas-fir and tanoak also occur. The park extends from the ocean to approximately 8 km inland. On the slopes, the understory is varied with redwood oxalis, sword fern, littleleaf Oregon grape, both species of redwood ivy, and both red and black huckleberry. Red huckleberry typically grows on old stumps in shady sites. Along the creek, deer fern, lady fern, sedges, and many species in the Saxifragaceae are abundant, including miterwort, bishop's cap, fringecups, sugar scoop, piggyback plant, and boykinia. Alum root grows on the canyon walls.

Several perennial herbs in the lily family are commonly seen during the spring months in the canyon. These include fairy false Solomon's seal, and slender false Solomon's seal. Slink pod and clintonia are more common on the flats and ridges. Trillium is common in both sites. The rare corn lily (false hellebore) occurs in moist sites.

Several grass species occur occasionally in ruderal sites in the redwood forest. Some occur in open "grassy areas or meadows" in the landscape mosaic. Vanilla grass is the most common grass, commonly occurring as an understory species growing through redwood leaf litter. The fungi and lichens of redwood forests, generally not included in floras, are listed in appendices 3.2 and 3.3, respectively.

Rare Redwood Forest Species

Although the redwood forest is not known for its high number of rare plant species, habitats such as open, wet meadows and streamsides associated with redwood forests in the landscape mosaic harbor many species listed as endangered, threatened, or imperiled by state or federal agencies. Some rare species, such as *Campanula californica* (swamp harebell), occur in the heart of the redwood forest—in small seeps and other vernally moist sites, a completely different habitat from that of its relative *Campanula prenanthiodes* (bluebell), which occurs on canyon walls. *Lilium maritimum* (coast lily) often occurs in sunny boggy areas with Labrador tea, but is also found in apparently dry sites underneath redwood and golden chinquapin. The small orchid *Listera cordata* (twayblade) occurs in shady sites in rich duff. *Veratrum fimbriatum* (corn lily or false hellebore) grows in wet sites in the shade of redwoods. Altogether, fifty-seven rare and endangered vascular plant species have been found in redwood forests and associated habitats in Mendocino, Humboldt, and Del Norte Counties (appendix 3.4). Another 134 rare species (ranked by the California Natural Diversity Data Base as imperiled at global or state scales) occur elsewhere in the redwood region in redwood forests or other habitats (appendix 3.5).

Redwood Canopy Communities

Redwood and other forests have been studied primarily at ground level. Lately, however, tropical and temperate forest ecologists have become interested in the biology of forest canopies. Little is known about the canopy biology of redwood forests compared to other old-growth conifer forests in the Pacific Northwest. For example, old-growth Douglas-fir forests in western Oregon and Washington have been the focus of considerable canopy research since the early 1970s (e.g., Denison 1973; Pike et al. 1975, 1977; McCune 1993; Sillett 1995). In 1996, S. Sillett initiated studies of redwood canopies. This work has documented high frequencies of trunk "reiterations" high in the canopy, that is, main trunks snapping and sprouting, forming new trunks (Sillett 1999). This research also has uncovered complex communities of organisms associated with these canopy structures. (Crown structural dynamics and canopies of individual redwood trees are discussed in more detail in chap. 4.)

Canopy Soil

Huge quantities of redwood foliage are shed each year. Much of this litter accumulates on the surfaces of large branches at the bases of reiterated trunks, where it decomposes, eventually forming an organic soil. A great deal of precipitation is required to fully hydrate thick canopy soils. For example, deep soil mats in the Atlas Tree at Prairie Creek Redwoods State Park were still dry 10 cm below the surface after three days of heavy rain. Once hydrated, however, these soils lose water very slowly and may never completely dry out. The thickest (ca. 1 m deep) soil mat in the Atlas Tree, which was located at the top of the broken main trunk, was still dripping water from its base a week after a three-week-long rain. Redwood trees send adventitious roots into deep soil layers within their own crowns, exploiting a source of water and nutrients far removed from the forest floor.

Examination of canopy soils in Prairie Creek Redwoods State Park is revealing a rich biota. Sporocarps of saprophytic fungi are frequently observed on this soil during the rainy season. Several species of vascular plants become rooted in the material, their growth contributing to soil formation. Animals collected in canopy soils include beetles, crickets, earthworms, millipedes, mollusks, and amphibians. The clouded salamander is often abundant in thick soil mats, especially those supporting colonies of leather fern. This salamander, which has been found nesting up to 40 m high in redwood canopies, appears to breed in canopy soil; egg masses and juveniles have been observed. Tall Sitka spruce trees in Prairie Creek Redwoods State Park exploit canopy soils beneath thick moss mats with adventitious roots emerging directly from branches high in their crowns. Like their terrestrial roots, Sitka spruce canopy roots are associated with ectomycorrhizal fungi.

Epiphytes

A wide variety of vascular plants occur as epiphytes in old-growth redwood forest canopies (Sillett 1999). These epiphytes include ferns, such as leather fern, licorice fern, sword fern; the "spike moss" festoon; shrubs, such as black huckleberry, red huckleberry, trailing black currant, hairyleaf manzanita, and salal; and trees, such as cascara, tanoak, California bay, Sitka spruce, western hemlock, Douglas-fir, and redwood itself. The licorice fern is less frequent on redwood than other conifers because it requires thick moss mats, which are less common on redwoods than on other trees. The leather fern can be abundant on redwood trees, its mats covering several square meters and containing more than 2,000 fronds (Sillett 1999). The sword fern is rarely encountered as an epiphyte, and then only on thick soil mats in lower layers of the canopy. Festoon occasionally occurs on redwood in high-rainfall areas of Jedediah Smith Redwoods State Park, where it forms mats up to 4 m long with its curtainlike stems hanging 2 m or more below tree branches. Several dozen salal and huckleberry shrubs can occur in a single tree crown, often growing from fissures in trunks lacking canopy soil (Sillett 1999). They probably rely on mycorrhizal fungi to supply water and nutrients from decaying wood. These ericaceous shrubs are an important resource for arboreal animals in redwood forests; they are pollinated by insects (e.g., bumblebees) in the spring and produce abundant crops of berries in the fall. Where canopy soils are thick, epiphytic western hemlock plants can attain several meters in height; however, no other epiphytic trees have been observed more than a meter tall.

Large redwoods on alluvial terraces support nonvascular epiphytes, including mosses, liverworts, cyanolichens, alectorioid lichens, and other green algal lichens (see box 3.3). Nevertheless, redwood's thin, peeling bark retards accumulation of epiphytes. Thick bryophyte mats and cyanolichens, for example, are much less common on redwood trees than on other conifers in Prairie Creek Redwoods State Park. Thick bryophyte mats occur occasionally on redwood trees, but only on large branches, where the rate of bark sloughing presumably is much slower. The heaviest loads of epiphytic bryophytes on redwoods occur in high rainfall areas of Jedediah Smith Redwoods State Park. Here, large redwood branches are often covered with thick bryophyte mats and associated licorice ferns. Cyanolichens are rarely, if ever, abundant on redwoods, though they may be prevalent on adjacent Sitka spruce trees. On the other hand, alectorioid lichens (e.g., *Alectoria* and *Usnea*) and other green algal lichens (e.g., *Hypogymnia* and *Sphaerophorus*) are often abundant on redwood trees, perhaps even more than on other conifers.

Conclusions

The "redwood forest" is not monolithic. We are only beginning to appreciate the great variety of forest associations in which redwood occurs over its range. Regional patterns in these communities suggest three major types (sections)—northern, central, and southern redwood forests—which reflect major environmental and genetic (see chap. 4) gradients related to latitude and history. Within each of the three sections, landscape-level controls, especially those associated with parent material and longitude (i.e., distance from coast), appear to be

Box 3.3. Fungal and Lichen Diversity of Redwood Forests

Before 1986, amateur and professional mycologists identified 127 fungi species associated with redwood (Bonar 1942, 1965; Cooke and Harkness 1884–1886, 1886; Ellis and Harkness 1881; Farr et al. 1989; Fritz and Bonar 1931; Harkness 1884–1886a,b; Largent 1985, 1994; Phillips and Harkness 1884–1886; Smith 1947, 1951, 1952). Today, the number of fungal species known to be associated with redwood has risen to 320 (appendix 3.2). All major taxonomic groups of fungi are involved, though they are not equally represented. Fungi in the division Basidiomycota are best represented.

Collecting of fungi in redwood forests before 1989 occurred irregularly, almost always in the fall, and did not involve establishment of permanent plots. At that time, J. Langenheim established permanent plots throughout the range of redwood to study endophytic fungi. Nineteen species were identified from leaves and branchlets of young (1–10 years) trees (Espinosa-Garcia and Langenheim 1990; Espinosa-Garcia et al. 1996; Rollinger and Langenheim 1993). D. Largent in 1993 established three plots in Prairie Creek, one plot in Jedediah Smith Redwoods State Park, and one plot in Redwood National Park (Little and Largent 1993). To date, 295 species of fungi have been associated with redwood in these plots (Largent 1998).

To better understand fungal diversity in redwood forests, the following must be accomplished: (1) establish permanent plots throughout the range of redwood; (2) survey dying vascular plants, dung and insect frass, the rhizosphere surrounding plant roots, the fruiting bodies of fungi, the remains of invertebrates, mammals, and birds, and the surfaces of bryophytes, lichens, soil, and vascular plant roots and stems (Rossman 1994); (3) investigate the fungi of forest canopies; (4) document differences in species composition by season, substrate, and habitat over time (Frankland 1998); and (5) study collections deposited in botanical collections not cited in publications.

Lichens—fungi with either a symbiotic blue-green bacterium or a green alga as their photosynthetic partner—are also abundant in redwood forests. *Nostoc* is the most common blue-green symbiont, but 90 percent of lichen algae are green (Chlorophyta). *Trebouxia* or *Trentepohlia* are the most common green algae symbionts. *Trebouxia* is rarely found in the free-living state (Hale and Cole 1988).

Lichens have three common growth forms: a flat crustose form normally found on bark or rocks, a fruticose form that hangs from branches or grows up from the soil, and a foliose form that is leaflike and clings to branches, bark, and sometimes rock. A fourth type, squamulose, consists of small lobes that lack a lower cortex (outer skin) and rhizines (multicellular, rootlike structures) (Hale 1979). By far the most common lichen on redwood bark is the light green *Leproloma membranacea*. It covers the lower parts of the tree, giving the distinct impression that the trees have been washed with a green silt. A list of 93 species of lichens was compiled for the redwood forest along the Mendocino Coast (appendix 3.3) important. At a more local level, the distribution of redwood associations correlates well with topographic variation.

Additional studies in other areas of redwood's range are needed to characterize forest variation at the association level, especially because redwood appears more sensitive to microsite variation than most of its associated tree species (see chap. 4). The distribution of understory plants varies tremendously along environmental gradients, but complete floras are available for few sites. Better management and conservation of redwood forests will follow from more complete understanding of the natural history and ecology of these forests.

Not much is known about the canopy biology of redwood forests, but information is accumulating rapidly with recent studies. Old redwood canopies host a compositionally and structurally diverse biota, including not only nonvascular epiphytes but many vascular plants, including trees. Many species of invertebrates and vertebrates (e.g., clouded salamander) are found in canopy soils as well, forming a little-understood community.

Appendix 3.1. Preliminary Checklist of Vascular Plant Species Found in the Redwood Forest at Van Damme State Park, Mendocino County, California.

Compiled by Teresa Sholars.

Species	Common Name	Family
Abies grandis	grand fir	Pinaceae
Acer macrophyllum	bigleaf maple	Aceraceae
Achlys triphylla	vanilla leaf	Berberidaceae
Actea rubra	baneberry	Ranunculaceae
Adenocaulon bicolor	trailmarker	Asteraceae
Adiantum aleuticum	five-finger fern	Pteridaceae
Agrostis idahoensis	bent grass	Poaceae
Alnus rubra	red alder	Betulaceae
Anagalis arvensis	scarlet pimpernel	Primulaceae
Anaphalis margaritacea	pearly everlasting	Asteraceae
Anthoxanthum odoratum	sweet vernal grass	Poaceae
Aquilegia formosa	columbine	Ranunculaceae
Aralia californica	spikenard	Araliaceae
Arctostaphylos columbiana	hairy manzanita	Ericaceae
Asarum caudatum	wild ginger	Aristolochiaceae
Athyrium filix-femina	lady fern	Aspidaceae
Baccharis pilularis	coyote bush	Asteraceae
Bellis perennis	English daisy	Asteraceae
Berberis nervosa	littleleaf Oregon-grape	Berberidaceae
B. darwinii	barberry	Berberidaceae
Blechnum spicant	deer fern	Blechnaceae
Boykinia elata	boykinia	Saxifragaceae
Bromus vulgaris	brome	Poaceae
Calamagrostis bolanderi	reed grass	Poaceae
Calypso bulbosa	calypso	Orchidaceae
Campanula prenanthoides	bluebell	Campanulaceae

Species	Common Name	Family
Cardamine oligosperma	yellow cress	Brassicaceae
C. californica	toothwort	Brassicaceae
Carex obnupta	sedge	Cyperaceae
Castilleja wightii	Indian paintbrush	Scrophulariaceae
Ceanothus thrysiflorus	blue blossom	Rhamnaceae
Chimaphila umbellata	pipsissewa	Ericaceae
Chrysosplenium glechomaefolium	golden saxifrage	Saxifragaceae
Chrysolepis chrysophylla	golden chinquapin	Fagaceae
Claytonia perfoliata	miner's lettuce	Portulacaceae
C. siberica	Indian lettuce	Portulacaceae
Cirsium vulgare	thistle	Asteraceae
Clintonia andrewsiana	clintonia	Liliaceae
Conium maculatum	poison-hemlock	Apiaceae
Coptis laciniata	goldthread	Ranunculaceae
Corallorhiza maculata	spotted coralroot	Orchidaceae
C. striata	striped coralroot	Orchidaceae
Coronopus didymus	wart-cress	Brassicaceae
Corylus cornuta	hazel	Betulaceae
Cynoglossum grande	hound's tongue	Boraginaceae
Cyperus bipartitus	nut-sedge	Cyperaceae
Dactylis glomerata	orchard grass	Poaceae
Dicentra formosa	bleeding heart	Fumariaceae
Disporum hookeri	fairy-bells	Liliaceae
Disporum smithii	fairy-bells	Liliaceae
Equisetum telmateia	horsetail	Equisetaceae
Epilobium brachycarpum	fireweed	Onagraceae
Erechtites minima	Australian fireweed	Asteraceae
Euonymus occidentale	burning bush	Celastraceae
Euphorbia peplus	petty spurge	Euphorbiaceae
Festuca californica	California fescue	Poaceae
F. elmeri	fescue	Poaceae
F. idahoensis	blue bunchgrass	Poaceae
Festuca occidentalis	western fescue	Poaceae
Fragaria vesca	strawberry	Rosaceae
Garrya elliptica	silk tassel	Garryaceae
Gaultheria shallon	salal	Ericaceae
Geranium molle	cranesbill	Geraniaceae
Glyceria elata	fowl manna grass	Poaceae
Gnaphalium chilense	cudweed	Asteraceae
G. purpureum	cudweed	Asteraceae
Goodyera oblongifolia	rattlesnake plantain	Orchidaceae
Hedera helix	English ivy	Araliaceae
Hemitomes congestum	gnome plant	Ericaceae
Heracleum lanatu	cow parsnip	Apiaceae
Heteromeles arbutifolia	toyon	Rosaceae
Heuchera pilosissima	alum root	Saxifragaceae
Hierochloe occidentalis	vanilla grass	Poaceae
Holcus lanatus	velvet grass	Poaceae
Holodiscus discolor	cream bush	Rosaceae
Hypochoeris radicata	cat's ear	Asteraceae
Hypericum anagalloides	tinker's penny	Hypericaceae
Iris douglasiana	iris	Iridaceae

(continues)

Appendix 3.1. *(continued)*

Species	Common Name	Family
Juncus bufonius	toad rush	Juncaceae
J. effusus	green rush	Juncaceae
Juncus patens	blue-green rush	Juncaceae
Lamiumpurpureum	henbit	Lamiaceae
Lathyrus vestitus	pea	Fabaceae
Leontodon leysseri	hairy hawkbit	Asteraceae
Lithocarpus densiflora	tanoak	Fagaceae
Listera cordata	twayblade	Orchidaceae
Lonicera hispidula	twining honeysuckle	Caprifoliaceae
Lysichiton americanum	yellow skunk-cabbage	Araceae
Madia madioides	tarweed	Asteraceae
Maianthemum dilatatum	false lily-of-the-valley	Liliaceae
Marah oreganus	wild cucumber	Cucurbitaceae
Melilotus albus	white melilotus	Fabaceae
M. officinalis	yellow melilotus	Fabaceae
Mitella caulescens	mitrewort	Saxifragaceae
M. ovalis	bishop's cup	Saxifragaceae
Myosotis latifolia	forget-me-not	Boraginaceae
Myrica californica	wax-myrtle	Myricaceae
Navarretia squarrosa	navarretia	Polemoniaceae
Oenanthe sarmentosa	oenanthe	Apiaceae
Osmorhiza chilensis	sweet-cicely	Apiaceae
Oxalis oregana	redwood oxalis	Oxalidaceae
Pedicularis densiflora	Indian warrior	Scrophulariaceae
Petasites palmatus	coltsfoot	Asteraceae
Pinus muricata	bishop pine	Pinaceae
Pityrogramma triangularis	gold-backed fern	Polypodiaceae
Poa annua	annual bluegrass	Poaceae
Polygala californica	milkwort	Polygalaceae
Polygonum aviculare	common knotweed	Polygonaceae
Polypodium californicum	California polypody	Polypodiaceae
P. glycyrrhiza	licorice fern	Polypodiaceae
P. scouleri	leather fern	Polypodiaceae
Polystichum munitum	sword fern	Aspidaceae
Prunella vulgaris	selfheal	Lamiaceae
Pseudotsuga menziesii	Douglas-fir	Pinaceae
Pteridium aquilinum	bracken fern	Pteridaceae
Pyrola picta	pyrola	Ericaceae
Ranunculus californicus	buttercup	Ranunculaceae
R. repens	buttercup	Ranunculaceae
R. uncinatus	buttercup	Ranunculaceae
Rhamnus purshiana	cascara	Rhamnaceae
Rhododendron macrophyllum	rhododendron	Ericaceae
Ribes menziesii	canyon gooseberry	Grossulariaceae
Ribes sanguineum	red-flowering current	Grossulariaceae
R. bracteosum	stink current	Grossulariaceae
Rosa gymnocarpa	wood rose	Rosaceae
R. nutkana	Nutka rose	Rosaceae
R. parviflorus	thimbleberry	Rosaceae

Species	Common Name	Family
R. discolor	Himalaya blackberry	Rosaceae
Rubus leucodermis	black raspberry	Rosaceae
R. spectabilis	salmonberry	Rosaceae
R. ursinus	blackberry	Rosaceae
Rumex obtusifolius	bitter dock	Polygonaceae
R. crispus	curly dock	Polygonaceae
Salix scouleriana	Nuttall willow	Salicaceae
S. lucida	shining willow	Salicaceae
Sambucus racemosa	red elderberry	Caprifoliaceae
Sanicula crassicaulis	snakeroot	Apiaceae
Satureja douglasii	yerba santa	Lamiaceae
Scirpus microcarpus	bulrush	Cyperaceae
Scoliopus bigelovii	slink pod	Liliaceae
Scrophularia californica	figwort	Scrophulariaceae
Senecio jacobaea	tansy ragwort	Asteraceae
S. mikanioides (Delaireria odorata)	Cape ivy (German ivy)	Asteraceae
Senecio vulgaris	groundsel	Asteraceae
Sequoia sempervirens	redwood	Taxodiaceae
Smilacina racemosa	false Solomon's seal	Liliaceae
S. stellata	slender false Solomon's seal	Liliaceae
Sonchus asper	sow-thistle	Asteraceae
S. oleraceus	sow-thistle	Asteraceae
Stachys chamissonis	hedge-nettle	Lamiaceae
S. ajugoides	hedge-nettle	Lamiaceae
Stellaria media	chickweed	Caryophyllaceae
Synthyris reniformis	snow queen	Scrophulariaceae
Taraxacum officinalis	common dandelion	Asteraceae
Tellima grandiflora	fringecups	Saxifragaceae
Thermopsis macrophylla	false lupine	Fabaceae
Tiarella trifoliata	sugar-scoop	Saxifragaceae
Tolmiea menziesii	piggy-back	Saxifragaceae
Torreya californica	nutmeg	Taxaceae
Toxicodendron diversilobum	poison-oak	Anacardiaceae
Trientalis latifolia	star flower	Primulaceae
Trifolium dubium	clover	Fabaceae
T. repens	white clover	Fabaceae
Trillium ovatum	trillium	Liliaceae
Tsuga heterophylla	western hemlock	Pinaceae
Umbellularia californica	California bay	Lauraceae
Urtica dioica	stinging nettle	Urticaceae
Vaccinium ovatum	black huckleberry	Ericaceae
V. parvifolium	red huckleberry	Ericaceae
Vancouveria hexandra	redwood ivy	Berberidaceae
V. planipetala	redwood ivy	Berberidaceae
Veronica americana	brooklime	Scrophulariaceae
Veratrum fimbriatum	false hellebore	Liliaceae
Viola glabella	creeping violet	Violaceae
V. sempervirens	redwood violet	Violaceae
Whipplea modesta	modesty	Saxifragaceae

Appendix 3.2. Fungi Associated with Redwood.

Compiled by David Largent.

KEY. *Division or subdivision:* a = Ascomycota; b = Basidiomycota, m = Myxomycotina, o = Oomycota, z = Zygomycota. *Type of fungus:* e = endophyte, ec = ectomycorrhiza, p = pathogen, s = saprobe, en = endomycorrhiza, l = basidiolichen. *Substrate:* ba = bark, dl = dead leaves, dw = dead wood, ib = inside branchlets, og = on ground

Agaricus diminutivus (b, s, og)
A. semotus (b, s, og)
Albatrellus avellaneus (b, s, og)
A. caeruleoporus (b, s, og)
A. cristatus (b, s, og)
Alboleptonia sericella (b, s, og)
A. sericella var. lutescens (b, s, og)
Aleuria aurantia (a, s, og)
Amphisphaeria wellingtoniae (a, s, dw)
Amylostereum chailletii (b, s, dw)
Armillariella mellea (b, p, root rot)
A. ostoyae (b, p, root rot)
Asterophora lycoperdoides (b, p, on fruiting body)
Aureobasidium pullulans (d, s, wood in use)
A. sp. (d, e, ib)

Bolbitius vitellinus (b, s, og)
Botryosphaeria dothidea (a, e/p, stem)
Botrytis cinerea (d, e/p, ib)
B douglasii (d, p, seedling blight)
Brachysporiella tubinata (d, s, dw/leaves)

Callistosporium luteoolivaceum (b, s, og)
Camarophyllus foetens (b, s, og)
C. graveolens (b, s, og)
C. niveus (b, s, og)
C. pratensis v. pratensis (b, s, og)
C. russocoriaceus (b, s, og)
Caulorhiza umbonata (b, s, on roots)

Cercospora sequoiae (d, p, needle blight)
Ceriporiopsis rivulosa (b, p, w ring rot)
Chaetasbolisia falcata (d, p, on leaves)
Chloroscypha chloromela (a, e/p, branchlet blight)
Claudopus byssisedus (b, s, dw or ground)
Clavaria purpurea (b, s, og)
C. vermicularis (b, s, og)
Clavulina cinerea (b, s, og)
C. cristata (b, s, og)
Clavulinopsis corniculata (b, s, og)
C. helveola (b, s, og)
C. laeticolor (b, s, og)
Clithris sequoiae (a, p, on twigs)
Clitocybula aveillaneialba (b, s, dw)
Clitopilus prunulus (b, s, og)
Collybia butryacea (b, s, og)
C. confluens (b, s, og)
C. dryophila (b, s, og)
C. umbonata (b, s, og)
Colpoma angustata (a, s, dw)
Coniochaetae sp. (b, e, ib)
Conocybe tenera (b, s, og)
Conoplea fusca (d, e, ib)
Coprinus comatus (b, s, og)
C. micaceus (b, s, og)
Corticium sanguineum (b, s, dw)
C. umbrinum (b, s, dw)
Crepidotus applanatus (b, s, dw)
C. pubescens (b, s, dw)
C. sp. (b, s, dw)

C. variabilis (b, s, dw)
Cryptosporiopsis abietina (d, e/p, ib)
Cucurbitaria cononillae (a, e/p, ib)
Cystoderma fallax (b, s, og)
Cytispora pinastri (a, p, needle blight)
Cytolepiota sp. (b, s, og)

Dacrymyces palmatus (b, s, dw)
Didymium squamulosum (b, s, on leaves)
Diplomitoporus lenis (b, s, dw)
D. lindbladii (b, s, dw)

Entoloma bloxami (b, ec, og)

Fomitopsis meliae (b, s/p, wood in use)

Galerina allospora (b, s, om)
G. filiformis (b, s, om)
G. perangusta (b, s, om)
Gelatinosproium sp. (d, e/p, ib)
Geniculosporium sp. (d, e, ib)
Gloeophyllum carbonarium (b, s, dw)
G. odoratum (b, s, dw)
G. protractum (b, s, dw)
Glomus caledonium (z, en, m)
G. convolutum (z, en, m)
G. fasciculatum (z, en, m)
G. macrocarpum (z, en, m)
G. radiatum (z, en, m)
Gymnopilus rufescens (b, s, dw)
G. sanguinea (b, s, dw)
G. sp. (b, s, dw)
G. ventricosum (b, s, dw)
Gymnopus villosipes (b, s, dw)

Hemimycena sp. (b, s, dw)
Heterobasidion annosum (b, p, root rot)
Hohenbuehelia geogenum (b, s, dw)
Hygrocybe acutoconicus var. *microsporus* (b, s, og)

H. aurantiosplendens (b, s, og)
H. cantharellus (b, s, og)
H. coccineus var. *coccineus* (b, s, og)
H. coccineus var. *umbonatus* (b, s, og)
H. conicus (b, s, og)
H. flavescens (b, s, og)
H. helobius (b, s, og)
H. laeta (b, s, og)
H. laetissimus (b, s, og)
H. marchii (b, s, og)
H. minutulus (b, s, og)
H. mollis (b, s, og)
H. moseri (b, s, og)
H. psittacinus var. *psittacinus* (b, s, og)
H. puniceus (b, s, og)
H. singeri (b, s, og)
H. sp. (b, s, og)
H. subvitellinus (b, s, og)
H. virescens (b, s, og)
Hygrophoropsis aurantiacus (b, s, og)
Hymenochaete tabacina (b, s, dw)
Hymenoscyphus sphaerophoroides (a, s, dl)
Hypholoma capnoides (b, s, dw)
H. dispersum (b, s, og)
H. fasiculare (b, s, dw)
H. sp. (b, s, dw)
Hypocrea rufa (a, s, dw)
Hypoxylon bipapillatum (a, e, ib)
H. thouarsianum (a, s, dw)

Inocephalus concavus (b, s, og)
I. cystomarginatus (b, s, og)
I. minimus (b, s, og)
I. rhombisporus (b, s, og)

Lacrymaria velutina (b, s, og)
Laetiporus sulphureus (b, s, dw)
Lasiosphaeria canescens (a, s, dw)
Lepiota atrodisca (b, s, og)

L. brunneodisca (b, s, og)
L. clypeolaria (b, s, og)
L. cristata (b, s, og)
L. flammeatincta (b, s, og)
L. sp. (b, s, og)
L. nuda (b, s, og)
Leptonia acuto-umbonata (b, s, og)
L. anatina (b, s, og)
L. asprella (b, s, og)
L. atrosquamosa (b, s, og)
L. caesiocincta (b, s, og)
L. carnea (b, s, og)
L. carneogrisea (b, s, og)
L. chalybaea (b, s, og)
L. chalybaea var. *chalybaea* (b, s, og)
L. chalybaea var. *squamulosipes* (b, s, og)
L. decolorans forma cystidiosa (b, s, og)
L. decolorans forma decolorans (b, s, og)
L. diversa (b, s, og)
L. exalbida (b, s, og)
L. exilis (b, s, og)
L. foliocontusa var. *caeruleotincta* (b, s, og)
L. foliocontusa var. *discolor* (b, s, og)
L. foliocontusa var. *foliocontusa* (b, s, og)
L. formosa (b, s, og)
L. formosa var. *formosa* (b, s, og)
L. formosa var. *microspora* (b, s, og)
L. fuligineomarginata (b, s, og)
L. fuscata (b, s, og)
L. gracilipes (b, s, og)
L. lividocyanula (b, s, og)
L. microspora (b, s, og)
L. mougeotii (b, s, og)
L. nigrosquamosa var. *californica* (b, s, og)
L. parva (b, s, og)
L. pigmentosipes (b, s, og)

L. scabrosa (b, s, og)
L. separata (b, s, og)
L. serrulata (b, s, og)
L. sodalis (b, s, og)
L. sp. (b, s, og)
L. strictipes (b, s, og)
L. subnigra (b, s, og)
L. subrubinea (b, s, og)
L. subviduense var. *marginata* (b, s, og)
L. subviduense var. *subviduense* (b, s, og)
L. trichomata (b, s, og)
L. turci (b, s, og)
L. viridiflavipes (b, s, og)
L. xanthochroa (b, s, og)
L. yatesii (b, s, og)
Leptostroma sequoiae (d, p, twig blight)
Leucoagaricus rubrotinctus (b, s, og)
L. sp. (b, s, og)
Leucogyrophana mollusca (b, s, dw)
Lycoperdon perlatum (b, s, og)
Lyophyllum decastes (b, s, og)
L. sp. (b, s, og)

Macrophoma sp. (d, p, twig dieback)
Marasmiellus candidus (b, s, dw)
Marasmius androsaceus (b, s, og)
M. calhouniae (b, s, og)
M. fuscopurpurea (b, s, og)
M. plicatulus (b, s, og)
Melanotus phillipsii (b, s, dw)
Meruliporia incrassata (b, s/p, heartrot)
Micromphale sequoiae (b, s, on needles and stems)
Morchella conica (s, s, bw)
Mycena albissima (b, s, dw)
M. clavata (b, s, dw)
M. epipterygia (b, s, og)
M. filopes (b, s, og)

M. fusco-ocula (b, s, dl)
M. glaucopus (b, s, og)
M. haematopus (b, s, og)
M. iodiolens (b, s, og)
M. maculata (b, s, og)
M. oregonensis (b, s, og)
M. paucilamellata (b, s, on twigs)
M. pura (b, s, og)
M. rugulosiceps (b, s, og)
M. sanguinolenta (b, s, og)
M. scabripes (b, s, og)
M. sp. (b, s, og)
Mycosphaerella sequoiae (a, p, needle
   blight)
Myxomphalia maura (b, s, og)
Myxotrichum ochraceum (a, s, dw)

Naucoria sp. (b, s, og)
Nodulisporium sp. (d, e, ib)
Nolanea ameides (b, s, og)
N. bicoloripes (b, s, og)
N. clandestina var. *oculobrunnea* (b, s,
   og)
N. hebes (b, s, og)
N. hirtipes (b, s, og)
N. lucida (b, s, og)
N. minuto-striata (b, s, og)
N. odorata (b, s, og)
N. proxima forma inodorata (b, s, og)
N. pseudostrictia (b, s, og)
N. pusillipapillata (b, s, og)
N. sericea (b, s, og)
N. sp. (b, s, og)
N. staurospora (b, s, og)
N. staurospora var. *incrustata* (b, s, og)
N. strictia (b, s, og)
N. undulata (b, s, og)

Oligoporus sequoiae (b, p, heartrot)
Omphalina epichysium (b, s, dw)
O. ericetorum (b, s, og)
Oncospora abietina (d, s, dw)

Panus conchatus (b, s, og)
Paraleptonia scabrulosa (b, s, og)
Paxillus atrotomentosus (b, s, dw)
P. panuoides (b, s, dw)
Pestalotiopsis funerea (d, e/p, needle
   blight)
Pezicula livida (a, p, bark canker)
Peziza fusca ? (a, s, og)
P. gemmea ? (a, s, dl)
P. molesta ? (a, s, redwood chips)
P. pithya ? (a, s, bark)
P. ustorum ? (a, s, og)
Phanerochaete burtii (b, s, dw)
P. sanguinea (b, s, dw)
Phellinus ferruginosus (b, s, dw)
P. viticola (b, s, dw)
Phellodon atrata (b, s, og)
Phlyctaena sp. (d, e/p, ib)
Pholiota astragalina (b, s, dw)
P. malicola (b, s, og)
P. sequoiae (b, s, dw)
P. sublateritium (b, s, dw)
P. terrestris (b, s, og)
Phomopsis occulta (d, e, ib)
P. occulta (d, s, db)
P. sp. I (d, e/p, ib)
P. sp. II (d, e/p, ib)
Phyllosticta sp. (d, p, needle blight)
Phyllosticta-like sp. (d, e, ib)
Phymatotrichopsis omnivora (a, p, root
   rot)
Physalospora sp. (a, p, twig dieback)
Phytoconis ericetorum (b, l, lichen)
Pithya cupressina (a, s, dl)
Pleurocybella porrigens (b, s, dw)
Pleuroplaconema sp. (b, e, ib)
Pleurotellus sp. (b, s, dw)
Pleurotus ostreatus (b, s, dw)
Pluteus cervinus (b, s, dw)
P. salicinus (b, s, dw)
P. sp. (b, s, dw)
Podostroma alutacea (a, s, og)

Psathyrella hydrophila (b, s, og)
P. longistriata (b, s, og)
P. sp. (b, s, og)
Pseudohydnum gelatinosum (b, s, dw)
Pulcherrinum caeruleum (b, s, og)
Pyrenochaeta sp. (d, e, ib)
Pyronema omphalodes (a, s, dw)
Pythium sp. (o, p, root rot)

Ramularia lactea (b, e, dw)
Rarmariopsis kunzeii (b, s, og)
R. sp. (b, s, og)
Rectipilus fasiculata (b, s, dw)
Rhodocybe hondensis (b, s, og)
R. nuciolens (b, s, og)
R. trachyospora (b, s, og)
Rosellinia mutans (a, s, dw)

Schizophyllum commune (b, s, dw)
Schizoporia flaviporia (b, s, dw)
S. paradoxa (b, s, dw)
Seiridium juniperi (d, e/p, ib)
S. sp. (d, p, twig and branch canker)
Serpula hexagonoides (b, s, dw)

Skeletocutis amorpha (b, s, dw)
Spathularia flavida (a, s, og)
Sphaeria confertissima (b, s, c)
Stereum hirsutum (b, s, dw)
S. ostrea (b, s, dw)
Stictis radiata (a, s, dw)
S. versicolor (a, s, ba)
Stropharia ambigua (b, s, og)

Torula herbarium (d, e, ib)
Trametes versicolor (b, s, dw)
Tremella mesenterica (b, s, dw)
Trichaptum abietinum (b, s, dw)
Tricholomopsis decorata (b, s, dw)
T. rutilans (b, s, dw)
Trichopilus jubatus (b, s, og)

Verpa digitaliformis (a, s, og)

Xeromphalina campanella (b, s, dw)
X. cauticinalis (b, s, og)
X. fulvipes (b, s, dw)
X. orickiana (b, s, dw)

Appendix 3.3. Lichens of the Central Redwood Forest Region.

Compiled by Teresa Sholars.

KEY. bk = bark, bt = broadleaf trees, c = conifers, cbk = conifer bark, cbr = conifer branches, fp = fence posts, h = humus, l = logs, m = mosses, o = oaks, r = rocks, rw = rotting wood, s = soil, t = trees, tb = tree bases, tt = tree trunks, w = wood

Fruticose species

Alectoria imshaugii (cbk)
Alectoria sarmentosam (c)
Bryoria furcellata (cbk)
Cladina portentosa ssp. *pacifica* (s)
Cladonia cervicornis ssp. *verticillata* (s)
Cladonia chlorophoraea (r, s, tb)
Cladonia crispata (m, s)
Cladonia fimbriata (tb)
Cladonia macilenta (fp, tb, w)
Cladonia pyxidata (bk, m)
Cladonia subsquamosa (m, s, tb, w)
Cornicularia californica (cbk, cbr)
Evernia prunastri (fp, l, t)
Letharia columbiana (cbk)
Letharia vulpina (cbk)
Niebla cephalota (r, w)
Ramalina farinacea (cbk, cbr)
Ramalina menziesii (c, o)
Sphaerophorus globosus (c, s)
Usnea arizonica (bt, o)
Usnea californica (cbk)
Usnea rubicunda Vulpicida canadensis
  (cbk)

Crustose species

Arthonia cinnabarina
Arthothellium sp.
Candelaria concolor var. *effusa*
Caloplaca holocarpa (bk)
Dibaeis baeomyces (s)
Dimerella lutea (m, tt)
Diploschistes scruposas (r, s)
Graphis striatula
Lecidea sp.
Leproloma membranacea (bk, h, s)

Mycoblastus sanguinarius
Ochrelechia subpallescens (bk)
Pertusaria amara (bk)
Pertusaria pustulata (obk)
Thelomma californica (fp)

Foliose species

Cavernularia lophyrea
Flavoparmelia caperata (tt)
Flavopunctelia flaventior (o)
Hypogymnia imshaugii (c)
Hypogymnia inactive (cbk, cbr)
Hypogymnia occidentalis (c)
Hypogymnia physodes (bt)
Hypogymnia tubulosa (c)
Leptochidium albociliatum (m)
Leptogium palmatum (m, r, s)
Leptogium saturninum
Leptogium tenuissimun (s)
Lobaria pulmonaria (bt, m, o, s)
Lobaria scrobiculata (tt)
Massalongia carnosa (m)
Melanelia elegantula (tr, t)
Melanelia subaurifera (bt, o)
Melanelia subolivacea (cbk, cbr)
Nephroma helveticum (m, t)
Nephroma laevigatum (r, t)
Nephroma resupinatum (r, t)
Parmelia saxatilis (r, t)
Parmelia sulcata (r, t)
Parmelina quercina (o)
Parmotrema arnoldii (t)
Parmotrema crinitum (cbk)
Peltigera spp.
Peltigera canina (l, s)
Peltigera collina (h, m, r, s, tb)

Peltigera polydactyla (m, s)
Peltigera venosa (s)
Phaeophysciaciliata (btbk)
Physcia (o)
Physcia adscendens (r, t)
Physcia aipolia (t)
Physcia dubia
Physcia stellaris (obk)
Physcia tenella (bt, o)
Physconia distorta (bt, o)
Platismatia glauca (c)
Platismatia herrei (cbk, cbr)
Platismatia stenophylla (c)
Pseudocyphellaria anomola
Psuedocyphellaria anthrapsis (bt, o)

Sticta fulginosa (m, r, t)
Sticta limbata (bk, m)
Tuckermanopsis ornata (cbr)
Umbilicaria phaea (r)
Xanthoparmelia coloradensis
  (r, t)
Xanthoparmelia plittii
Xanthoria candelaria (bt, r)
Xanthoria elegans (r)
Xanthoria fallax (bk)
Xanthoria polycarpa (obr)

Squamulose species
Pannaria conoplea (bk, s)
Pilophoron aciculare (bk, r, s)

Appendix 3.4. Rare and Endangered Vascular Plants of the Redwood Region Found in Mendocino, Humboldt, and Del Norte Counties. Adapted from Skinner and Pavlik 1994.

KEY. *California Native Plant Society (CNPS) status:* 1A = presumed extinct in California; 1B = rare, threatened, or endangered in California and elsewhere; 2 = rare, threatened, or endangered in California but more common elsewhere; 3 = need more information; 4 = of limited distribution

California Natural Diversity Data Base (CNDDB) status: G1 = fewer than 6 viable element occurrences (EOs) or fewer than 1,000 individuals or less than 2,000 acres; G2 = 6–20 EOs or 1,000–2,000 individuals or 2,000–10,000 acres; G3 = 21–100 EOs or 3,000–10,000 individuals or 10,000–50,000 acres; G4 = apparently secure, yet factors exist to cause some concern; G5 = demonstrably secure

T codes signify status of taxonomic subcategory (subspecies or variety); *S codes* signify status within state (CA), with number after decimal point indicating level of threat (.1 = highest, .3 = lowest); X = extinct; H = historical, not seen for ≥ 20 years; range of values (e.g., S2S3) and ? signifies uncertainty about appropriate status

Federal and state status: FE = federal endangered, FT = federal threatened, FPE = federal proposed endangered, FPT = federal proposed threatened, FC = federal candidate, FSC = federal special concern; CE = California endangered, CT = California threatened, CP = California protected, CS = California sensitive, CR = California rare

CNPS habitat codes: BgFns = bogs and fens; BUFrs = broadleaved upland forest; CBScr = coastal bluff scrub; CCFrs = closed-cone coniferous forest; CmWld = cismontane woodland; Chprl = chaparral; CoDns = coastal dunes; CoPrr = coastal prairie; CoScr = coastal scrub; Medws = meadows and seeps; MshSw = marshes and swamps; NCFrs = North Coast coniferous forests; RpFrs = riparian forests; VnPls = vernal pools

Species	Status	Habitat code
Arctostaphylos mendocinensis (pygmy manzanita)	1B, G1/S1?	CCFrs
Boschniakia hookeri (small groundcone)	2, G5/S1S2	NCFrs
Calamagrostis bolanderi (Bolander's reed grass)	4, G3/S3.3	BgFns, CCFrs, Medws, MshSw
Calamagrostis crassiglumis (Thurber's reed grass)	2, G3/S1.1, FSC	CoScr, MshSw
Calamagrostis foliosa (leafy reed grass)	4, G3/S3.2, CR	CBScr, NCFrs

(continues)

Appendix 3.4. *(continued)*

Species	Status	Habitat code
Campanula californica (swamp harebell)	1B, G2/S2.2, FSC	BgFns, CCFrs, CoPrr, Medws, MshSw
Carex californica (California sedge)	2, G5/S2?	BgFns, CCFrs, CoPrr, Medws, MshSw
Carex livida (livid sedge)	1A, G5/SH	BgFns, MshSw
Ceanothus gloriosus var. *gloriosus* (Point Reyes ceanothus)	4, G5T3/S2.2	CBScr, CCFrs, CoDns, CoScr
Cupressus goveniana ssp. *pigmaea* (pygmy cypress)	1B, G2T1/S1.2, FSC	CCFrs
Cypripedium fasciculatum (clustered lady's slipper)	4, G4/S3.2, FSC	NCFrs
Cypripedium montanum (mountain lady's slipper)	4, G4/S3.2	BUFrs
Epilobium oreganum (Oregon fireweed)	1B, G2/S2.2, FSC	BgFns
Epilobium septentrionale (Humboldt County fuchsia)	4, G3/S3.3	BUFrs, NCFrs
Erigeron biolettii (streamside daisy)	3, G3/S3?	BUFrs, CmWld, NCFrs
Gentiana setigera (Mendocino gentian)	1B, G2/S1.2, FSC	Medws
Glyceria grandis (American manna grass)	2, G5/S1.3?	BgFns, Medws, MshSw
Hackelia amethystina (amethyst stickseed)	4, G3/S3.3	Medws
Horkelia bolanderi (Bolander's horkelia)	1B, G1/S1.2, FSC	Medws
Juncus supiniformis (hair-leaved rush)	4, G4?/S3.3	MshSw
Lasthenia burkei (Burke's goldfields)	1B, G1/S1.1, FE, CE	Medws, VnPls
Lilium kelloggii (Kellogg's lily)	4, G3/S3.3	NCFrs
Lilium maritimum (coast lily)	1B, G2/S2.1, FSC	BUFrs, CCFrs, CoPrr, CoScr, NCFrs
Lilium rubescens (redwood lily)	4, G3/S3.2	Chprl
Limnanthes bakeri (Baker's meadowfoam)	1B, G1/S1.1, FSC	Medws, MshSw
Linanthus acicularis (bristly linanthus)	4, G3/S3.2	Chprl, CmWld, CoPrr
Listera cordata (heart-leaved twayblade)	4, G5/S3.2	BgFns, NCFrs

Species	Status	Habitat code
Malacothamnus mendocinensis (Mendocino bush mallow)	1A, GX/SX, FSC	CmWld
Melica spectabilis (purple onion grass)	4, G5/S3.3	Medws
Microseris borealis (northern microseris)	2, G5?/S1.1	BgFns, Medws
Pinus contorta ssp. *bolanderi* (Bolander's beach pine)	1B, G5T3/S3.2, FSC	CCFrs
Piperia candida (white-flowered rein orchid)	4, G3/S3.3	NCFrs
Piperia michaelii (purple-flowered piperia)	4, G3?/S3.2	CBScr, CCFrs, CmWld
Pityopus californicus (California pinefoot)	4, G4/S3.2	BUFrs, NCFrs
Pleuropogon californicus var. *davyi* (Davy's semaphore grass)	4, G5T3/S3.3	CmWld, Medws
Pleurogpogon hooverianus (north coast semaphore grass)	1B, G1/S1.1, FSC	BUFrs, Medws, NCFrs
Pleuropogon refractus (nodding semaphore grass)	4, G4/S3.2?	Medws, NCFrs
Pogogyne douglasii ssp *parviflora* (Douglas' pogogyne)	3, G?/T3?Q/S3.2	Chprl, MshSw, VnPls
Potamogeton epihydrus ssp. *nuttallii* (Nuttall's pondweed)	2, G5T5Q/S2.2?	MshSw
Ranunculus lobbii (Lobb's aquatic buttercup)	4, G4/S3.2	CmWld, NCFrs, VnPls
Rhynchospora alba (white beaked-rush)	4, G5/S3.3	BgFns, MshSw
Sanguisorba officinalis (great burnet)	2, G5?/S2.2	BgFns, BUFrs, Medws, MshSw, NCFrs, RpFrs
Sidalcea calycosa ssp. *rhizomata* (Point Reyes checkerbloom)	1B, G5T2/S2.2	MshSw
Sidalcea malachroides (maple-leaved checkerbloom)	1B, G2?/S2.2	BUFrs, CoPrr, NCFrs
Sidalcea malvaeflora ssp. *patula* (Siskiyou checkerbloom)	1B, G5T1/S1.1, FSC	CBScr, CoPr, NCFrs
Sidalcea oregana ssp. *eximia* (coast checkerbloom)	1B, G5T1/S1.2	Medws, NCFrs
Sidalcea oregana ssp *hydrophila* (marsh checkerbloom)	1B, G5T2/S2?	Medws, RpFrs
Veratrum fimbriatum (fringed false-hellebore)	4, G3/S3.3	CoScr,Medws,NCFrs
Viola palustris (marsh violet)	2, G5/S1S2	BgFns, CoScr
Wyethia longicaulis (Humboldt County wyethia)	4, G3/S3.3	BUFrs, CoPrr

Appendix 3.5. Vascular Plant Species Besides Those Listed in Appendix 3.4, Which Are Listed as G2 or G1 at a Global Scale or S2 or S1 at a State Scale by the California Natural Diversity Data Base, and Which Occur in the Redwood Region. Code Definitions as in Appendix 3.4.

Species	Status
Abronia umbellata ssp. *breviflora* (pink sand-verbena)	G5T2/S2.1, FSC
Acanthomintha duttonii (San Mateo thorn-mint)	G1/S1.1, FE, CE
Agrostis blasdalei (Blasdale's bent grass)	G2/S2.2, FSC
Agrostis clivicola var. *punta-reyesensis* (Point Reyes bent grass)	G3T1/S1.2, FSC
Arabis koehleri var. *stipitata* (Koehler's stipitate rock cress)	G3T3/S1.3
Arabis macdonaldiana (Mcdonald's rock cress)	G2/S2.1, FE, CE
Arctostaphylos andersonii (Santa Cruz manzanita)	G2/S2?, FSC
Arctostaphylos bakeri ssp. *bakeri* (Baker's manzanita)	G2T2/S2.1, FSC, CR
Arctostaphylos bakeri ssp. *sublaevis* (The Cedars manzanita)	G2T2/S2.2
Arctostaphylos canescens ssp. *sonomensis* (Sonoma manzanita)	G3T2/ S2.2
Arctostaphylos cruzensis (Arroyo De La Cruz manzanita)	G2/S2.2, FSC
Arctostaphylos edmundsii (Little Sur manzanita)	G2/S2.2, FSC
Arctostaphylos glutinosa (Schreiber's manzanita)	G2/S2.2, FSC
Arctostaphylos pajaroensis (Pajaro manzanita)	G2/S2.1, FSC
Arctostaphylos pallida (pallid manzanita)	G1/S1.2, FT, CE
Arctostaphylos pumila (sandmat manzanita)	G2/S2.2, FSC
Arctostaphylos silvicola (Bonnie Doon manzanita)	G2/S2.2, FSC
Arctostaphylos stanfordiana ssp. *decumbens* (rincon manzanita)	G3T1/S1.1
Arctostaphylos stanfordiana ssp. *raichei* (Raiche's manzanita)	GT32?/S2?, FSC
Arctostaphylos virgata (Marin Manzanita)	G2/S2.2
Arenaria paludicola (marsh sandwort)	G1/S1.1, FE, CE

Species	Status
Asplenium trichomanes ssp. *trichomanes* (maidenhair spleenwort)	G5T5/S2.3
Astragalus agnicidus (Humboldt milk-vetch)	G1/S1.1, FSC, CE
Bensoniella oregona (bensoniella)	G2/S2.2, FSC, CR
Blennosperma nanum var. *robustum* (Point Reyes blennosperma)	G4T1/S1.2, FSC, CR
Calochortus raichei (The Cedars fairy-lantern)	G1/S1.2, FSC
Calochortus tiburonensis (Tiburon mariposa lily)	G1/S1.2, FT, ST
Cardamine nuttallii var. *gemmata* (yellow-tubered toothwort)	G5T2?/S2.2, FSC
Carex albida (white sedge)	G1/S1.1, FE, CE
Carex leptalea (flaccid sedge)	G5/S2?
Carex praticola (meadow sedge)	G5/S2S3
Castilleja ambigua ssp. *humboldtiensis* (Humboldt Bay owl's-clover)	G4T2/S2.2, FSC
Castilleja mendocinensis (Mendocino Coast Indian paintbrush)	G2/S2.2, FSC
Ceanothus confusus (Rincon Ridge ceanothus)	G5T2Q/S2.1, FSC
Ceanothus divergens (Calistoga ceanothus)	G2/S2.2, FSC
Ceanothus ferrisae (coyote ceanothus)	G1/S1.1, FE
Ceanothus masonii (Mason's ceanothus)	G1/S1.3, FSC
Ceanothus sonomensis (Sonoma ceanothus)	G2/S2.2, FSC
Chlorogalum pomeridianum var. *minus* (dwarf soaproot)	G5T1/S1.2
Chorizanthe howellii (Howell's spineflower)	G1/S1.2, FE, CE
Chorizanthe pungens var. *hartwegiana* (Ben Lomond spineflower)	G1G2T1/S1.1, FE
Chorizanthe pungens var. *pungens* (Monterey spineflower)	G2T2/S2.2, FT
Chorizanthe robusta var. *hartwegii* (Scott's Valley spineflower)	G2T1/S1.1, FE
Chorizanthe robusta var. *robusta* (robust spineflower)	G2T1/S1.1, FE
Chorizanthe valida (Sonoma spineflower)	G1/S1.1, FE, CE

(continues)

Appendix 3.5. *(continued)*

Species	Status
Cirsium fontinale var. *fontinale* (fountain thistle)	G2T1/S1.1, FE, CE
Cirsium hydrophilum var. *vaseyi* (Mt. Tamalpais thistle)	G1T1/S1.2, FSC
Cirsium loncholepis (La Graciosa thistle)	G2/S2.1, FPE
Clarkia franciscana (Presidio clarkia)	G1/S1.1, FE, CE
Clarkia imbricata (Vine Hill clarkia)	G1/S1.1, FE, CE
Collinsia corymbosa (Round-headed Chinese houses)	G1/S1.2
Cordylanthus tenuis ssp. *capillaris* (Pennell's bird's-beak)	G4T1/S1.2, FE, CR
Cupressus abramsiana (Santa Cruz cypress)	G1/S1.1, FE, CE
Cupressus goveniana ssp. *goveniana* (Gowen cypress)	G2T1/S1.2, FPT
Cupressus macrocarpa (Monterey cypress)	G1/S1.2, FSC
Delphinium bakeri (Baker's larkspur)	G1/S1.1, FPE, CR
Delphinium hutchinsoniae (Hutchinson's larkspur)	G2/S2.2, FSC
Delphinium luteum (yellow larkspur)	G1/S1.1, FPE, CR
Dirca occidentalis (western leatherwood)	G2G3/S2S3
Empetrum nigrum ssp. *hermaphroditum* (black crowberry)	G5T5/S2?
Erigeron angustatus (narrow-leaved daisy)	G2/S2.2?
Erigeron serpentinus (serpentine daisy)	G1/S1.3
Erigeron supplex (supple daisy)	G1/S1.2, FSC
Eriogonum kelloggii (Kellogg's buckwheat)	G1/S1.2, FC, SE
Eriogonum nervulosum (Snow Mountain buckwheat)	G2/S2.2, FSC
Eriogonum nortonii (Pinnacles buckwheat)	G2/S2.3
Eriogonum nudum var. *paralinum* (Del Norte buckwheat)	G5T4?/S2?
Eriogonum pendulum (Waldo buckwheat)	G4/S2.2

Species	Status
Eriophyllum latilobum (San Mateo woolly sunflower)	G1/S1.1, FE, CE
Erysimum ammophilum (Coast wallflower)	G2/S2.2, FSC
Erysimum menziesii ssp. *eurekense* (Humboldt Bay wallflower)	G2T1/S1.1, FE, CE
Erysimum menziesii ssp. *menziesii* (Menzies's wallflower)	G2T2/S2.1, FE, CE
Erysimum teretifolium (Santa Cruz wallflower)	G2/S2.1, FE, CE
Erythronium hendersonii (Henderson's fawn lily)	G4/S1.3
Fritillaria liliacea (fragrant fritillary)	G2/S2.2, FSC
Fritillaria roderickii (Roderick's fritillary)	G1Q/S1.1, CE
Galium hardhamiae (Hardham's bedstraw)	G2/S2.3
Hesperolinon congestum (Marin western flax)	G2/S2.1, FT, CT
Holocarpha macradenia (Santa Cruz tarplant)	G1/S1.1, FPT, CE
Horkelia marinensis (Point Reyes horkelia)	G2/S2.2, FSC
Horkelia tenuiloba (thin-lobed horkelia)	G2/S2.2
Lasthenia conjugens (Contra Costa goldfields)	G1/S1.1, FE
Lathyrus palustris (marsh pea)	G5/S2S3
Layia carnosa (beach layia)	G1/S1.1, FE, CE
Legenere limosa (legenere)	G2/S2.2, FSC
Lessingia arachnoidea (Crystal Springs lessingia)	G1/S1.2, FSC
Lessingia micradenia var. *micradenia* (Tamalpais lessingia)	G2T1/S1.1, FSC
Lewisia oppositifolia (opposite-leaved lewisia)	G4/S2.2
Lilium occidentale (western lily)	G1/S1.2, FE, CE
Lilium pardalinum ssp. *pitkinense* (Pitkin marsh lily)	G4T1/S1.1, FE, CE
Limnanthes vinculans (Sebastopol meadowfoam)	G2/S2.1, FE, CE
Linanthus jepsonii (Jepson's linanthus)	G1/S1.2

(continues)

Appendix 3.5. *(continued)*

Species	Status
Lomatium martindalei (Coast Range lomatium)	G5/S2.3
Lupinus tidestromii (Tidestrom's lupine)	G2/S2.1, FE, CE
Lycopodiella inundata (bog club-moss)	G4?/S1?
Lycopodium clavatum (running-pine)	G5?/S2S3
Monotropa uniflora (Indian-pipe)	G5/S2S3
Montia howellii (Howell's montia)	G2?/SH, FSC
Navarretia leucocephala ssp. *bakeri* (Baker's navarretia)	G3T2/S2.2
Navarretia rosulata (Marin County navarretia)	G2?/S2?
Oenothera wolfii (Wolf's evening-primrose)	G2/S1.1, FSC
Pedicularis dudleyi (Dudley's lousewort)	G2/S2.2, FSC
Pentachaeta bellidiflora (white-rayed pentachaeta)	G1/S1.1, FE, CE
Phacelia argentea (sand dune phacelia)	G2/S1.1, FSC
Phacelia insularis var. *continentis* (North Coast phacelia)	G2T1/S1.2, FSC
Pinus radiata (Monterey pine)	G1/S1.2, FSC
Piperia yadonii (Yadon's rein orchid)	G1/S1.1, FPE
Plagiobothrys diffusus (San Francisco popcorn-flower)	G1/S1.1, FSC, CE
Polygonum hickmanii (Scotts Valley polygonum)	G1/S1.1?
Polygonum marinense (Marin knotweed)	G1Q/S1.1, FSC
Potamogeton foliosus var. *fibrillosus* (fibrous pondweed)	G5T2T3/S1S2
Puccinellia pumila (dwarf alkali grass)	G4?/S1.1?
Raillardiopsis muirii (Muir's raillardella)	G2/S2.3
Rhynchospora californica (California beaked-rush)	G1/S1.1, FSC
Rhynchospora globularis var. *globularis* (round-headed beaked-rush)	G5T5?/S1.1
Sanicula maritima (adobe sanicle)	G2/S2.2, FSC

Species	Status
Saxifraga nuttallii (Nuttall's saxifrage)	G4?/S1.1
Sedum eastwoodiae (Red Mountain stonecrop)	G1/S1.2, FC
Silene campanulata ssp. *campanulata* (Red Mountain catchfly)	G5T1/S1.2, FC, CE
Stebbinsoseris decipiens (Santa Cruz microseris)	G2/S2.2, FSC
Streptanthus albidus ssp. *albidus* (Metcalf Canyon jewel-flower)	G2T1/S1.1, FE
Streptanthus albidus ssp. *peramoenus* (most beautiful jewel-flower)	G2T2/S2.2, FSC
Streptanthus batrachopus (Tamalpais jewel-flower)	G1/S1.2, FSC
Streptanthus howellii (Howell's jewel-flower)	G2/S1.2
Streptanthus morrisonii (subspecies of jewel-flower)	G2Q/S2, FSC (?)
Streptanthus niger (Tiburon jewel-flower)	G1/S1.1, FE, CE
Thermopsis robusta (robust false lupine)	G2Q/S2.2
Thlaspi californicum (Kneeland Prairie pennycress)	G1/S1.1, FPE
Trientalis arctica (arctic starflower)	G5/S1.2
Trifolium amoenum (showy Indian clover)	G1/S1.1, FE
Trifolium buckwestiorum (Santa Cruz clover)	G1/S1.1
Vaccinium scoparium (little-leaved huckleberry)	G5/S2.2?
Viola langsdorfii (Langsdorf's violet)	G4/S1.1
Viola primulifolia ssp. *occidentalis* (western bog violet)	G4T2/S2.2, FSC

Chapter 4

REDWOOD TREES, COMMUNITIES, AND ECOSYSTEMS: A CLOSER LOOK

John O. Sawyer, Stephen C. Sillett, William J. Libby,
Todd E. Dawson, James H. Popenoe, David L. Largent,
Robert Van Pelt, Stephen D. Veirs Jr., Reed F. Noss,
Dale A. Thornburgh, and Peter Del Tredici

Chapter 3 provided an overview of the variation in the forests where redwood is found throughout its range and summarized the physical characteristics of the regional environment. Chapter 3 also reviewed the flora of the redwoods (including fungi and lichens) and the animal and plant assemblages of the amazing redwood canopy. In this chapter, we take a closer look at redwood forests, describing the life history and architecture of redwood trees; redwood genetics; relations to environmental factors, such as temperature, moisture, and soil; and disturbance regimes. We also summarize what is known about the life history of other tree species that commonly coexist with redwoods. The closer we look at redwoods, the more extraordinary they appear.

Author contributions: Sawyer, lead author and organizer; Sillett, redwood canopies and editing throughout; Libby, genetics; Dawson, fog relations and redwood height; Popenoe, soils; Largent, role of fungi; Van Pelt, canopy architecture; Veirs, fire and stand dynamics; Noss, section introductions, box on old growth, and general writing and editing; Thornburgh, stand dynamics of second growth; Del Tredici, ligno-tubers (burls).

Life History

Redwood is an evergreen, highly productive tree with many individuals exceeding 100 m in height. Crowns of mature trees are pyramidal, with horizontal or drooping branches. Old trees often have broad crowns with a complex structure. Trunks may be straight or tapering and buttressed at the base. The bark is dark cinnamon red (often gray on large trees), fibrous, and thick (30 cm). Leaf morphology varies between two extremes. Leaves in the exposed upper crowns of tall trees are upright and awl-like (sun leaves), while leaves in the sheltered lower crown are linear and spreading (shade leaves). Branchlets are persistent for three years, at which time they are shed in the autumn.

Stem fasciation, a flattening of the stem due to linked meristems, is known in redwood (Becking 1970; Roy 1972). The root system is composed of wide-spreading lateral roots with no taproot (Lindquist 1974). Redwood is an extremely long-lived tree, with a moderately high tolerance to shade. In addition, redwood normally endures fire and flood (Agee 1993). Genetically, redwood is a highly variable species, with biologically meaningful adaptations evident in local populations (Libby et al. 1996); this variability should be considered in conservation planning and restoration activities (Millar and Libby 1989, 1991).

Substrate Relations

Redwood trees occur on a variety of geologic substrates and on a wide range of soil types, which have a plethora of physical and chemical properties and degrees of development. The soils found under redwood are described in a general way in local soil surveys published by the Natural Resources Conservation Service (formerly the Soil Conservation Service) and in the soil-vegetation surveys co-published by the California Department of Forestry and U.S. Forest Service. More detailed descriptions of the soils, including laboratory data, were reported by Waring and Major (1964), Begg et al. (1984), Popenoe (1987), and Popenoe et al. (1992). The National Resources Conservation Service has posted series descriptions (NRCS 1998a) and laboratory data on the internet (NRCS 1998b).

Although redwood is naturally absent from soils derived purely from ultramafic parent materials (i.e., those with high levels of exchangeable magnesium in relation to calcium and potassium), it does grow in soils derived from a mixture of these rocks and other parent materials. Suitable soils range from sandy loam to clay. Although growth rates are highest in continuously moist, well-aerated soils, redwood tolerates some drainage impairment, at least during winter (Popenoe et al. 1992; Vasey 1970) unless the soils are heavy clays (Marden 1993).

The Seed Stage

Redwood is a monoecious species with pollen (male) and seed (female) cones developing on separate branches of the same tree. Seed cones are receptive to pollen from November through March, the period when pollen cones are mature and shedding pollen. Although pollen is shed during the rainy season, rainless periods suitable for pollen transfer often occur during intervals between precipitation events. Seed set varies among years and is strongly influenced by that season's precipitation pattern. Seed cones are terminal and 10–30 mm long. They mature in the next rainy season, opening from September to March. As with pollen, seeds are shed when cones open during the rainy season on dry, bright days. Open cones may persist until branchlets are shed (Olson et al. 1990; Veirs 1996).

Redwood trees begin to produce seeds when 5 to 15 years old (Boe 1968). Trees produce abundant seed crops each year (Becking 1996). Seed viability is thought to increase initially with tree age, with trees approximately 250 years old producing seeds with the highest viability. Trees more than 1,000 years old produce viable seeds, but the percentage of viable seeds is lower (Roy 1966). Stands in the central redwood forests of Mendocino County are known to produce few or no cones in many years (Olson et al. 1990).

Redwood seeds are winged, small, and light, with about 265,000/kg. Because they fall rapidly (average = 2.6 m/second, range = 1.5–6.2 m/second), most seeds do not travel far from source trees (Olson et al. 1990). Seed viability for redwood is considered low and variable among locations (Ornduff 1998), but this result may reflect biases of collection techniques. Seeds collected from tree-tops by helicopter show more than 90 percent viability. Many ground-collected seeds, on the other hand, are empty, filled with tannin, or damaged by fungi. Sound seed estimates are low (1–32 percent), based on X-ray examination (Roy 1966). Seed viability also is fleeting, as seeds are susceptible to ultraviolet light. Stored seeds become nonviable in five or fewer years (Fritz and Redelius 1966).

The Seedling or Sprout Stage

Redwood seeds have the ability to germinate soon after they are shed from the parent tree. Full germination takes from three to six weeks. Warm, moist mineral soil provides the best seedbed, but seeds germinate and seedlings establish in duff, on logs, in debris, and under plants as long as adequate soil water is available. Seedlings can endure considerable surface heat (up to 60°C) if their roots have a source of water. Favorable water relations (i.e., humid air, moist soil) and warm temperatures lead to successful germination, but seedlings are extremely susceptible to infection by damping-off fungi in leaf-litter (Davidson 1971; Olson et al. 1990) as well as consumption by banana slugs, bush rabbits, and nematodes (Snyder 1992). Mean annual leaf-litterfall is high (3,120–3,740 kg/ha/yr) for two central redwood stands and even higher (3,940–4,690 kg/ha/yr) for two northern redwood stands (chap. 3, Pillars and Stuart 1993), when compared to other temperate forests (Bray and Gorham 1964). Thus, safe sites for redwood seedling establishment (i.e., those lacking thick accumulations of leaf-litter) are often scarce.

Redwood's ability to resprout vigorously after cutting is highly unusual for a conifer. Cut stumps often are encircled by more than 100 sprouts. This fast-growing swarm of new green stems and leaves sustains the stump-root system (Cole 1983). In addition, each sprout soon develops its own root system, and in a remarkably short time, the dominant sprouts create circles of new trees around a stump (Olson et al. 1990). The sprouts arise from a large "burl," or lignotuber, located at the base of the tree. Seedlings develop two buds at the junction of the stem and root, which divide to develop a basal burl as the seedling grows (Cooper 1965; Rydelius and Libby 1993). Recent anatomical studies have shown that the development of lignotubers by redwood seedlings is part of their normal ontogeny from detached meristems located in the axils of the two cotyledons (Rydelius and Libby 1993; Del Tredici 1998, 1999). Within four months of germination, each axillary cotyledonary meristem gives rise to a large central bud with two or more collateral accessory buds. As seedlings age, bud and cortex proliferation phenomena associated with lignotuber formation spread up the stem to include axillary meristems located immediately above the cotyledonary node.

Simmons (1973) found that less than 2 percent of 1-year-old seedlings had basal burls, whereas almost 75 percent of 6-year-old seedlings had them. In a greenhouse setting, all seedlings with bud collars survived the flames of a propane torch by growing at least one new leader. Because their basal burls are small, seedlings and saplings produce few sprouts.

Redwood may be propagated by cuttings with no special treatment. Rooting in excess of 90 percent is obtained routinely when using cuttings from young plants. Cuttings from older trees are more difficult to root (Olson et al. 1990). In a nursery setting, a single seedling can produce one million cuttings in three years (Libby and McCutchan 1978). Tissue-created plants (explants) also have produced successful redwood individuals (Olson et al. 1990). Simpson Timber Company in Korbel, California, grows nursery stock using plant-tissue culture techniques in combination with rooted cutting strategies.

Sprouts have been observed developing on broken branches that have fallen from the canopy and lodged in the ground. Trees can also sprout from fallen logs. Roy (1966) reported that cuttings from treetops pushed into soil develop roots and grow into healthy trees without special treatment. Indeed, determining whether a small redwood plant in the forest is a seedling or a sprout can be difficult. In addition, it is common to see individual plants that have been damaged several times only to start a new leader (i.e., to reiterate) once again.

Seedlings often become established in the northern redwood forests on fallen logs (Bingham 1984). Bingham sampled three sections of slope forest in Prairie Creek Redwoods State Park and estimated 56 redwood seedlings/ha on logs. In this way, redwood mimics Sitka spruce and western hemlock of the temperate rain forests from Oregon to Alaska, which both use logs as nurseries.

The Tree Stage

As a redwood tree grows, its ability to sprout from lignotubers continues until the tree reaches advanced age (Neal 1967). Lignotubers eventually form massive, basal swellings that are covered with leafy shoots and/or suppressed shoot buds. Vigorous trees may produce many spouts. Very old trees often produce few or no basal sprouts when damaged (Rydelius and Libby 1993). In response to a variety of environmental factors, redwood also will produce induced lignotubers on the trunks of mature trees, as well as on the "layered" lateral branches of young trees, at the point where they come in contact with the soil. Arborists use the ability of redwoods to sprout from axillary buds along the length of the trunk and branches to trim and shape ornamental redwood trees. Recovery of a tree's crown after fire is commonly noticed in the field as well, and can be attributed to sprouting of axillary buds (fig. 4.1).

From an ecological perspective, lignotubers function as sites for (1) production and storage of suppressed shoot buds that can sprout following injury to the primary stem; (2) storage of carbohydrates and mineral nutrients, which may allow for the rapid growth of suppressed buds following stress or damage; and (3) generation of roots, which increase the stability and vigor of both young and old trees. For trees growing on steep slopes, the lignotuber functions as a kind of clasping organ, which anchors the plant to the substrate. As important as lignotuber sprouting is for the perpetuation of mature trees, however, it is even

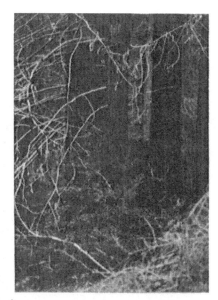

Figure 4.1. Sprouting by a redwood after fire.

more important for the survival of sprouts that are struggling to establish themselves in dense shade or on exposed slopes.

Despite the prevalence of sprouting in redwoods, little is known about the factors affecting frequency of sprouting. More than 90 percent of the cut stumps sprouted in a 40-year-old stand (Lindquist 1979), but only 30 percent of the stumps sprouted in an old-growth stand (Powers and Wiant 1970). A stand near Humboldt State University cut by hand in 1890, yarded by oxen, and burned several times produced sprouts on 71 percent of the stumps, but less than 1 percent of these developed on the tops of stumps. Veirs (1996) reported that 63 percent of the residual dead trees/snags had rings of sprouted trees around their bases. Surveys of old-growth stands in parks and preserves indicate frequent occurrences of "cathedral trees" (i.e., clumps of large trees derived from basal sprouts of a "mother" tree). Veirs (1996), however, argued that less than 10 percent of the trees in northern redwood forests are of sprout origin, and that sprouts are more common in drier, warmer sites, where fires are more severe and regeneration by seedlings more difficult.

Juvenile redwood trees usually grow rapidly in full sunlight in moist soil conditions. Established seedlings commonly grow 50 cm in their first year. Saplings from either seed or sprout sources can grow more than 2 m in a single growing season. Generally, sprouts outstrip seedlings when young, but as trees of seed origin exceed 40 years, they outstrip trees of sprout origin (Olson et al. 1990). Shading commonly limits sprout height at about 4 m, and high sprout density reduces individual stem vigor (Griffith 1992).

Redwood trees are long-lived and continue to increase in size for centuries. The oldest verified redwood tree is at least 2,200 years old (Fritz 1957), but many redwood trees may be much older. Diameter growth can be rapid (2.5 cm annually) or extremely slow (1 mm annually), depending on conditions. Old-growth stands typically have a range of ages and a mixture of vigorously growing and slow-growing trees (Olson et al. 1990). Although no generally accepted definition of old-growth redwood forests is available, their characteristics are becoming well known (box 4.1).

The tallest known living redwood tree is 112.0 m tall. The tallest historically reported redwood was Eureka Tree, which was cut in 1914 and measured at 115.8 m tall. The accuracy of this measurement, however, is not known. Douglas-fir and *Eucalyptus regnans* trees exceeding this height probably existed until the end of the nineteenth century (Carder 1995). The tallest known living tree that is not a redwood is Doerner Fir, a 100.3 m tall Douglas-fir in Coos County, Oregon.

Redwood is a distinctive species, in that the tallest trees are so similar in height. There are 15 known living trees more than 110 m tall (table 4.1) and 47 trees more than 105 m in height. Other species often have one tree that is much taller than its closest rival. The tallest known living redwood tree grows in a

Box 4.1. What Are Old-Growth Redwoods?

Conservation strategies for redwood include protecting the existing old-growth stands and allowing as much as possible of the second and third growth to regain old-growth conditions. To meet the requirements of the U.S. Endangered Species Act for listed and candidate species associated with old redwoods forests, such as the marbled murrelet and northern spotted owl, private landowners are developing silvicultural systems intended to hasten the development of old-growth or late-seral conditions in managed stands. Pacific Lumber, for example, has developed a Habitat Conservation Plan (HCP) and Sustained Yield Plan (SYP) which, among other things, seek to create and maintain "multistoried, uneven-aged, late-seral forest habitat." This plan proposes the use of selection harvesting to "enhance the growth of a few large trees while creating and maintaining special habitat elements including decadent trees, snags, downed logs, and other woody material" (Pacific Lumber Company 1998). Conservationists, however, claim that the silvicultural treatments proposed in the HCP/SYP will not produce or sustain true old-growth conditions.

For conservationists to evaluate the ability of management options to meet the needs of species associated with old-growth redwoods, some understanding of the specific habitat requirements of the species is desirable, as well as a general understanding of the characteristics of the old-growth forest. Detailed habitat requirements are known for only a few, well-studied species. For example, nests of marbled murrelets in California are found exclusively on old-growth trees above 30 m on large limbs with high vertical cover (see chap. 5). For lesser-known species, the old-growth forest as a whole must serve as a surrogate for their needs. Unfortunately, as noted by Hunter (1989), "There is no generally accepted or universally applicable definition of old growth." Similarly, Tuchmann et al. (1996) stated that "specifying exact age ranges for late-successional and old-growth forests is impossible because of variations in climate, soil quality, disturbances, and numerous other factors."

Despite the lack of a universal definition for old-growth forests—redwoods or otherwise—the general characteristics of these forests are well known to ecologists. As Hunter (1989) pointed out, the core of a conceptual definition is that "old-growth forests are relatively old and relatively undisturbed by humans." In the Pacific Northwest, the planning team for the Northwest Forest Plan, as a "general rule," defined late-successional (late-seral) forests as those with trees at least 80 years old and old-growth forests as a "subset of late-successional forests with trees 200 years or older" (Tuchmann et al. 1996).

The working definitions of the Northwest Forest Plan are consistent with others that have been offered for Douglas-fir and other coniferous forests in the Pacific Northwest and northern coastal California. Franklin (1982) specified age ranges for several seral stages in these forests: herb and shrub, 30 years; young forest, 30–100 years; mature forest, 100–200 years; old growth, 200–800 years; and climax forests, more than 800 years. Morrison (1988) recognized three categories of old growth: (1) "Classic" old growth—stands meeting all minimum old-growth criteria in which at least 8 trees per acre exceed 300 years in age or 40 inches in

(continues)

Box 4.1. (*continued*)

diameter; included in this grouping are stands that may be considered "super old growth," with trees exceeding 700 years in age or 72 inches in diameter; (2) "Early" old growth—stands meeting all minimum old-growth criteria in which at least 8 trees per acre exceed 200 years in age or 32 inches in diameter; these stands may include some older trees but in insufficient numbers to qualify as classic old growth; and (3) "Mature"—stands where more than 20 trees per acre exceed 80 years in age or 21 inches in diameter; included in this category are stands that fail to meet one or more of the minimum old-growth criteria.

Another way to characterize old-growth forests relates to their history (F. Euphrat, unpub.). Again, three types can be recognized: Type I: stands that have never been logged (i.e., primary forests); Type II: stands that have been logged, but retain significant structure associated with old trees and have a relatively continuous canopy; and Type III: stands that have been logged, but retain significant structure associated with old trees and have a discontinuous canopy.

Most definitions of "old growth" take more than tree age into consideration. Among the frequently noted characteristics are deep, multilayered canopies; a range of tree ages and sizes (consisting of both rapid-growing and slow-growing individuals); abundant shade-tolerant species; numerous large, standing snags and downed logs in various size and decay classes; and abundant tree cavities (Franklin et al. 1981; Franklin 1982; Old-Growth Definition Task Group 1986; Morrison 1988; Norse 1990). These structural characteristics are important in determining the biological diversity of redwood forests, as in other forest types. For example, the complex communities of redwood canopies, described in chapter 3, are not found in young forests. Similarly, species that rely on snags and coarse woody debris, such as many salamanders, also do not occur or persist in young forests, unless considerable woody debris ("biological legacies") remains from the old growth that formerly occupied the site. The California slender salamander is often significantly more abundant in mature and old-growth redwood forest compared to young forest (see chap. 5). Furthermore, most of the 320 species of fungi found in redwood forests (see chap. 3, appendix 3.2) depend on coarse woody debris—large, downed logs that retain moisture in the dry season. These structures are supplied in abundance and perpetuity only in true late-seral forests.

Variation in site class (i.e., suitability of conditions for tree growth) has an influence on the age at which a stand develops old-growth structural characteristics. For Douglas-fir–tanoak forests in California, Beardsley and Warbington (1996) considered old-growth conditions to begin at 180 years on high site classes, 240 years on medium site classes, and 300 years on low site classes. Specific to redwood, Helms (1995) noted that "most existing stands on alluvial flats are about 800 years old, although redwood may grow vigorously for 2,000 years." Bingham and Sawyer

(1991) characterized redwood stands in the Angelo Reserve (central redwood forests section) as young at 40–100 years, mature at 100–200 years, and old-growth at 200–560 years. Helms (1995) agreed that 100-year-old stands of redwood are young growth: "Young growth stands at 100 years have yields ranging from 10,600 cubic feet per acre on low-index sites to 51,080 cubic feet per acre on high-index sites."

Although redwood is a relatively fast-growing tree, its long life-span and potentially huge final size suggest that true late-seral and, especially, old-growth conditions are slow to develop. Old-growth redwood stands are composed of very large, tall trees. Natural, old-growth stands of redwoods also typically existed in patches larger than the remnants left behind when the surrounding landscape is logged. As noted by Morrison (1988), "Old-growth stands less than 80 acres in size are not viewed as viable old-growth units because external influences can easily penetrate and because they are vulnerable to disturbances such as windthrow." Most forest ecologists agree that small patches, especially when isolated, cannot be considered viable forest ecosystems (Noss and Cooperrider 1994).

Today, with some 96 percent of the original, old-growth redwood forest already destroyed by logging (U.S. Fish and Wildlife 1997) and much of the remaining forest highly fragmented, many old-growth species may have difficulty surviving. Today's landscape matrix serves as habitat for early successional and generalist species but generally not for old-growth species (Diaz and Bell 1997). Some second-growth stands have begun to regain old-growth characteristics; others retain these characteristics as residual features of the former, old-growth stands. In some instances, second-growth redwood stands that retain residual old-growth components are suitable habitat for such typically late-seral species as the northern spotted owl, red tree vole, flying squirrel, and fisher (Noon and Murphy 1997; and see chap. 8, this vol.). Nevertheless, species adapted to the unique environment of old-growth forests, such as canopy alectorioid lichens and other nonvascular epiphytes, as well as many invertebrates, cannot persist in young forest patches (chap. 8). Furthermore, many older second-growth stands have been recently cut, making this stage rarer than old growth in some areas.

Conservation of old-growth redwoods requires maintaining and, where possible, restoring relatively large areas of very old trees. Knowledge of how to restore or create old-growth redwood forests is in its infancy. Thinning can increase the rate of recovery of old-growth forest conditions in young stands (see chap. 3 and 8), but it is too early to say how long it takes to reestablish a complete old-growth community. If it is indeed fog that allows redwoods to achieve their great heights (see Box 4.2), and if complex redwood canopies are needed to capture fog water from the atmosphere, then loss of old-growth redwoods over large landscapes may make it difficult to reestablish redwoods on heavily cutover lands.

Table 4.1. Diameter at Breast Height (dbh), Height, and Trunk Wood Volume of the Tallest and Largest Known Living Redwood Trees.

TALLEST LIVING REDWOODS	Trunk dbh (m)	Total Height (m)	Trunk Wood Volume (m³)	Location
Mendocino Tree	3.11	112.0		(not given—site is fragile)
Harry Cole Tree	4.94	111.7		Redwood National Park
Paradox Tree	3.78	111.6		Humboldt Redwoods SP
National Geographic Tree	4.30	111.4		Redwood National Park
Federation Giant	4.57	110.8		Humboldt Redwoods SP
Swamp Tree	3.03	110.7		Montgomery Woods SR
Pipe Dream Tree	4.27	110.6		Humboldt Redwoods SP
Twin Tower Tree	4.57	110.5		Redwood National Park
Lost Hope Tree	5.03	110.4		Humboldt Redwoods SP
Alice Rhodes Tree	3.35	110.4		Humboldt Redwoods SP
Rockefeller Tree	4.05	110.3		Humboldt Redwoods SP
John Muir Tree	4.00	110.3		Humboldt Redwoods SP
Chevron Tree	3.29	110.2		Humboldt Redwoods SP
Godwood Creek Giant	7.44	110.0	726	Prairie Creek Redwoods SP
Harriet Weaver Tree	4.05	110.0		Humboldt Redwoods SP
Redwood Creek Giant	5.20	109.8	745	Redwood National Park
Montgomery Giant	5.24	109.4		Montgomery Woods SR
Randy Stoltmann Tree	3.26	109.2		Humboldt Redwoods SP
Libby Tree	4.27	109.2		Redwood National Park
Dome Top Tree	3.40	109.1		Humboldt Redwoods SP
Rifle Tree	3.41	109.1		Humboldt Redwoods SP
Dr. A. C. Carder Tree	4.15	109.0		Humboldt Redwoods SP
New Hope Tree	4.84	108.8		Jedediah Smith Redwoods

LARGEST LIVING REDWOODS (by wood volume)	Trunk dbh (m)	Total Height (m)	Trunk Wood Volume (m³)	Location
Del Norte Titan	7.23	93.7	1045	Jedediah Smith Redwoods SP
Cal-Barrel Giant	6.40	93.6	1019	Prairie Creek Redwoods SP
Lost Monarch	7.68	97.7	989	Jedediah Smith Redwoods SP
Howland Hill Giant	6.04	100.3	951	Jedediah Smith Redwoods SP
Sir Isaac Newton Tree	6.58	94.8	949	Prairie Creek Redwoods SP
Terex Titan	6.46	79.9	910	Prairie Creek Redwoods SP
Adventure Tree	5.15	103.9	898	Prairie Creek Redwoods SP
El Viejo del Norte	6.74	98.7	891	Jedediah Smith Redwoods SP
Bull Creek Giant	6.80	101.2	879	Humboldt Redwoods SP
ARCO Giant	7.35	80.8	869	Redwood National Park
Newton B. Drury Tree	6.07	77.4	843	Prairie Creek Redwoods SP

LARGEST LIVING REDWOODS (by wood volume)	Trunk dbh (m)	Total Height (m)	Trunk Wood Volume (m³)	Location
Westridge Giant	5.91	72.2	837	Prairie Creek Redwoods SP
Big Tree	7.22	87.2	810	Prairie Creek Redwoods SP
Elk Tree	5.85	95.4	791	Prairie Creek Redwoods SP
Atlas Tree	7.10	88.6	791	Prairie Creek Redwoods SP
Giant Tree	5.15	108.2	787	Humboldt Redwoods SP
Browns Creek Tree	6.40	73.5	781	Prairie Creek Redwoods SP

Note: Measurements of trunk wood volume are available only for trees exceeding 700 m³. SP = State Park. SR = State Reserve.

Source: Height and volume measurements were recorded by Michael Taylor, R. Van Pelt, and/or S. Sillett between 1995 and 1998 using a tripod-mounted Criterion laser surveyor. Sillett confirmed ground-based measurements for National Geographic Society Tree and Paradox Tree by lowering a weighted measuring tape from the top to the ground along the trunk. Both methods agree within a few centimeters.

ravine, far inland, surrounded by oak woodland. The moist ravine bottom, while small, supports many redwood trees more than 100 m tall. The most favored environment for tall trees seems to be the alluvial flats along Bull Creek in Humboldt Redwoods State Park, where more than 60 percent of the tallest known redwood trees occur (table 4.1).

Redwood ranks second to giant sequoia in trunk wood volume among the world's living trees (table 4.2). The following historical evidence suggests that redwood trees were once larger than giant sequoias. The Fieldbrook Stump in Humboldt County, California, is 9.8 m in diameter at a point 1.5 m from the ground (some observers, however, believe this stump to be two trees). By comparison, the General Sherman Tree in Sequoia National Park, California, is 9.0 m in diameter. More convincing is the Maple Creek Tree, which was logged in 1926. This tree had a log-by-log account published in the October 1926 Timberman (p. 109). While only 6.6 m diameter at the base, this tree had such a slight taper that by 60 m off the ground it was still 4.6 m in diameter (see figure 4.2 for the general relationship between diameter and height above ground). The tree's height was estimated to exceed 93 m and trunk volume to be 1,794 m³. By comparison, the largest living giant sequoia is 1,487 m³, and the largest known living redwood tree is 1,037 m³; other big tree species have much smaller volumes (table 4.2).

Table 4.2. Trunk Wood Volume, Diameter at Breast Height (dbh), and Height of Largest Living Trees for Species with Individuals Exceeding 300 m³ Trunk Wood Volume.

Species	Trunk Wood Volume (m³)	Trunk dbh (m)	Height (m)	Location
Giant sequoia	1,487	8.53	83.8	Sequoia National Park
Redwood	1,045	7.23	93.7	Jedediah Smith Redwoods SP
Western red-cedar	501	6.15	48.5	Olympic National Park
New Zealand kauri	480	4.39	51.5	Waipoua Forestry Sanctuary
Douglas-fir	349	4.23	62.5	Port Renfrew, British Columbia
Sitka spruce	337	4.55	75.6	Olympic National Park
Formosan cypress	316	5.82	49.4	Mount Morrison, Taiwan

Note: R. Van Pelt recorded the measurements using a tripod-mounted Criterion laser surveyor, except for the New Zealand kauri (Burstall and Sale 1984) and the Formosan cypress (Carder 1995).

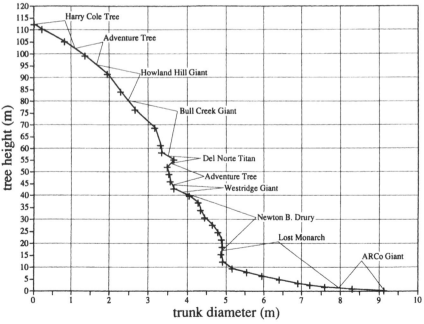

Figure 4.2. Maximum trunk diameters for living redwood trees at heights above ground. Identities of the trees (see table 4.1) are indicated at the appropriate heights. R. Van Pelt reported the following measurements, using a tripod-mounted Criterion laser surveyor. S. Sillett measured trunk diameters for Del Norte Titan, Lost Monarch, Atlas Tree, and Redwood Creek Giant with a fiberglass tape.

Crown Structure

Young redwood trees exhibit the classic conifer architecture: one large, central, vertically oriented trunk supporting numerous, smaller, horizontally oriented branches. Tall trees growing in sheltered environments may retain this simple architecture well into old age. Paradox Tree is a prime example. This 111.6 m tree, which emerges 10 m above the surrounding forest canopy, has one main trunk and 167 fairly horizontal branches exceeding 5 cm basal diameter. There is little correlation between branch basal diameter and height; the largest branches occur between 85 and 100 m above the ground (fig. 4.3). The considerably smaller branches in the lower portion of the crown are sprouts that arose from fissures in the bark of the trunk receiving high light levels. Nearly all of the original branches in the lower crown have been lost by self-pruning. The top of the crown is fully exposed, with ample evidence of wind damage. No large branches occur above 105 m (fig. 4.3), and many stubs and broken branches occur in the top 15 m of the crown. Foliage is evenly distributed around the crown above 100 m, and many branches, including the topmost, bore female cones in 1996.

Many large redwood trees display a highly complex crown structure. Main trunks of tall redwoods often snap and resprout; even branches can sprout new

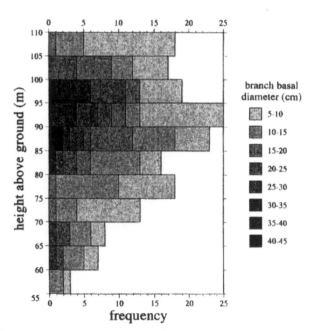

Figure 4.3. Frequency distribution of branch size classes by crown height in Paradox Tree, Humboldt Redwoods State Park (see text for details).

erect stems that can be regarded as new trunks. Such trunks and their crowns are, in effect, reiterations of the tree's architectural model. (The term *leader replacement* also has been used for this process.) Highly reiterated crowns resemble forest stands more than individual trees because each reiterated trunk supports its own branch systems.

As with many species, the rate of height growth in redwood trees diminishes with age. Height and trunk diameters are strongly correlated ($R^2 > 0.9$) for small trees, but the correlation becomes weaker as trees get larger and, presumably, older (fig. 4.4). Old trees probably invest proportionally more energy in lateral crown expansion and stabilization than in height growth compared to young trees (fig. 4.5). In general, crowns of large trees in Prairie Creek Redwoods State Park are more highly reiterated than crowns of large trees in Humboldt Redwoods State Park. The higher frequency of reiterations in Prairie Creek Redwoods State Park may be a consequence of its closer proximity to the ocean, which could lead to more frequent and severe windstorms, resulting in crown breakage and subsequent reiteration.

The cause of reiterations in redwood trees remains unclear, but their distribution appears to represent a growth response to an increased availability of light. The distribution of reiterations also reflects the disturbance history within a

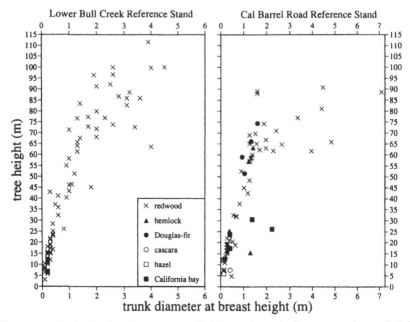

Figure 4.4. Relationship between tree height and trunk diameter at breast height in two 0.5-ha reference stands containing old-growth redwood forest (see text for details).

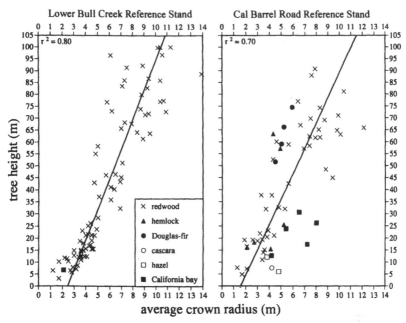

Figure 4.5. Relationship between tree height and crown radius in two 0.5-ha reference stands containing old-growth redwood forest (see text for details).

tree's crown. Thus, trees with highly reiterated crowns have complex, individualized structure. Consider the crown of a redwood tree 82 m tall (fig. 4.6). This tree has 62 reiterated trunks, including several that arise from the broken tops of other trunks. The largest reiteration arises from the main trunk at 60 m and is nearly 1 m in basal diameter. It supports nine additional reiterated trunks, including one that arises from the end of another trunk at 74 m. Two reiterations represent fusions of two or more trunks (see right side of fig. 4.6 around 60 m and left side around 50 m). In both cases, a large reiteration arises from the main trunk and then rejoins it several meters higher in the crown. The cluster of four broken tops between 53 and 56 m demonstrates a recurring pattern of crown damage. Several large, dead branches have become lodged here, including some that have fused with living branches. Such fusions presumably originate as wound responses to prolonged wind-induced rubbing that exposes cambium.

Trunk-to-trunk, trunk-to-branch, and branch-to-branch fusions are common in redwood trees. Because individual trunks in a large tree crown move independently in the wind, fusions may stabilize the crown, perhaps reducing the probability of breakage during storms. Regardless of their effects on crown stability, fusions divert the flow of water and nutrients through a tree's vascular

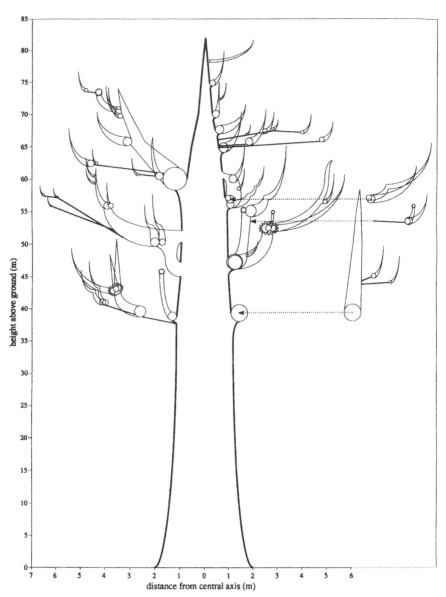

Figure 4.6. Crown structure of a redwood 82 m tall in Prairie Creek Redwoods
State Park. Circles denote trunk bases. Circle diameters are proportional to trunk
basal diameters and are drawn to the same scale as the x-axis. Note that the scale
of the x-axis is expanded to illustrate crown structure more clearly. Horizontal
lines denote branches. Only branches giving rise to reiterated trunks are shown.
The main trunk is assumed to be vertical, and all distances are calculated with
reference to its presumed central axis. Broken trunks are indicated by circles with
jagged edges.

tissues. The effectiveness of this diversion is best illustrated by an example. A 30-m tall reiteration emerges from a tall redwood tree's main trunk at 60 m. Most of the reiteration's branches are oriented away from the main trunk, which has a dense crown, but a few of its branches rub against the main trunk. One of these became completely fused with the main trunk. The branch was unable to withstand the force generated by trunk movement during a storm and was severed from the reiterated trunk because the fusion was stronger than the branch's attachment to its trunk of origin. Amazingly, however, the branch remains alive and vigorous.

Although many large redwood trees have complex structure, few have crowns as massive and sprawling as Atlas Tree, which is the fifteenth largest known living redwood (table 4.1). Fire scars are present in Atlas Tree up to 70 m. Its crown has 21 reiterated trunks, but 85.1 percent of its trunk wood volume occurs in the main trunk, which is more than 7 m in diameter at the base and still 3 m in diameter at 55 m (fig. 4.7). Between 55 and 60 m, the main trunk is broken and gives rise to five major reiterations, the largest 2 m in basal diameter and 28 m tall. This trunk gives rise to two other reiterations. One strange reiteration is forked near the base, and both resulting trunks have broken tops (left side of fig. 4.7 around 60 m). The taller of these trunks ends in a massive array of steeply downward-sloping branches, which are covered by an enormous leather fern mat. Most of the branches in Atlas Tree hang down around the periphery of the crown like a gigantic curtain shading the interior of the tree. Several horizontally oriented branches, however, support reiterated trunks. The largest of these emerges from the main trunk at 25 m, and the trunk it supports is 0.7 m in basal diameter and 17.5 m tall. R. Van Pelt's detailed sketch of Atlas Tree captures much of the structural complexity of its remarkable crown (fig. 4.8).

The largest known living redwood tree, which was discovered in May 1998 by S. Sillett and M. Taylor, is the Del Norte Titan (table 4.1). Three regions of negative taper can be found on the main trunk of this impressive tree: between 25 and 30 m, 50 and 55 m, and 60 and 65 m (fig. 4.9). At 55 m, the main trunk is still 3.7 m diameter, but quickly diminishes to 2.4 m after three major reiterations emerge. Its 43 reiterated trunks account for 9.5 percent of the tree's total wood volume. The largest reiterated trunk is 1.6 m basal diameter and 47.7 m tall. It is fused to the main trunk by a 1.4 m–diameter branch at 45 m. Both the main trunk and the largest reiterated trunk are thicker above the fused branch than below it. Another large trunk emerges from the main trunk at the same location as the largest reiterated trunk. The crotch formed by these three trunks holds a layer of canopy soil 2 m deep. Several broken reiterated trunks occur between 59 and 67 m (see left side of main trunk in fig. 4.9). This area of the crown is loaded with decaying wood, soil, and vascular epiphytes. Dozens of steeply downward-sloping branches arise from the left side of the largest reiter-

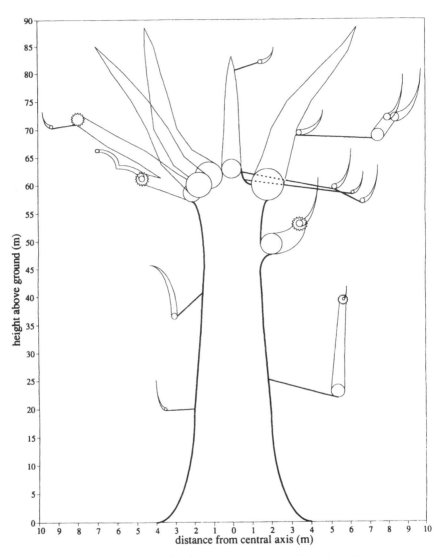

Figure 4.7. Crown structure of Atlas Tree, Prairie Creek Redwoods State Park. Circles denote trunk bases. Circle diameters are proportional to trunk basal diameters and are drawn to the same scale as the x-axis. Note that the scale of the x-axis is expanded to illustrate crown structure more clearly. Horizontal lines denote branches. Only branches giving rise to reiterated trunks are shown. The main trunk and the centrally located reiterated trunk are assumed to be vertical, and all distances are calculated with reference to their presumed central axis. Broken trunks are indicated by circles with jagged edges. (See text for details.)

Figure 4.8. Scale drawing of Atlas Tree.
(Drawing by R. Van Pelt)

ated trunk (not shown in fig. 4.9). Some of these branches are 30 m long and
fuse with each other. The upper crown of the Del Norte Titan is declining; the
main trunk is rotten and hollow above 85 m, with several large, dead branches.
Nevertheless, several smaller branches and a reiterated trunk arising from a
branch near the top are quite vigorous.

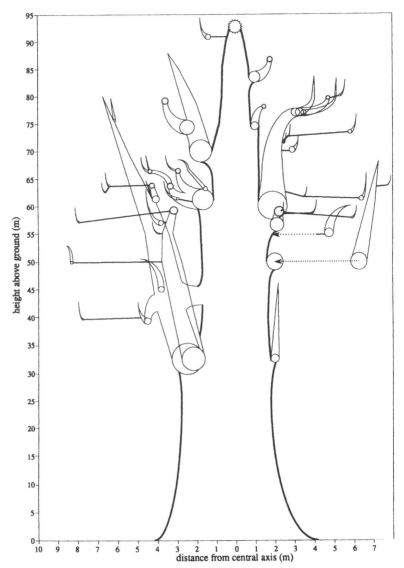

Figure 4.9. Crown structure of Del Norte Titan, Jedediah Smith Redwoods State Park. Circles denote trunk bases. Circle diameters are proportional to trunk basal diameters and are drawn to the same scale as the x-axis. Note that the scale of the x-axis is expanded to illustrate crown structure more clearly. Horizontal lines denote branches. Only branches giving rise to reiterated trunks are shown. The main trunk and the centrally located reiterated trunk are assumed to be vertical, and all distances are calculated with reference to their presumed central axis. Broken trunks are indicated by circles with jagged edges. (See text for details.)

Environmental Relations

Factors such as light, temperature, moisture, and soil conditions are critical to the survival of plants. The distribution of redwood, as with other tree species, reflects adaptations of the species to a particular set of conditions. The relationship of redwood to fog, although long noted, is finally receiving serious scientific attention.

Light Requirements

Redwood is considered shade-tolerant to very shade-tolerant, as it can grow in surprisingly dim parts of forests (Baker 1949). In addition, redwood plants that have been growing slowly in the shade for many years can grow rapidly when shade and root competition are removed (Fritz 1931; Roy 1966; Olson et al. 1990). Large trees also can accelerate growth when released from shade. One 1,000-year tree went from growing 12 rings/cm to 2.4 rings/cm when supplied with higher levels of light (Fritz 1957).

Not all ecologists regard redwood as highly shade tolerant; they emphasize that redwood grows best in well-illuminated environments, not under dense shade. Olson et al. (1990) reports that a one-year-old sprout may be 2 m tall, but a 100-year-old suppressed tree may be less than 10 m tall. W. Libby describes (pers. obs.) an 80-year-old selectively cut stand where most stump sprouts grew vigorously until they reached heights of about 4 m, then remained at about that height (nevertheless, two stems were growing vigorously in heavy shade well above that height). D. Thornburgh (pers. obs.) reported that in five-year-old selective cuts, grand fir, Sitka spruce, and western hemlock were 3 m in height and were outgrowing redwood. Studies are needed in a variety of forest conditions to better understand the role of light and the interaction of available light, moisture, and nutrients in redwood's growth and ecology.

Temperature Requirements

In a greenhouse setting, the best growth of redwoods, in dry weight, is attained at moderate temperatures (19°C day, 15°C night) with little or no thermoperiod (Hellmers 1963, 1966). Laboratory studies (Anekonda et al. 1994) suggest that temperature for peak metabolic activity varies for redwood plants in different parts of the species' range. In the field, redwood trees grow under many temperature regimes, but forests dominated by redwood are associated with soils that have an isomesic regime (i.e., mean annual temperature from 8 to 15°C). Here the difference between summer and winter surface soil temperatures ranges from about 3 to 5°C (Hauxwell et al. 1981).

Soil Requirements

The principal soils under northern and central redwood forests are those on alluvium associated with streams (Udifluvents), on steep slopes with high organic content (Haplohumults), and on broad ridges with marked horizon development (Palehumults). Redwoods occur locally in areas of moderately impaired drainage. Some soils have high organic content and little horizon development (Hapludalfs and Eutrochrepts). (See Popenoe 1987; Soil Survey Staff 1998, and NRCS 1996 for definition and description of soils.) Soil pH in redwood forests ranges from about 5.0 to 6.5. Properties of the principal soils reflect the influence of intense leaching, which occurs during mild winters with high precipitation.

Little soil information has been collected from southern redwood forests, perhaps because of their small extent. Ben Lomond and Gamboa (classified by Cook [1978] as Haploxerolls; NRCS 1996, 1998a) are the only soil series mapped by the National Resources Conservation Service as associated with redwood. These soils are well drained, with pH from 6.0–7.5. In northern Monterey County, California, Borchert et al. (1988) found redwood growing on well-drained, loamy, gravelly, neutral soils. Properties of these soils reflect the influence of less intense leaching than those farther north.

Regional Soil Moisture Patterns

Redwood trees can grow under many soil moisture regimes, but forests dominated by redwood are associated with soils that are dry (water potential > 1,500 kPa) for less than 30 days in most years (i.e., an udic moisture regime). In addition, forests dominated by redwood are not associated with soils that flood or puddle during the active growing season (i.e., an aquic moisture regime). Redwood does, however, tolerate some drainage impairment (Popenoe et al. 1992; Vasey 1970).

Throughout the range of redwood, the patterns and extent of soils suitable for its establishment and success vary greatly. Vegetation patterns, therefore, show striking differences regionally and locally (Zinke 1988). For example, soils meeting redwood's requirements of temperature, moisture, and oxygen are extensive in the northern redwood forest region. Here, most topographic settings support redwood forests, so they are widely distributed across the landscape. Because annual precipitation is greater in the north than in the south and the summer dry season is relatively short, the udic moisture regime extends from the coast eastward beyond the eastern range of redwood into the Douglas-fir–tanoak forests (Lewis 1982; Sandek 1984). Soils associated with certain marine terraces that experience alternating prolonged water saturation lack redwood forests and instead support beach pine, bishop pine, red alder, Sitka spruce, or willow.

Soils meeting redwood's requirements are less extensive in the central redwood forest region. The high terraces associated with the Eel and other rivers support magnificent redwood-dominated forests. In the more extensive upland situations, however, soils with suitable moisture regimes to support redwood are less extensive than in the northern redwood region (Zinke 1988). Forest composition changes accordingly as redwood mixes with other tree species (i.e., California bay, Douglas-fir, madrone, and tanoak) and eventually disappears.

In southern redwood forests, with less annual precipitation and a longer summer dry season, soils meeting redwood's requirements are relatively limited in extent. Here, redwood stands are regularly described as "riparian" because redwood grows near creeks or in canyon bottoms and ravines (Zinke 1988). In upland conditions, depending on topography and parent material, soils rapidly approach ustic or even xeric moisture regimes. Vegetation is typically chaparral, coastal scrub, and grassland.

Fog

Redwood commonly grows in areas characterized by cool fog during the summer (Cannon 1901; Cooper 1917; Byers 1953; Oberlander 1956; Parsons 1960; Myers 1968; Freeman 1971; Azevedo and Morgan 1974; Harris 1987). Fog forms on days when a warm layer of air caused by a southward and descending migration of the central Pacific high-pressure cell meets with cold water upwelling along the coast. Vertical mixing of moist eddies of air lifts water off the Pacific Ocean and moves it inland, where it bathes plants in coastal plains and hills (Myers 1968). Fog is heaviest from 0700 to 1100 hours and least prevalent at 1500 hours (Harris 1987).

Fog is known to be an important source of water for California's major hydrological catchments containing redwood forests (Cannon 1901; Cooper 1917; Byers 1953; Oberlander 1956; Parsons 1960; Freeman 1971; Azevedo and Morgan 1974; Harris 1987; Dawson 1996, 1998). For decades, a "fog lore" developed that included numerous claims about how fog serves as a potential source of water for plants (Byers 1953; Oberlander 1956; Parsons 1960; Freeman 1971; Azevedo and Morgan 1974; Harris 1987). Early investigations in the northern and central redwood forests showed that soil moisture is higher around trees or in forest stands where the canopy stripped fog from the air mass (Oberlander 1956; Parsons 1960; Azevedo and Morgan 1974). In one such study, Freeman (1971) showed that fog drip from redwood foliage could be equivalent to 50 mm/tree/day or more. Recent studies have boosted these early estimates and have shown that fog inputs on a per-tree basis can exceed 100 mm/day during summer, when rainfall is essentially absent but when plants' demand for water is high (Dawson 1996, 1998). These studies show that fog contributes 25–50 percent of total water input each year, and that this input is due to the presence of trees. When trees are absent, input declines by some 30 percent.

Several early papers written about redwood and fog suggested that moisture input from fog might reduce plant moisture stress (Cannon 1901; Monteith 1963; Kerfoot 1968), enhance growth (Becking 1968), or even improve the nutrient status of forest soils (Cooper 1917; Stone 1957; Azevedo and Morgan 1974; Harris 1987). The importance of fog water input in the water relations of redwood trees, as well as many understory species, has been further supported by recent studies. Analysis of hydrogen and oxygen isotopes showed that redwood and other plants can absorb fog directly through the leaves and that this unique water source is particularly important for young saplings and understory species, especially in years of below-normal rainfall (Dawson 1993, 1996, 1998). Fog may even explain, in large part, why redwood trees are able to grow so tall (box 4.2).

Box 4.2. How Does Water Get to the Tops of Tall Trees?

How do redwoods get to be 112 m tall? To answer this question we need to appreciate the challenge of getting water, perhaps the most important resource limiting plant growth, to the tops of these giants. In simple terms, the challenge that redwood trees face is building wood that will efficiently transport water from the soil into the roots and eventually to the growing apices tens of meters above the ground, while minimizing the negative effects of gravity and friction on water transport. At sea level, atmospheric pressure will permit water to rise only 10.3 m within a capillary tube evacuated and closed at its upper end. Water drawn up a capillary above this height will begin to experience negative pressures within the water column, allowing dissolved air bubbles to come out of solution until air fills part of the space between the walls of the capillary, and the water column breaks. In short, air bubbles and the process by which they interrupt water flow break the water column (Tyree and Sperry 1989).

Properties of wood such as the size and structure of the water conducting tissues, the properties of the material dissolved in water, and the water stored in wood permit water to rise higher than that in a capillary tube. These factors, however, are still not enough to fully explain how water gets to the top of a tree 112 m tall.

Plant physiologists have long known that hydrostatic pressures generated by plant root systems, or within stems, can account for only a very small fraction of the forces needed to "lift" water within a tree. We also know that water is not physically "pumped" against an uphill gradient inside the water-conducting tissues of plants. Therefore, water moving from the soil, into the roots, and eventually into conducting tissues on its way to the leaves must, essentially, be "pulled" along this path.

One long-standing and popular explanation for how water reaches such great heights in trees was advanced by Dixon and Joly (1894) and called the "cohesion theory." Although recently there have been some challenges to the theory, and other factors may indeed be important (Zimmerman et al. 1993, 1994; Canny 1997a,b), evidence suggests that cohesion offers a partial explanation for how water ascends to the tops of tall trees (Pockman et al. 1995; Holbrook et al. 1995; Sperry et al. 1996).

The cohesion theory has three main elements. For sap to reach heights within trees that exceed 10 m, there must be (1) a "driving force" for the movement of water against the forces of gravity, negative pressure, and friction; (2) adhesion of water to tissue surfaces provided by the unique hydrogen-bonding properties of water molecules and their attraction to other surfaces; and (3) cohesive or bonding forces generated by the attraction of water molecules to each other.

Water can be pulled out of tree leaves through the soil-plant-atmospheric continuum by the evaporative power of the atmosphere because the pathways through which water flows in the wood prevent the water column from breaking, and because water has unique properties. When the minute pores (stomata) embedded within the leaf surface open, water within the leaf will vaporize and move out. Movement takes place because air inside the leaf is saturated with water vapor, but air outside of the leaf, even in humid environments, is less saturated. Water vapor will move through air from a relatively "wetter" to a relatively "drier" region. For the leaf to remain well hydrated, water leaving the leaf must be replaced by water within the leaf and stem tissues. Because water has high cohesion, as one bit of water exits, another bit of water is pulled up along the pathway to replace it. This process is what allows water to get to the tops of redwood trees.

Although redwoods grow in a Mediterranean climate, with wet winters and dry summers, they rarely experience severe water deficits. During the summer, when rainfall is extremely low, fog serves as an additional water source (Dawson, 1996, 1998), so redwoods have water available year-round. Moreover, when fog is present, it significantly reduces the atmospheric water stress commonly imposed by low humidity. The potential for water within the water transport tissue to break is dramatically relieved, keeping the conducting pathway intact. In addition, modest amounts of fog water can be absorbed directly by leaves, lessening the water stress. For these reasons, redwood may be better able to avoid the hydraulic limitations to height that are faced by species growing in other climates (Ryan and Yoder 1997). The direct influence of fog on the redwoods may reduce water stress, enhance growth, and may even influence long-term survival (Dawson 1996; Dawson et al. submitted).

The results from these studies demonstrate strong interactions among plants, fog, human, and climatic factors. The occurrence or lack of El Niño, fire, floods, windthrow events, climate change, and logging (e.g., reduction in forest patch size) can have important ecological consequences for fog-inundated forests and their hydrology (Bruijnzeel 1991; Dawson 1998). Ongoing research is exploring the influence of fog on plant water and carbon and nutrient balance, the role of fog in tree regeneration, and the importance of fog in biogeochemical and hydrological cycles.

Site Productivity and Mycorrhizae

Redwood forests can produce biomass throughout the year (Waring and Franklin 1979). Among the factors influencing site productivity, water is typically first and oxygen is second in importance. Nitrogen can affect redwood productivity, too. Other essential nutrients are usually abundant (table 4.3).

Redwood, like most other conifers, relies upon fungal associations called mycorrhizae to capture soil nutrients and water. Roots of almost all vascular plants are connected to the hyphae of fungi in the soil, a mutually beneficial relationship between the fungus and the plant. In redwoods, the type of relationship is described as endomycorrhizal because the fungus forms bladderlike sacs and highly branched units inside the cells of the redwood root's cortex (Gerdemann 1968; Kough et al. 1985; Newman and Reddell 1987; Afek et al. 1994). Five species of *Glomus* have been collected under redwood trees (Trappe 1998) and are putative fungal partners in endomycorrhizal associations with redwood.

The hyphae that form the endomycorrhizae extend throughout the substrate, absorb nonmobile nutrients (P, Zn, Cu) from the soil, and translocate these nutrients rapidly to the plant, thereby extending the nutrient-absorbing surfaces of roots (Smith and Read 1997). Endomycorrhizae also increase drought resistance of trees. In return, the fungi receive carbohydrates from the plant. Exudates and hyphae of mycorrhizal fungi provide an energy source for bacteria, protozoans, arthropods, and microfungi in the soil.

Mycorrhizal fungi are obligate aerobes, so whatever soil conditions favor redwood's mycorrhizae will also favor redwood. Nitrogen can be limiting in soils without an A horizon (the surface layer rich in organic material) (Pritchett 1979), such as fresh alluvium and landslide scars, which tend to be quickly colonized by seedlings including redwood, blue blossom, and red alder. Alluvium also may bury old A horizons, which retain their organic matter and nitrogen and remain full of actively growing redwood roots.

Table 4.3. Nutrient Levels for the Upper Meter of Soil Dominated by Northern and Central Redwood Forests.

Element	Northern Forests	Central Forest
Organic C	180,000	128,000
Total N	11,000	6,500
Exchangeable Ca	6,400	5,600
Exchangeable Mg	1,600	1,300
Exchangeable K	1,600	550

Source: Data for northern redwood forests from Popenoe (1987); data for central redwood forests from Zinke et al. (1996).

Genetics

Population geneticists find that each species usually has a genetic architecture, or structure, different from those of related species and of unrelated species sharing the same environment (Libby and Critchfield 1987). Conservation planning increasingly includes an attempt to maintain appropriate levels of genetic variation in species and populations. Such planning should be based on an understanding of how genetic variation is distributed and organized within and among populations of species. Common-garden studies often are used to perceive genetic structure. In these experiments, replicated clones of different individuals within families, different families within stands, and stands from different locations and environments are planted. These plantings are replicated at each of several different locations or environments, and the plants' development is followed through time. We summarize below several aspects of redwood's genetic structure: within- and among-population genetic variation, heritability, nonadditive genetic variation, and genotype—environment interactions. An accessible review of these concepts is available in Silverton and Lovett Doust (1993).

Provenance Variation

"Provenance" (i.e., the place where a population has recently evolved) variation implies that populations from different places are adapted to different environments. Terpene evidence suggests that the southern redwood forests were founded by a colonizing population different from those that founded the redwood forests north of San Francisco, whereas the northern and central redwood forests were founded by two other, incompletely mixed populations (Hall and Langenheim 1987). Provenance variation is crucial to consider when planning a conservation or restoration strategy for redwoods because mixing genes from different populations could result in reduced adaptation to local conditions (i.e., outbreeding depression).

Although the populations were founded independently, sufficient time has elapsed for most or all of the current native redwood populations to adapt to local environmental conditions (Anekonda et al. 1994). Several common-garden studies conducted in or near the native redwood range support this hypothesis. In these studies, growth of redwood trees near each location tended to be somewhat (but not substantially) better than that of trees from more distant populations (Millar et al. 1985). Within each region, redwoods from the more favorable (usually lower-elevation) environments generally grew better in the common-garden environments than did other redwoods from that region (Anekonda 1992; T. S. Anekonda, unpub. data), at least during the first few years. Nevertheless, although there were nontrivial differences in performance, redwoods from all parts of the native range survived and grew well in each of these common-garden locations in or near the native range. This finding suggests a need for caution in restoration activities with redwood. Redwoods from populations that are not locally native will survive if planted, will grow, and will probably contaminate the genetic structure of the local population, which is the object of the conservation or restoration interest.

Kuser et al. (1997) reported the results of studying 180 redwood clones from 90 locations in the natural range at four common-garden locations in the United States, two in France, and one each in Spain, Britain, and New Zealand. Early results indicate that, at those locations more distant and different from redwood's natural range, many or even most redwood plants from the less-appropriate parts of the redwood range died or failed to thrive. Thus, although redwood populations do not show the large adaptive differences well known in such wide-ranging species as Douglas-fir (Ching and Bever 1960; Rowe and Ching 1973) and western hemlock (Kuser and Ching 1980), origin differences are important to consider when growing redwood outside of its native range.

Heritability

A good way to understand the heritability (degree of genetic control) of traits is to study clones. Natural redwood clones, such as stump sprouts, are effectively studied when scattered on hillsides transected by roads. This approach has proven useful in the hills southeast and east of Santa Cruz, California. Human-created clones also can be studied (Kuser et al. 1984). Contiguously planted redwood clones differ substantially from each other in such traits as foliage color, branch length and angle, crown density, bark color and texture, trunk straightness, and overall growth. When these clones are propagated, they differ in speed of rooting, percentage of rooting, and root-system size and configuration. Substantial differences in physiological traits such as cold/frost tolerance (A. Bailly, AFOCEL, France, 1991, pers. comm. to W. Libby) and biosynthetic metabolism (Anekonda 1992) are also evident.

Given these observations on contiguously planted clones, it is surprising that cloned redwoods planted noncontiguously have such low calculated heritabilities compared to other conifers (Anekonda 1992). The likely explanation for these results is that redwood is unusually responsive to environmental variation; that is, traits are modified by the environment. During computation of heritablity, the effects of microsite variation can be reduced by using the performance of nearest neighbors to adjust the data. When this was done for redwood, estimated heritabilities approximately doubled for most traits (Anekonda and Libby 1996).

Nonadditive Genetic Variation

No available studies directly estimate nonadditive genetic variation in redwood. The presence of inbreeding depression in redwood, however, is good indirect evidence that some nonadditive variation is operative. Whereas the hexaploid nature of redwood might suggest inbreeding depression is infrequent (Libby et al. 1981), three studies have found a consistent pattern of inbreeding depression (Libby et al. 1981; Kirchgessner and Libby 1985; W. J. Libby et al., unpub. data). Under "near-ideal" laboratory or nursery conditions, redwood shows little response to inbreeding, as inbred offspring and outcrossed offspring grow similarly. Nevertheless, under "stress conditions," as when trees compete in field conditions, noninbred redwoods grow much larger than inbred redwoods. This outcome has important implications for breeding and deployment strategies for plantations, as well as for understanding the success of seedlings and clones under conditions of natural regeneration.

Among-Population and Within-Population Genetic Variation

Genetic differences among stands of redwoods within the same physiographic region are generally present for most traits, but they make up a relatively small percentage of the species' total genetic variation. The largest component of redwood's genetic variation for most traits resides within stands of trees (Anekonda 1992). We caution that this generalization is based on studies of very young trees, and that genetic differences in adaptive traits among the different physiographic regions of origin may, over time, become more important as the trees grow and encounter infrequent but strong selective events on each site (Millar and Libby 1989).

Redwood is unusual among conifers in being hexaploid (Saylor and Simons 1970). It has 66 chromosomes (sometimes more), whereas most conifers have from 20 to 24. Because redwood is a hexaploid, it is possible to have much allelic variation within a single individual (i.e., alleles are alternative forms of the same gene). For example, single (triploid) gametes have been shown to have one, two, or even three different alleles of a gene coding for a particular enzyme. The resulting individual-tree enzyme variability made it possible, within four self-pollinated families, to distinguish ten siblings from each other in each of these very closely related families (Rogers 1994). In short, redwoods have enormous within-family genetic variability, and we now have the tools to find and characterize it.

The considerable hexaploid-level genetic variability within individual redwood trees is enhanced substantially by maturational (i.e., developmental genetic) gradients within each tree (Tufuor 1973). In addition, a mosaic of endophytic fungi living in the cells of redwood leaves (Espinosa-Garcia and Langenheim 1990) make large redwoods biologically diverse organisms. Various organisms attempting to live on or attack redwoods encounter this variation. Thus, a large redwood tree, or a clonal clump of redwood sprouts, may present as much operational diversity to its potential pests and pathogens as does a whole field of genetically variable herbs.

Within four natural stands studied in Humboldt Redwoods State Park, the pattern of enzyme variation indicated only weak neighborhood aggregations of closely related redwood trees. Clusters of redwoods, sometimes called "fairy rings," are usually thought to be clonal. The fairy rings within these stands, however, often proved to have other seedlings or clones sharing the ring with the resident clone. In some instances, the resident clone had a linear string of clonal members extending out about 30 m from a fairy ring, indicating that perhaps branches driven into the ground when one of the live stems crashed in a storm had rooted near and along its length (Rogers 1994). One might expect that unusually well-adapted clones would occupy large areas, but no extensive clones were located in the stands studied.

Genotype-Environment Interactions

When genotype-environment interactions are present, the relative performance of some trait or set of traits changes among environments. Some clones are better adapted to one environment, while other clones are better adapted to another environment. This sort of genotype-environment interaction is relatively high at the clonal level for redwoods across physical distances of less than 3 km (Anekonda 1992) and at the population level across such great environmental differences as those between Great Britain and California's Sacramento Valley (Kuser et al. 1997).

Major Coexisting Tree Species

Several tree species commonly intermingle with redwoods in stands, sometimes as codominants. Some knowledge of the life histories of these species is necessary to understand potential competitive relationships of redwood with other tree species. Below we summarize what is known about four common tree species within the redwood forests.

Western hemlock is a highly shade-tolerant, relatively short (usually > 60 m) tree common in the northern redwood forests. Its viable seeds are plentiful, and seedlings can establish both on organic and mineral substrates. Rotting logs are favored seedbeds. Western hemlock responds well to release in canopy gaps after languishing in the shade for years. Shallow roots and thin bark make it sensitive to windthrow and fire. Many damaging insects and fungi are associated with western hemlock, and they can cause high mortality. Old trees rarely exceed 700 years in age, but trees older than 500 years are common (Packee 1990).

Douglas-fir is an intermediately shade-tolerant tree in California. It is common in forests throughout the range of redwood. As with western hemlock, viable seeds are abundant, and seedlings can establish on organic or mineral substrates. Douglas-fir responds to release from shade when young. The thick bark of older trees withstands fire well. Many species of insects and fungi are associated with Douglas-fir, but they mainly attack younger trees. Old-growth stands tend to have more predaceous insects, which may control herbivorous insects

(Schowalter 1989). Old trees rarely exceed 1,000 years in age, and trees older than 500 years are uncommon in California (Hermann and Lavender 1990).

California bay and tanoak are very shade tolerant, relatively short (> 30 m) hardwood trees common in many stands throughout the range of redwood. They reproduce by seed and especially from sprouts arising from basal burls. Stems are easily killed by fire, but plants respond by creating a new set of sprouts. California bay is considered allelopathic. California bay and tanoak do not have the longevity of their conifer associates (Stein 1990; Tappeiner et al. 1990).

The abundance of these highly shade-tolerant species in old-growth stands of redwood has caused some ecologists to speculate about redwood's fate in the absence of major disturbances (Cooper 1965; Daubenmire and Daubenmire 1975). These ecologists have hypothesized that shade-tolerant species will replace redwood in time without substantial stand disturbance. High numbers of seedlings and saplings, however, do not necessarily allow one to predict future stand composition. Survivorship patterns and the availability of safe sites for seedling establishment under old-growth conditions are unknown for redwood and its associates (box 4.3). A better understanding of these aspects of redwood ecology is clearly needed to allow ecologists to fully evaluate redwood's fate in stands with low levels of disturbance.

Box 4.3. Stand Dynamics of Redwood Forests

Redwood forests vary from pure stands found on alluvial terraces to scattered individual redwood trees in favorably moist microsites in upland vegetation dominated by other species. At the southern extreme of its range, in Monterey County, redwood stands on coastal slopes are composed of many small trees that sprouted after fires on dry, low-rainfall sites where the likelihood of seedling establishment is very low. In the northern part of the range, most redwood trees in old-growth forests are of seedling origin (Veirs 1982).

Fire frequency and intensity increase with distance from the Pacific Coast and elevation. Fire-return intervals of up to 500–600 years are found in the moistest sites near the coast. Brown (1991) reports three or more stand-altering fires per century in the drier, warmer inland sites. Because lightning is rare along the coast, and fire conditions are rarely extreme, more severe fires were probably blown into the redwoods from their eastern margins, as suggested by Greenlee (1983) for the southern redwood forests.

Douglas-fir apparently benefits from fire, because this relatively drought-tolerant species thrives in the open, sunny conditions following fires that remove understory and varying amounts of canopy foliage. The proportion of Douglas-fir in the northern redwood forests increases with increasing elevation and distance from

(continues)

Box 4.3. (*continued*)

the coast. Fires in the northern stands kill thin-barked smaller trees of all species, but they rarely kill larger, thick-barked redwood and Douglas-fir. Stand-opening fires favor Douglas-fir establishment, whereas longer fire-free intervals favor establishment of the more shade-tolerant redwood, tanoak, and western hemlock in favorable understory microsites. Such microsites include fallen logs, root wads, almost any woody debris for western hemlock, and deep litter for the large-seeded tanoak. Repeated fires with short return intervals favor redwood because of the young tree's ability to sprout (Rydelius and Libby 1983).

In old-growth northern redwood forests not severely modified by fire, the average age of a redwood is about 600 years; a few individual trees exceed 1,500 years (Veirs 1982). Approximately 2.5 trees/ha must reach canopy status per century to maintain the age structure of these stands. Standing dead redwoods are rare; killing diseases are unknown for redwood past the seedling stage. Several visits to a site last burned in 1974 revealed that a likely cause of death in large overstory trees was girdling caused by fire killing the cambium of the tree at ground level. Girdling kills some of the roots when healing growth is inadequate to reconnect enough stem and root system. With less water transport to the crown, subsequent growth rates are lower, and the probability of surviving another fire is reduced.

Mortality from windthrow of overstory trees occurs when soils are saturated by winter rains. Fire frequency and intensity are low on well-watered and shaded alluvial stream deposits. Fresh sediments under an existing canopy can be colonized by redwood seedlings. Only a few seedlings receive enough light to survive, and still fewer trees ever reach the canopy. In Tall Trees Grove on Redwood Creek in Redwood National Park, many seedlings established after the 1964, 1973, and 1975 floods; some of them have since been partly buried by fine sediments deposited during subsequent floods.

Douglas-fir is rarely found on alluvial sites and is better represented on upper slopes. Absence of other nearby Douglas-fir, infrequent seed years, relatively heavy seeds, and relative shade intolerance reduce the probability of seedlings establishing in alluvial sites. When changing flow patterns erode stream banks, portions of streamside groves can be washed away during floods. Alder, black cottonwood, redwood, and bigleaf maple establish quickly on the freshly deposited alluvium and eventually dominate the forest in one or more combinations. Douglas-fir is less likely to win the seedling sweepstakes following floods.

The timing of seed fall and alluvial sediment deposition may be important in redwood stand dynamics. Late-season flooding could wash away seeds while depositing fresh sediment. Whereas repeated fires can kill upland redwoods, elevated water tables caused by heavy sediment loads in stream channels (Nolan and Marron 1995) can kill streamside redwoods (Figure 4.9). Unlike willow, maple, and Sitka spruce, redwood is intolerant of prolonged flooding of its root zone. Where stream beds are elevated by extremely high sediment loads, which raise adjacent groundwater levels, redwood roots are deprived of oxygen, water transport to their foliage slowly fails, and the trees die over a few years. When trees on toe-slopes are buried by mudflows, the results are the same. The more complete the lack of oxygen, the faster trees die, sometimes within one season.

In the central redwood forests, fire frequencies are higher and redwoods are more confined to canyon bottoms and low north-facing slopes. Successful redwood regeneration by seed is probably less likely because of lower total annual precipitation, and the proportion of trees sprouted from fire-killed stems is likely to be greater. Douglas-fir is also more prevalent in such stands. On the Los Posadas State Forest, some 120 km from the ocean, young redwoods are rare. However, seedlings were observed in the mineral soil of a dirt road berm, having established after a year of unusually high rainfall.

In the southern redwood forests of the Big Sur coast, Monterey County, redwoods are limited to small valley bottom stands surrounded by grassland and chaparral. Trees of both seedling and sprout origins occur in moist shaded sites. On slopes well above the valley bottoms, dense stands of young stems of sprout origin are found where successful seedling establishment is highly unlikely. These stands may be post-Pleistocene relicts of a wetter climate, surviving frequent fires by sprouting and resprouting from basal burl tissue, persisting as clones for many centuries. Data obtained following the July, 1985, Rat Creek fire suggest a very short fire interval followed by nearly three fire-free decades (Veirs, unpub.).

In the northernmost redwood stands, the absence of multiple-stemmed trees, snags, and logs suggests the possibility that these are newly established stands, representing a northward migration of redwood into southwestern Oregon. Redwood is damaged by hard freezes; its northern range may have been reduced during Pleistocene glacial periods, with recovery occurring today.

In Humboldt County, prairies disturbed by livestock and abandoned in the early 1900s have been invaded by Douglas-fir. As the Douglas-fir stand begins to open after 50–75 years, redwood and tanoak are becoming established. Redwoods also occur on streamside landslide scars where proximity of seed sources, relatively high moisture levels, and some shading facilitated their establishment. In the absence of fire, redwoods will probably increase their range and establish in the shade of Douglas-fir and other trees. Douglas-fir will disappear.

With increased fire and disturbance, redwood trees will have greater mortality, seedling success will decline, and competitors with greater drought resistance and higher light requirements will be more successful. Park managers must identify the natural character of the redwood forests under their care and work to maintain natural stand-shaping conditions for their perpetuation.

Disturbance Regimes

Ecologists consider disturbance a critical factor in forest ecology (Pickett and White 1985). A disturbance breaks the dominance of a site by established individuals, allowing room for new individuals of the same or different species to grow. Disturbances that are intermediate in frequency or intensity tend to promote maximum diversity in many systems because they prevent dominance by a few, highly competitive species but are not so severe as to allow only the most disturbance-tolerant species to survive (Connell 1978; Huston 1979). Different species have evolved different ways of avoiding, tolerating, or exploiting disturbances. Variation in disturbance regimes in relation to climate, topography, and the characteristics of particular plant species (e.g., flammability) have considerable influence on the nature of the vegetation in any area.

Redwood evolved along with fire, windstorms, floods, and potentially damaging animals and fungi. The life history characteristics of redwoods reflect adaptations to these factors, and their effects are evident in stand dynamics (box 4.3). Numerous studies have found that basal sprouting in redwood, for example, is closely associated with the occurrence of fire. Such findings are consistent with studies in other Mediterranean-type climates, which report the occurrence of lignotuber-producing angiosperms in habitats where fire, or other types of recurring disturbance, is common.

Today, the dominant element of the disturbance regime for redwood forests is usually logging and other human disturbances, which are discussed in subsequent chapters. Redwood has some resiliency to this new disturbance regime. The remarkable ability of redwood trees to resprout from basal lignotubers, regardless of age, is an important part of the species' persistence in the face of extensive clear-cutting. Throughout much of its natural range, logging has transformed redwood into a clonally reproducing species that spreads by means of a massive, underground lignotuber. One report describes a colony of forty-five redwood trunks that formed a third-generation fairy ring 17 m by 15 m across, whereas another illustrates a lignotuber exposed by erosion that was 12.5 m across and weighed 475,000 kg.

Fire

Casual surveys of redwood stands in parks and preserves throughout the species' range present abundant evidence of past fires. Trunks and snags are often black with charcoal. In many examples, trees are sufficiently burned out to create "goose pens." Trees surrounding a burned skeleton are commonly seen; they are apparently sprouts brought on by a fire in the remote past. New branches and entire crowns can grow from buds activated when the cambium is damaged by fire. Besides the ability of redwoods to sprout from lignotubers after fire, the bark is thick and fibrous and resists ignition.

A fire-history study in an old-growth stand in Redwood National Park revealed that the effects of surface fires in 1894 and 1974 on stand structure were inconsequential; however, competition among trees was apparently reduced (Abbott 1987). The width of growth rings increased after the fires. Most scarring from the 1974 fire occurred on the upslope side of trees, whereas basal spouting was induced on the downslope side. Understory trees present before the fire with a dbh of less than 20 cm were top-killed, but most of these subsequently resprouted.

Nives (1989) documented fire behavior on the forest floor in a northern redwood stand. She found fuel depth and load the best predictors of fire intensity and duration. The influences of elevation and topographic position, and their related weather, affected relative humidity and fuel moisture and, therefore, ease of ignition and rates of spread. Test fires burned with low intensity and a low rate of spread, changing stand characters little.

Fire-return intervals vary widely among redwood forests, depending on several factors. In northern redwood stands, Veirs (1980a) used stand-age structure distributions to suggest that the fire-return interval varied along an ocean to inland gradient. Coastal sites experienced a fire-return interval of 250 to 500 years, intermediate sites 150 to 200 years, and inland sites 33 to 50 years. Using cross-sections to create a master chronology from two intermediate sites, Brown and Swetnam (1994) concluded that the fire-return interval was an astonishingly low seven years.

Using basal sprouts and fire scars, Stuart (1987) estimated presettlement fire-return intervals of 26 years for stands at Humboldt Redwoods State Park in the central redwoods section. Other studies (Fritz 1931; Jacobs et al. 1985) derived similar intervals. Finney and Martin (1992) found intervals of 6 to 23 years in stands at Annadel State Park, Sonoma County, where individual redwood trees grow with Douglas-fir and hardwoods. In southern stands, Greenlee and Langenheim (1990) found fire-return intervals of 17 to 175 years. In general, fires occurring at shorter intervals are expected to be less intense.

The wide range of estimated fire-return intervals for redwood forests suggests that fire has variable influences on redwood forest composition and structure. Fire may have favorable or detrimental consequences on the success of redwood, as well as that of associated tree species (box 4.3). Very frequent fires may favor tree species other than redwood (Veirs 1975). For example, small western hemlocks were abundant under a redwood overstory where fire had recently occurred (Veirs 1980b).

Wind and Flood

The Pacific Coast of northern California is subject to regular windstorms. High winds are therefore routine events in redwood forests, killing individual trees or small clumps of trees. Besides inducing tree-falls, severe winter storms in redwood forests induce "reiteration-falls" as sections of complex crowns are lost. Detached reiterated trunks destroy areas of the crown as they fall, and broken remnants of the crown respond to these disturbances by reiterating (fig. 4.7). Coarse woody debris from wind damage is frequently retained in complex crowns, providing platforms for debris accumulation. This debris, as well as scars from torn-out trunks and broken branches, provide entry points for wood-decay fungi. The resulting rotten hollows are colonized by terrestrial vascular plants, especially shrubs (see chap. 3).

Floods are also routine throughout the range of redwood. The most celebrated are those on alluvial flats along the Eel River in the central redwood forests (Becking 1968; Stone and Vasey 1968; Zinke 1988). Here, nearly pure stands of massive redwood show little butt swell characteristic of upland trees. Instead, the parallel trunks stick directly out of the alluvium. These floods have toppled hundreds of trees and buried others several meters deep. After a flood, redwood stems and roots sprout new roots into the newest layer of alluvium, which makes an excellent seed bed (Stone and Vasey 1968). After the 1966 flood, a carpet of seedlings established; thirty years later, patches of seedlings grow in more lighted areas, some only a few centimeters tall and others exceeding 2 m in height (Sugihara 1996).

Animal Damage

Redwood leaves, branchlets, roots, and wood are seldom severely damaged by nonhuman animals, presumably because of the presence of volatile essential oils and tannins. Despite this chemical protection, many insects infest redwood, although none are capable of singularly killing mature trees (Roy 1966). Cones and seeds are attacked by the cone moth and rounded borer; foliage is attacked by aphids, black araucaria, green chafer, redwood leaf beetle, redwood mealybug, redwood scale, and a spider mite; buds and shoots are attacked by the tip moth; twigs are attacked by anthaxia, bark beetle, black-horned borer, cypress puto, cypress twig borer, redwood bark beetle, roundheaded borer; and large branches and trunks are attacked by the bark moth, blackhorned borer, carpenter ants, California prionus, cypress puto, dampwood termite, large cedar borer, powderpost beetle, redwood bark beetle, roundheaded borer, sapwood borer, small cedar borer, spined woodborer, and western horntail (Olson et al. 1990; Roy 1966; Synder; see Species List).

Mammals also attack individual redwood trees, often damaging and sometimes killing them. Black bears strip the bark of small trees and eat the inner bark (cambial layer), whereas black-tailed deer browse redwood sprouts (Becking 1968; Olson et al. 1990; Roy 1966). Bears, which reputedly cause the greatest damage, concentrate their attacks on young trees, feeding most heavily on stems 25–50 cm dbh (Giusti 1990). Bear damage occurs primarily after the animals emerge from their dens in late winter and early spring, a time when sap flows are high.

Ecological Roles of Fungi

Almost all portions of a redwood tree are utilized as an energy source by fungi. Different fungi are found on living and dead cones, living and dead needles and branchlets, living and dead bark, heartwood of living trees, and dead wood in logs and branches, roots, and humus. Fungal hyphae, threadlike portions of a fungus collectively referred to as the mycelium, ramify throughout the substrate. Hyphae secrete enzymes that degrade the substrate into soluble compounds that are absorbed through the hyphal walls. The processes of enzymatic degradation and nutrient absorption allow fungi to function ecologically as partners in mutualistic associations (e.g., mycorrhizae; see earlier section on site productivity), as endophytes (i.e., fungi that cause symptomless infections in apparently healthy plant tissue), as saprobes (i.e., fungi obtaining energy from dead material), as pathogens (i.e., fungi causing disease in the parasitized host), and as food sources for other organisms.

Endophytic fungi of redwood function either as mutualists or as pathogens. The most abundant redwood endophyte, a species of *Pleuroplaconema*, is a mutualist (Rollinger and Langenheim 1993). Fungi that are endophytic mutualists stimulate tree growth or provide a competitive advantage by preventing herbivory and controlling pathogens (Espinosa-Garcia et al. 1996). Eleven of nineteen known endophytic species are either pathogenic or belong to genera with pathogenic representatives.

Two hundred sixty-eight species of saprobic fungi are associated with redwood (chap. 3; appendix 3.2). Fruiting bodies of mushrooms and coral fungi are abundant, especially in northern redwood forests during the wettest months of the year. They probably are responsible for the relatively fast decomposition rate of redwood leaf-litter (Pillers and Stuart 1993). In addition, saprobic fungi contain enzymes that can degrade cellulose, lignin, and chitin in the wood of snags and logs. Well-rotted coarse woody debris, particularly in the form of a brown cubical rot prevalent in redwood logs, retains water, which serves as a reservoir for use by fungi and redwood roots during times of drought. Several saprobic fungi, such as angel's wings, produce fruiting bodies that are consumed by deer and humans (O'Dell et al. 1996).

Redwood leaves, branchlets, roots, and wood are notoriously decay-resistant, presumably resulting from the presence of volatile essential oils (monoterpenoids and sequiterpenoids) and tannins (Espinosa-Garcia et al. 1996). Despite this chemical protection, at least twenty species of fungi are pathogens of redwood trees. These fungi cause bark canker, branchlet blight, leaf blight, root rot, rot of living heartwood, stem canker, and twig dieback (see Species List). Pathogenic fungi contribute to a forest's structural diversity by killing trees that later become snags and logs, creating important wildlife habitats (O'Dell et al. 1996).

The fruiting bodies and hyphae of many fungi in a redwood forest are an important food source for deer, elk, bear, small mammals, and insects. Some small mammals, such as the California red-backed vole, rely on these fungal structures for most of their food supply. Fly larvae, colembolids, other insects, and millipedes also consume fungal structures (O'Dell et al. 1996).

Conclusions

Redwood's ecological niche varies throughout its range, but some generalizations are possible. Seedling establishment occurs under many circumstances, particularly where water conditions are favorable and soil surfaces lack organic material. Sprouts are an important method of reproducing new trees. Redwood trees have few dangerous enemies; once established, they can grow for a very long time with the possibility of becoming among the tallest and most massive trees on earth. The presence of fog, which reduces moisture stress and allows the water transport system to operate to great heights, appears to be a key factor in allowing redwoods to grow so tall.

Crowns of ancient redwood trees are structurally complex, as they frequently develop multiple trunks by reiteration. Branch-to-branch, branch-to-trunk, and trunk-to-trunk fusions within complex tree crowns divert the flow of water and nutrients through the tree and may serve to stabilize the crown during storms. Crown complexity of old trees also promotes biotic diversity in redwood canopy communities.

Compared to other conifers, redwood has relatively high genetic variability. Modest genetic variation exists among regions, and modest to higher levels occur among stands within a region. Genotype-environment interactions, on both individual and population levels, is substantial over large environment gradients. Redwood shows little effect of inbreeding under favorable conditions, but strong inbreeding depression can occur in stressful environments, with individuals showing reduced growth.

Chapter 5

TERRESTRIAL FAUNA
OF REDWOOD FORESTS

*Allen Cooperrider, Reed F. Noss, Hartwell H. Welsh Jr.,
Carlos Carroll, William Zielinski, David Olson,
S. Kim Nelson, and Bruce G. Marcot*

The redwood forests support an unknown but large number of invertebrate species and, over all seasons, more than two hundred species of vertebrates (A. Cooperrider, unpub. comp.). Beyond this, knowledge of the redwoods fauna is surprisingly limited. Despite the enormous public attention that has been focused on redwood forests in recent years, comparatively little has been written about the ecology of their terrestrial fauna. One explanation for this deficiency is that few animal species are endemic to the redwood region and no known species is restricted to redwood stands. Although the biogeography of redwood forests and associated fauna has not been well studied, the evidence suggests that most vertebrate species opportunistically adapted to life in the redwoods relatively recently, rather than having coevolved with redwoods over a long period of geologic time. Vertebrate species lists from redwood and Douglas-fir forests in the redwood region show few differences (table 5.1), with a nearly 95 percent

Author contributions: Cooperrider, lead writer and organizer; Noss, writing and editing throughout; Welsh, amphibians and reptiles; Carroll and Zielinski, forest carnivores; Olson, invertebrates; Nelson, marbled murrelet; Marcot, editing throughout and contribution to table on endemics.

Table 5.1. Similarity of Terrestrial Vertebrate Fauna of Redwood and Douglas-fir Forests.

| | Number of Native Species | | |
	Redwood Forest	Douglas-fir Forest	Species in Common
Amphibians	18	18	16
Reptiles	16	18	15
Birds	105	110	101
Mammals	63	66	61
Total	202	212	193

Note: Data are derived from the California Wildlife Habitat Relationships (California WHR) system (Mayer and Laudenslayer 1988; Zeiner et al. 1988, 1990a,b). Nonnative species have been excluded, but no effort was made to revise or correct the WHR (i.e., several reviewers pointed out errors in the WHR, which we will assume balance each other in species numbers).

overlap in species presence; furthermore, the species that account for the 5 percent dissimilarity are not species characteristic of one type or the other. Nevertheless, the redwood forests contain some of the older and more sensitive vertebrate taxa that characterize the temperate rain forests of the Pacific Coast.

Vertebrate Distributions

Most of this chapter focuses on vertebrates because information on invertebrates in redwood forests is extremely limited. Invertebrates are addressed, however, as the first of three "case studies" of redwoods fauna later in this chapter. Below we review the biogeographic history and current biogeography of terrestrial vertebrates in redwoods.

Historical Biogeography

The amphibians and reptiles (herpetofauna) of the temperate rainforests of the Pacific Northwest, inclusive of the redwoods, appear to have derived from at least three historical sources. The oldest group probably evolved *in situ* from an Eocene or earlier origin, at least 50 million years ago; some of the progenitors of these species (e.g., the salamander genera *Aneides* and *Plethodon*) had a transcontinental distribution that was fragmented by tectonic uplift and mountain-building during the Miocene (5–20 million years ago; Welsh 1990). Lineages with this history have been referred to as the Western American Complex of the Old Northern Element (Savage 1960). Other examples include the giant salamanders, the torrent salamanders, salamanders of the genus *Ensatina*, the tailed frog, and a reptile, the western pond turtle.

A second component of the Pacific Northwest herpetofauna had a relatively recent Asian origin, having diffused across the Bering land bridge during the Pleistocene about one million years ago (the Holarctic Element; Savage 1960). Members of this group include the red-legged frogs (two subspecies) and the western toad. Lineages of the Western American Complex of the Old Northern Element and the Holarctic Element have been described as temporally distinct components of a Pacific Northwest track (Welsh 1988).

The third group of Pacific Northwest herpetofauna is the Madrean Track (*sensu* Welsh 1988). This track derives from the south, with forms radiating northward during the Pliocene-Pleistocene from earlier origins in the Sierra Madre Occidental region of Mexico. It consists of species that diffused both north and west into present-day Arizona and California, and then south onto the Baja California peninsula (Welsh 1988). Madrean Track forms in the Pacific Northwest include the salamander genera *Batrachoseps* and *Hydromantes*, the frog genus *Pseudacris* (formerly *Hyla*), and most of the genera of snakes and lizards that occur in the redwood zone (e.g., *Sceloporus, Elgaria, Eumeces,* and *Thamnophis;* see complete list in Welsh 1988). Several snake and lizard lineages are members of the North American Desert and Plains Track (Welsh 1988) and are at the extremes of their distributions in the mesic Pacific Northwest (e.g., gopher snake, common kingsnake).

The mammals of the redwood region also have diverse origins. Some groups, including the marsupials (now extinct in the region, except for the Virginia opossum, an introduced species from eastern North America), multituberculates (extinct egg-laying mammals, probably ancestral to the monotremes), and some placental mammals, apparently evolved in North America or other portions of Laurasia from reptile-like progenitors; they were well established as a diverse assemblage across much of North America by the early Cretaceous, 135 million years ago (Eisenberg 1981). Other mammal groups immigrated from Eurasia at various times. By the late Cretaceous, 70 million years ago, North America had been invaded by several families of Asian placental and multituberculate mammals. Other early mammal groups apparently migrated directly through a Europe–North America connection about 50 million years ago, when Asia and Europe were separated (Hallam 1994); many of these archaic groups were lost in a late Eocene extinction. The largest extinction of North American mammals occurred during the Miocene, about 6 million years ago, during a period of rapid climate change.

During the Pliocene, about 5–3 million years ago, beavers (*Castor*) and bears (*Ursus*) arrived from Eurasia, probably crossing the Bering Strait. Also during the Pliocene, porcupines arrived from South America via the newly created Isthmus of Panama (Hallam 1994). Interchange between North America and Eurasia continued across the Bering land bridge during the Pleistocene. Elephants (e.g., *Mammuthus*) arrived in North America about 1.6 million years ago, and the first wave of humans (*Homo*) arrived at least 25,000 years ago

(Hallam 1994). Elk, members of the genus *Martes* (marten, fisher), and some other mammals also arrived from Eurasia and have close relatives in the Old World today. Many mammals of the Pacific states, as elsewhere in North America, went extinct during the Pleistocene, especially toward the end of the epoch (Ingles 1965). The late-Pleistocene extinctions have been attributed to overkill by humans (Martin 1984), but climatic changes certainly played a role (Webb and Barnosky 1989). Many mammals found in the redwood region today appear to be relatively recent (post-Pleistocene) arrivals to the West Coast.

The origin of redwood forest birds is even more complex and difficult to generalize about because of the small fossil record and birds' great mobility and wide-ranging migratory habits. Most of the modern families of birds in North America and Europe were there by the early Oligocene, some 35 million years ago (Martin 1983). Some of the less-mobile birds of the redwood forest, such as the blue and ruffed grouse, are closely related to Eurasian species and have no South American counterparts. Thus, their origin probably parallels that of the mammals that show a circumboreal, Northern Hemisphere distribution. Most of the bird species in the redwoods today, however, are neotropical migrants— species that migrate from North America to tropical regions of Mexico and Central and South America. Many of them, such as the nine warblers (family Parulidae), are members of groups that are widespread in South America but not found in Eurasia, although their origins within the Americas are complex (Darlington 1957; Brown and Gibson 1983).

Current Biogeography

Among amphibians and reptiles, member lineages of the Pacific Northwest Track occur throughout the redwood zone. Several genera of amphibians have more than one species in the Pacific Northwest, and some (e.g., *Dicamptodon, Plethodon,* and *Aneides*) have more than one species in the redwoods. The single reptile member of this track, the western pond turtle, occurs in larger streams and rivers, as well as ponds, throughout the redwoods region and most of the Pacific Northwest (though it is in decline). Some member lineages of the Madrean Track (i.e., *Batrachoseps* and *Pseudacris regilla*) reach far northward and occur throughout the redwood forest. Many of the reptiles members of this track, and those of the North American Desert and Plains Track, are near or at their physiological limits in the cool, moist habitats that characterize these forests and are therefore patchy in occurrence and more likely to be found in ecotonal habitats, where they can bask to maintain critical physiological functions.

The birds and mammals of the redwood forest are, for the most part, much less responsive to the moisture regime of the habitat and are not restricted to moist habitats. Rather, most are widely distributed along the Pacific Coast, the western states, or the continent. In generalizing about the Pacific coastal rain forest, Bunnell and Chan-McLeod (1997) concluded that although the area is extremely rich in vertebrate diversity, relatively few species (41 species; 11 percent of the total) are restricted to the region. Of these 41 Pacific coastal rain-forest species, 21 are found in the redwood forest, which represents the southern terminus of the region. This includes 12 of the 21 amphibians, all 3 reptiles, and 8 of 11 mammals, but none of the 6 bird species. The bird species included in Bunnell and McLeod's categorization are all coastal species (five seabirds and one shorebird) rather than forest dwellers.

Endemism and Near-Endemism

No species or subspecies of vertebrate is known to be strictly endemic to the redwood forest. Several species and subspecies, however, are endemic to the redwood region and several more are near-endemic (table 5.2). The red-bellied newt is found largely in redwood forests, but also in other habitats (Stebbins 1985; Petranka 1998). Some of the species listed in table 5.2 have extremely limited distributions. For example, the Santa Cruz long-toed salamander is found in a small area of Santa Cruz County but has never been reported in redwood forest. Several families of amphibians are endemic to the temperate rain forests of the Pacific Northwest; they comprise a large part of the amphibian fauna of the redwoods. These families are the Ascaphidae, a frog family with a single species (the tailed frog), considered the most primitive extant frog on earth (Ford and Cannatella 1993); the Dicamptodontidae, the giant salamanders, which are the largest terrestrial salamanders in the world; and the Rhyacotritonidae, the very ancient and primitive torrent salamanders (Good and Wake 1992). This high endemism at the family level is unusual, normally found only at much larger geographic scales.

Six species of birds—blue grouse, Steller's jay, pygmy nuthatch, hermit thrush, hairy woodpecker, and red crossbill—have subspecies confined to the coastal coniferous forest, including the redwood region, but the species are much more widespread, occurring in montane forest as well (Small 1974). Because many ornithologists no longer recognize subspecies, and the American Ornithologists' Union no longer maintains a checklist of subspecies, these birds (if any were to qualify) are not listed in table 5.2. One mammal species and several subspecies are more or less confined to the redwood region.

Although it appears that no species or subspecies is restricted to redwood forests, many are found largely within the redwood region. The ultimate survival of several taxa is probably tied to the persistence of these forests. The temperate rain forest of the Pacific Coast, inclusive of the redwoods, contains many ancient, "long-branch" animal taxa indicative of *in situ* evolution over millennia. The fauna of late-seral (i.e., mature and old- growth) redwoods likely represents

Table 5.2. Terrestrial or Amphibious Vertebrate Species Endemic or Near-Endemic (i.e., > 75% of Total Range) to the Redwood Region.

Species	Endemism and Status	Range	Habitat	Source
AMPHIBIANS				
California giant salamander (*Dicamptodon ensatus*)	Endemic to redwood region	From Sonoma and Napa Counties south to Santa Cruz Co.	In and about permanent and semipermanent streams in mesic coastal forests	Good 1989, Petranka 1998
Santa Cruz long-toed salamander (*Ambystoma macrodactylum croceum*)	Endemic to redwood region; G5T1/S1, FE, CE, CP	Small portion of Santa Cruz and Monterey Counties, around Monterey Bay	Under debris near water; not reported in redwood forests	Smith 1978, Petranka 1998
California slender salamander (*Batrachoseps attenuatus*)	Near-endemic to redwood region	Extreme southwestern Oregon south in coast ranges to slightly south of San Francisco Bay; also in west foothills of Sierra Nevada	Under logs and rocks in forests and other habitats, including grasslands, chaparral, and oak woodlands	Leonard et al. 1993, Yanov 1980, Petranka 1998
Red-bellied newt (*Taricha rivularis*)	Endemic to redwood region and virtually to redwood forest	Northwestern California in Sonoma, Mendocino, and Humboldt Counties	Along clean, rocky streams with moderately fast flow; in redwood forests, but also Douglas-fir/hardwood and meadows	Stebbins 1985, Petranka 1998, H. Welsh pers. obs.
Del Norte salamander (*Plethodon elongatus*)	Near-endemic to redwood region; G3/S3, CSC, CP	Southwestern Oregon and northwestern California Coast Ranges	Rocky talus and outcrops within old redwood and Douglas-fir forests	Welsh 1990, Raphael 1988, Diller and Wallace 1994, Petranka 1998
California (southern) clouded salamander (*Aneides vagrans*)	Near-endemic to redwood region; recently split from A. ferreus	From the South Fork of the Smith River in extreme northwestern California to just north of San Francisco Bay	Coarse woody debris and large logs in coastal forests; highly arboreal, nesting up to >40 m in redwoods	Petranka 1998, T. Jackman pers. comm.
Black salamander (*Aneides flavipunctatus*)	Near-endemic to redwood region	Extreme southwestern Oregon south into Santa	Decaying logs or stumps or wet talus within mature	Leonard et al. 1993, Raphael 1988.

124

Species	Status	Distribution	Habitat	Reference
		Cruz and Santa Clara Counties, California	mixed evergreen and hardwood forests	Petranka 1998
REPTILES				
San Francisco garter snake (*Thamnophis sirtalis infernalis*)	Endemic to redwood region; subspecies of common garter snake; G5T2/S2, FE, CE, CSC, CP	Western portion of San Francisco Peninsula from about S.F. Co. line south along crest of hills at least to Crystal Lake and along coast to Point Ano Nuevo, San Mateo Co.	Ponds, marshes, roadside ditches, streams, meadows, city lots; near water, within coastal scrub forest and redwood forest	Bounty and Rossman 1995
Pacific Coast aquatic garter snake (*Thamnophis attratus attratus*)	Near-endemic to redwood region	From southwestern Oregonto Sonoma and Marin Counties, where it intergrades with T.a. hydrophilus	Primarily rivers and streams, but also other aquatic habitats	Rossman et al. 1996
Pacific Coast aquatic garter snake (*Thamnophis attratus hydrophilus*)	Near-endemic to redwood region	From Big Sur coast around San Simeon north to Sonoma and Marin Counties, where it intergrades with T.a. attratus	Primarily rivers and streams, but also other aquatic habitats	Rossman et al. 1996
Coast garter snake (*Thamnophis elegans terrestris*)	Near-endemic to redwood region; subspecies of western terrestrial garter snake	Coast ranges from southern Oregon to Point Conception, California	A variety of habitats, including grassland, brushland, woodland, and forest, from sea level to mountains	Stebbins 1966, Rossman et al. 1996
MAMMALS				
Fog shrew (*Sorex sonomae sonomae*)	Near-endemic to redwood region (barely); split from S. pacifica	From central Oregon Coast Range south to Marin Co.	Alder/salmonberry, riparian alder, and skunk-cabbage marsh habitats; less often in mature and immature coniferous forest	Maser et al. 1981, Carraway 1990

(continues)

Table 5.2. (*continued*)

Species	Endemism and Status	Range	Habitat	Source
Point Reyes mountain beaver (*Aplodontia rufa phaea*)	Endemic to redwood region; G5T2/S2, FSC, CSC	Small portion of Marin County (Point Reyes)	Coastal scrub	Steele 1989
Point Arena mountain beaver (*Aplodontia rufa nigra*)	Endemic to redwood region; G5T1/S1, FE, CSC	Small area north of Point Arena, Mendocino Co.	Coastal scrub thickets on north-facing slopes; also herbaceous and wooded areas	Steele 1986, 1989
Point Reyes jumping mouse (*Zapus trinotatus orarius*)	Endemic to redwood region; G5T2?/Q/S2?, FSC, CSC	Marin County (Point Reyes)	Riparian, grassland, and wet meadow habitats	Zeiner et al. 1990b
California tree vole (*Arborimus pomo*)	Near-endemic to redwood region; G4S3, FSC, CSC	Northern to central California Coast Range	Dense coastal forests of at least Douglas-fir; poorly known	Johnson and George 1991
Salt-marsh harvest mouse (*Reithrodontomys raviventris*)	Endemic to redwood region; G1G2/S1S2, FE, CE, CP	Wetlands of San Francisco Bay and its tributaries	Dense pickleweed salt marsh and marginal habitats	Zeiner et al. 1990b
Humboldt marten (*Martes americana humboldtensis*)	Endemic to redwood region; G5T2T3/S2S3, CSC	Coastal forests in northern California to near (or slightly beyond) Oregon border	Redwood and other conifer and mixed hardwood-conifer forests	Hall 1981

Note: No known vertebrate is strictly endemic to redwood forests, but at least one (red-bellied newt) is virtually so. Species that occur largely within the region but do not meet the near-endemic criterion include the tailed frog (*Ascaphus truei*), red-legged frog (*Rana aurora*), and Allen's hummingbird (*Selasphorus sasin*); their conservation within the region would account for a substantial portion of their overall distribution. Status information includes California Natural Diversity Data Base codes: G1 = less than 6 viable element occurrences (EOs) or less than 1000 individuals or less than 2000 acres; G2 = 6–20 EOs or 1000–2000 individuals or 2000–10,000 acres; G3 = 21–100 EOs or 3000–10,000 individuals or 10,000–50,000 acres; G4 = apparently secure, yet factors exist to cause some concern; G5 = demonstrably secure. T codes signify status of taxonomic subcategory (subspecies or variety); S codes signify status within state (CA); range of values (e.g., S2S3) and ? signify uncertainty about appropriate status. Federal and state status: FE = federally endangered, FT = federally threatened, FSC = federal special concern; CE = California endangered, CT = California threatened, CP = California protected, CSC = California special concern. This list is probably incomplete, because scientific information on the status and distribution of subspecies and races of vertebrates in the region is incomplete and under revision.

the more vulnerable end of a continuum of biological resilience in coastal temperate rain forests, with many species showing relatively narrow adaptations and tolerances.

Faunal Description

In this section, we review the extant vertebrate fauna of the redwoods. We provide more detail for amphibians and reptiles than for mammals and birds because the amphibians and reptiles of the redwoods have been better studied as a group in this habitat. Much more detail on one group of mammals (the forest carnivores) and one species of bird (the marbled murrelet) will be provided later in the case studies.

Amphibians

Amphibians display the widest range of natural history strategies of all vertebrates, including the typical biphasic aquatic larvae and terrestrial adult mode to strictly terrestrial and strictly aquatic modes. This range of strategies can be found among both the anurans (frogs and toads) and the salamanders. Except for the lungless salamanders of the family Plethodontidae, however, all of the amphibians of the redwood forest use aquatic habitats for reproduction. Consequently, the streams, wetlands, and riparian habitats of the redwoods are rich in amphibian life, potentially containing up to ten species during the breeding season (one toad, four frogs, and five salamanders). Though the adults of some species (e.g., tailed frog, foothill yellow-legged frog, and southern torrent salamander) are closely associated with riparian and aquatic habitats throughout their lives, adults of other species (e.g., Pacific tree frog, California and northern red-legged frogs, western toad, Pacific and California giant salamanders, northwestern salamander, and the rough-skinned and red-bellied newts) can be found throughout forest habitats. The adults of this latter group of more terrestrial species make annual or semiannual migrations to aquatic sites, such as ponds, marshes, or streams, during the breeding season but otherwise spend much of their lives in the upland forest. Many of these species use subterranean haunts, such as rodent burrows and root channels, for shelter during the day and forage on the forest floor at night. The Pacific tree frog is highly arboreal and can be found at any stratum of the forest from underground to the canopy.

The amphibian order Anura (frogs and toads) is represented by four families in the redwood forest: the Bufonidae, or true toads, with one species, the western toad; the Ranidae, or true frogs, with three species, the California and northern red-legged frogs and the foothill yellow-legged frog; the Hylidae, or treefrogs, with one extant species, the Pacific tree frog; and the Ascaphidae (or Leiopelmatidae) with a single extant species, the tailed frog. All of these anurans exhibit the classic biphasic life history with aquatic larvae and terrestrial or semiaquatic adults. The tailed frog is associated with old forests, in low-order, fast-flowing, shaded streams with cold, clear waters and low amounts of fine sediments and abundant clean, coarse sediments (Welsh 1990, 1993). The foothill yellow-legged frog is associated with larger streams with open canopies and abundant, abovewater rocky substrates for basking. The northern and California red-legged frogs and the western toad use primarily lentic (nonflowing) waters with vegetated banks for breeding, but are less permanently associated with stream-bank and riparian habitats than the tailed frog and foothill yellow-legged frog. Details of the larval biology of these anurans and those of the aquatic salamanders described below will be discussed in chapter 6.

The amphibian order Caudata (salamanders) is represented by five families in the redwood forest: the Ambystomatidae, or mole salamanders, with a single species, the northwestern salamander; the Dicamptodontidae, or giant salamanders, with the Pacific giant salamander (north of Sonoma Co.) and the California giant salamander (south of Sonoma Co.); the Plethodontidae, or lungless salamanders, with seven species; the Rhyacotritonidae, or torrent salamanders, with a single species, the southern torrent salamander; and the Salamandridae, the true salamanders or newts, with two species, the rough-skinned newt and the red-bellied newt. Except the Plethodontidae, these salamander families all have aquatic larvae and therefore use lentic or lotic aquatic habitats for reproduction. Torrent salamanders, a species associated with headwaters in mature forests (Welsh 1990), and paedomorphic individuals (individuals that retain larval characteristics such as gills when sexually mature) of the giant salamander, are highly aquatic and rarely leave the water. The northwestern salamander and the two newt species spend much of their adult lives in uplands habitats, usually seeking shelters in burrows and under logs during the day or during dry or cold periods.

The amphibians with the greatest representation in the redwoods are the lungless salamanders (family Plethodontidae). Four genera of plethodontid salamanders are found in redwood forests: the slender salamanders (genus *Batrachoseps*), ensatina (genus *Ensatina*), the climbing salamanders (genus *Aneides*), and the woodland salamanders (genus *Plethodon*). The California slender salamander occurs north of the Pajaro River in Santa Cruz County, whereas the Pacific slender salamander occurs south of the Pajaro River to San Luis Obispo County. Welsh and Lind (1991) reported ensatina, followed by the California slender salamander, the most abundant salamander species in the mixed conifer–hardwood forests of northwestern California. The California slender salamander is the most common salamander in the redwood forests, followed closely by ensatina (fig. 5.1). Several subspecies of *Ensatina*, including the Oregon salamander *(E. e. oregonensis)*, the painted salamander *(E. e. picta)*, and the yellow-eyed salamander *(E. e. xanthoptica)*, occur in the redwoods. Research from mixed conifer–hardwood forests of the eastern United States indicates that plethodontid salamander biomass can equal or surpass that of all other small vertebrates combined in forested habitats (Burton and Likens 1975). Population estimates equaling or surpassing those reported for the eastern United States have been reported for plethodontid salamanders in forested habitats of the Pacific Northwest (see Welsh and Lind 1992 and references therein). Although these studies were not conducted in redwood forest, there is every reason to believe that the numbers of plethodontid salamanders (especially the slender salamander and ensatina) in redwoods may equal or surpass those in other forest types of the Pacific Northwest. The California slender salamander has been found significantly more abundant in mature and old-growth redwood forest compared with young forest (ANOVA; F = 5.90, P = 0.013; fig. 5.2).

The genus *Aneides* includes two species in the redwoods, the terrestrial black salamander and the more arboreal clouded salamander. The clouded salamander frequents large conifer logs in the forests of northwestern California and is a good climber (Welsh and Lind 1991). Recently, it has been found nesting 40 m above the ground in a large, live redwood tree, using fern mats and other microhabitats in the canopy (Welsh and Wilson 1995; S. Sillett, pers. comm.). This is the only salamander outside of the New World tropics known to use forest canopy, and is therefore the focus of a new research effort.

The genus *Plethodon* is represented by two species in the redwoods, the Del

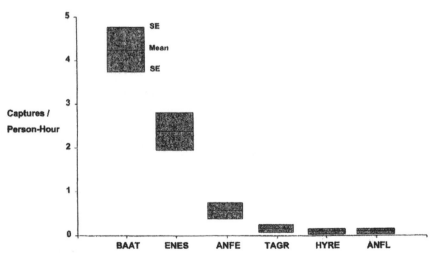

Figure 5.1. Relative abundance of the six most common amphibian species from eighteen redwood stands. Data were collected using time-constrained searches (Welsh 1987) in the springs of 1984–1986. BAAT = California slender salamander; ENES = ensatina; ANFE = clouded salamander; TAGR = rough-skinned newt; HYRE = Pacific tree frog; and ANFL = black salamander.

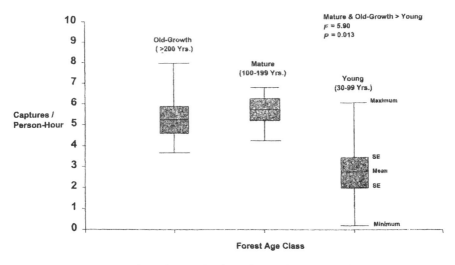

Figure 5.2. Captures of California slender salamanders (*Batrachoseps attenuatus*) by forest age class from eighteen redwood forest stands. Data were collected by time-constrained searches (Welsh 1987) in the springs of 1984–1986.

Norte salamander and Dunn's salamander. The Del Norte salamander, while occasionally using downed wood, is most often associated with rock substrates, such as talus, rock rubble, and rock outcrops (Welsh and Lind 1995). Based on data primarily from interior conifer-hardwood forests, Welsh (1990) described this species as associated with late-seral forests. Forest canopy closure was the best predictor of the presence of Del Norte salamanders, with sites supporting salamanders having a mean of 83 percent closure (Welsh and Lind 1995). On the other hand, Diller and Wallace (1994) did not find Del Norte salamanders restricted to late-seral stands of redwoods in north coastal California. The range of Dunn's salamander barely reaches south into the redwoods of extreme north-western California. This species is associated with damp rocky substrates, usually near stream channels.

Reptiles

Compared to amphibians, reptiles have lower overall diversity and lower relative abundances in the redwood forest. This pattern can be explained by physiological requirements. Amphibians require cool, moist habitats; many species avoid daylight or direct sunlight. In contrast, reptiles are often heliotherms, requiring direct sunlight or higher ambient temperatures to establish body temperatures sufficient to function on a daily basis. This being the case, the most likely places to find reptiles associated with redwood forest would be in clearings, along an edge with other habitat types, or along the margins of larger streams and rivers where adequate sunlight penetrates into the forest or onto riparian substrates. A single species of turtle (order Chelonia)—the western pond turtle—occurs in ponds or lakes and along streams and rivers in redwood forest. This turtle is associated with openings for basking adjacent to deep-water habitats, where it hides from predators (Reese and Welsh 1998). The western pond turtle uses upland forested habitats both for nesting and hibernation (Reese and Welsh 1997).

The other reptiles of the redwood forest belong to the order Squamata, the snakes and lizards. The lizards are represented by five species: western fence lizard, sagebrush lizard, northern alligator lizard, southern alligator lizard, and western skink. Of these, only the northern alligator lizard is likely to be encountered in the deep forest, where it forages for invertebrates in the litter and under cover objects on the forest floor. The other species of lizards might be encountered in clearings or along stream banks in open habitats or along forest edge where sunlight penetrates. The snakes that can be found in the redwoods are representative of three families: the Viperidae, represented by the western rattlesnake; the Boidae, represented by the rubber boa; and the Colubridae, which includes the remaining nine species: sharp-tailed snake, California kingsnake, western racer, gopher snake, ringneck snake, northwestern garter snake, common garter snake, western terrestrial garter snake, and aquatic garter snake. The ringneck snake, sharp-tailed snake, garter snakes (except the aquatic garter snake), and rubber boa all frequent forest habitats and are usually associated with logs and other cover on the forest floor, where they forage for small vertebrates or invertebrates. The gopher snake, kingsnake, rattlesnake, and western racer occur in relatively open, terrestrial habitats. Aquatic garter snakes are associated with wetlands. All these species could be encountered along the redwood forest edge or in ecotones with open, sunny areas.

Birds

Approximately 100 native bird species representing 28 families use redwood forests (A. Cooperrider, unpub. comp.). These figures includes species that use redwood forest for feeding, breeding (nesting), or cover in at least one season. Thus, it includes species such as the belted kingfisher, which are primarily associated with aquatic and riparian inclusions within the forest, but it does not include coastal or oceanic species unless, like the marbled murrelet, they make use of redwood forest for vital needs. The group with the most species (23) in redwood forests includes wood warblers, tanagers, sparrows, and blackbirds, most of them migratory and until recently lumped into the family Emberizidae.

Surprisingly, because birds are the easiest to observe of all forest vertebrates, the avifauna of the redwood forests has not been well studied. One general pattern has been noted by observers: moving away from the coast or from areas of hardwoods or mixed conifers into stands of pure redwood, the number of bird species declines precipitously. Perhaps the depauperate avifauna of redwood forests explains the relative lack of interest among ornithologists. As pointed out by Small (1974), "Where the forests are thick, pure, and extensive, the number of bird species sharply declines. In California only six species of birds nest primarily in this forest."

The six bird species noted by Small (1974) as characteristic of redwood forest are ruffed grouse, resident locally in riparian areas and canyons; Vaux's swift, which nests in burned and hollowed conifers, usually redwoods; the migratory race of Allen's hummingbird, a summer visitor and transient, found largely in open areas near forest edges; chestnut-backed chickadee, a resident from the Oregon border south to about Morro Bay in San Luis Obispo County, but not restricted to coniferous forest; rufous hummingbird, which nest only in the coastal forests of Del Norte County; and "probably the most characteristic bird of the coastal coniferous forest," the varied thrush, which nests "deep in the dark forest glades where it is resident the year around although in winter there is some movement to the south and towards less dense stands of forest" (Small 1974). The varied thrush is a fragmentation-sensitive species associated with large patches of late-seral redwood forest (Hurt and George, in prep.).

A search of Breeding Bird Censuses published in *American Birds*, 1980–1994 (i.e., censuses have not been published thereafter) revealed only two censuses (1981–1982) in a redwood forest: an old-growth Douglas-fir–redwood forest in Mendocino County. Thirty-nine territorial individuals of 20 species were located in the 10.22-ha plot in 16 site visits in 1981, and 33 territorial individuals of 18 species were recorded in 16 visits in 1982 (Judah 1983). Curiously, Small's (1974) "most characteristic bird," the varied thrush, was not present. Rather, the most common species were western flycatcher, brown creeper, Wilson's warbler, chestnut-backed chickadee, hermit thrush, Allen's hummingbird, hermit warbler, rufous-sided (spotted) towhee, common flicker, hairy woodpecker, and winter wren. This avifauna is typical of coniferous forests and other habitats in the Pacific Northwest. Northern spotted owls were also present in Judah's (1983) surveys in both years.

Unlike other vertebrates, many bird species use the redwood forest on a seasonal, as opposed to year-round, basis. Many birds, such as the hermit warbler and Vaux's swift, nest and raise young in the redwoods, then migrate south for the winter. Others, such as Townsend's warbler, nest farther north or at higher elevations and then winter in the redwoods. Even birds such as mountain quail, blue grouse, and the varied thrush, which are permanent residents in the redwoods region, often make seasonal, elevational movements.

Mammals

The redwood forests support 61 species of mammals, representing 19 families (A. Cooperrider, unpub. comp.). Four of these families—Vespertilionidae (insectivorous bats; 11 species), Sciuridae (squirrels and chipmunks; 8 species), Muridae (mice and voles; 13 species), and Mustelidae (the weasel family; 6 species)—make up more than half of the species present. Almost all of these species make use of the forest for breeding, feeding, and cover. Most of these species are permanent residents, except some of the bats, which make short- to long-distance migrations, and the ungulates (elk and mule deer), which may make moderate seasonal movements and use habitats besides redwood forest. Some of the bat species, such as the Yuma myotis and fringed myotis, raise young in caves and crevices.

Among the mammals that appear most closely tied to late-seral redwoods are the bats. Until recently, bats had been virtually unstudied in redwood forests. Recent research on the use by bats of basal hollows in old-growth redwoods confirmed that these hollows are important roost sites, as judged by the quantities of bat guano in hollows. Species of bats could not be identified by guano; however, three fecal morphotypes were identified (Gellman and Zielinski 1996). Further study determined that bat use of hollows in old-growth redwoods is greatest in small, residual stands of old growth on commercial forestland, as opposed to within larger areas of unfragmented old growth (Zielinski and Gellman 1999). Eight species of bats were mist-netted in the study areas, with the Yuma myotis most abundant. This research underscores the value of retaining remnant stands of old growth in managed forests.

The role of mammals in the functioning of the redwood ecosystem is not well understood. We expect that many of the relationships among species groups and ecosystem functions are common to many Pacific Coast forests. For example, Maser (1988) describes the complex interaction among northern flying squirrels, mycorrhizal fungi, and tree growth in Douglas-fir forests, a set of relationships apparently common to redwood forests (see chap. 4 on the ecological role of fungi).

Another way in which vertebrates can affect forests is through potential alteration of succession by herbivory. Foresters have been concerned that heavy browsing by ungulates might prevent regeneration of "desirable" tree species. In some forest regions, deer and elk are known to alter the species composition of forests, at least locally, by overbrowsing seedlings and saplings of favored species of trees and shrubs (e.g., see Alverson et al. 1988). Heavy browsing, in turn, can reduce densities of other animals, shrub-nesting birds, for example (Noss and Cooperrider 1994 and references therein). We found no documentation of regeneration of redwoods or associated woody species being affected significantly by browsing of native ungulates, although these animals occasionally cause damage to individual trees. Damage to regenerating forests by dusky-footed woodrats was investigated by Giusti (1999), who found that woodrats live "within clumps of sprouting redwoods, feeding heavily on the emergent sprouts." The extent of damage and its effect on stand regeneration were not established. Damage to trees by black bears, who strip bark to eat the sugar-rich cambium, apparently can be a greater problem, especially when bears emerge in late winter (Giusti 1990), but whether this damage has a significant effect on regeneration at a stand level remains an open question. Bears appear to do most damage to young, even-aged redwood stands, as they concentrate their feeding on stems 25–50 cm dbh (Giusti 1990).

Species Richness Patterns

One simple, though incomplete, expression of biodiversity is the number of species present in an area: species richness (Noss and Cooperrider 1994). Among temperate coniferous forests and other major ecosystems in North America, the redwood region does not stand out as extremely high in native species richness (Ricketts et al. 1999), though it is biologically distinct in other ways (see chap. 1). Late-successional redwood forests, in particular, have been characterized as relatively depauperate in vertebrate species. Few quantitative data are available to confirm this observation, however. We pose the following generalizations about species richness patterns of terrestrial vertebrates in redwood forests as working hypotheses, which can be verified or falsified by further research.

1. Vertebrate species richness is lowest in homogeneous stands of young or mature redwood-dominated forests. Species richness of vertebrates is generally higher in heterogeneous habitats than in more homogeneous ones, a phenomenon well documented for bird communities in many types of habitat (MacArthur et al. 1962; Roth 1976) but not well studied in redwood forests. Lower vertebrate species richness in homogenous redwood stands is predicted by the knowledge that many vertebrates require habitat features and resources, such as fruit and nut-bearing trees and shrubs, that tend to be scarce in pure stands of redwoods. The scarcity of these plants, and of herbaceous flora, is at least partially explained by low light levels, which also affect some vertebrates (e.g., heliothermic reptiles) directly. Furthermore, old-growth redwood forests, which like other old-growth forests can be expected to have higher rates of treefall gap formation (Clebsch and Busing 1989), are predicted to have greater vertebrate diversity than young (40–100 years) or mature (100–200 years) forests as a result of higher structural heterogeneity and light levels (see chap. 4, box 4.1). Light levels (solar radiation) near ground level decrease in old-growth redwood forests with increasing proximity to boundaries with young redwood plantations (Russell et al. In press).

2. Species richness of vertebrates is greater where redwood-dominated stands have inclusions of riparian habitat. This prediction is based on the addition

of species that depend on riparian or aquatic habitat for at least some of their life-history needs. Higher light levels and rates of gap formation (because of vulnerability to wind and bank undercutting along streams) in riparian areas will contribute to greater heterogeneity and species richness.

3. Species richness is greater where redwood forest includes codominants such as tanoak and Douglas-fir in the canopy. As in the first two generalizations above, this prediction follows from the presence in mixed forests of resources such as acorns, which some animal species rely on.

4. Species richness is greater where late-seral (mature or old-growth) redwood forest is replaced by a mixture of seral stages, from the herb-shrub through old-growth stages. This prediction, which has considerable potential for misinterpretation, follows from the increase in horizontal habitat hetero-geneity (among-habitat diversity) and edge effects when a landscape is com-posed of a mosaic of stands of different ages (Noss 1983). Foresters and wildlife managers have often claimed that logging and creation of edge effects is good for wildlife, but in fact, many native species are sensitive to fragmentation and increase in the amount of edge habitat (see review in Noss and Cooperrider 1994:196–203). For high species richness of native vertebrates to be sustained on a regional scale, effects of fragmentation must be moderated by maintaining patches of late-seral forest large and connect-ed enough for persistence of species that require such forest.

5. Species dependent on specific structural components of old-growth red-wood forests (e.g., snags, downed logs, complex canopies of live trees) will decline as these forests are replaced by simplified plantations; therefore, species richness or abundances are predicted to be lower in plantations com-pared to natural forests of any age (Hansen et al. 1991; Noss and Cooperrider 1994). Many salamanders, for example, are dependent on coarse woody debris (see above). Plantations and other young, managed forests that retain structural legacies will be expected to retain more species, at least temporarily (see chap. 8).

Habitat Relationships

In making decisions regarding land uses or conservation priorities, questions such as the following often arise:

- What species are (or would be expected to be) present in a given area?
- How significant is one area relative to others for supporting a given species or group of species?
- How might the species composition of an area—or the abundance of a par-ticular species of concern—change when habitat is changed?

Answering these questions would help ensure that limited resources for conservation activities, such as money for purchasing forestlands, are used most effectively, and that land management plans consider potential effects of habitat alteration on species composition. Ideally, a conservation decision-maker would have time and funding to conduct a thorough on-the-ground inventory of all areas for all species of concern. Typically, neither time nor money is available for such efforts, and some assessment must be made without a complete field inventory.

Habitat Relationships Models

To deal with questions such as these, wildlife biologists have developed several methods, termed *habitat evaluation techniques* or *habitat relationships models*, for evaluating areas generically. These techniques take advantage of the observation that presence and abundance of wildlife species depend on available habitat; thus, knowledge of habitat characteristics can be used to predict presence/absence and sometimes abundance of wildlife species. These techniques have been described and explored in detail in other publications (see Cooperrider 1986a and Cooperrider et al. 1986); the discussion here focuses on specific techniques developed for or in common use in the redwood region.

Virtually all habitat evaluation methods developed to date are for vertebrates, with the most detailed models for birds and mammals. This bias reflects the interests of wildlife agencies (funded largely by hunting and fishing licenses) in vertebrates and the lack of natural history information on most species of invertebrates. A common method of predicting presence/absence of vertebrates in an area is based on the vegetation type and seral stage. For example, violet-green swallows are largely dependent on older seral stages of forest for breeding and feeding habitat, whereas cliff swallows favor early seral forests or open grasslands. Thus, armed with knowledge of the vegetation types and seral stages present in a landscape, a biologist often can predict with reasonable accuracy which species are likely to occur there.

Predictions from habitat relationships models, however, often contain substantial errors of omission and commission. For example, an assessment of the reliability of the California Wildlife Habitat Relationships (WHR) model in four seral stages of mixed evergreen forest in northwestern California found that 14 percent of the species were observed but not predicted (an error of omission), whereas 11 percent of the species predicted to occur were not observed (an error of commission). Species richness among seral stages differed significantly from model predictions for birds but not for amphibians, reptiles, and mammals (Raphael and Marcot 1986). Similarly, gap analysis models, which use simple habitat relationships models to associate vertebrate distributions with vegetation maps derived from remote sensing, always include some errors of omission and commission—11 percent and 21 percent, respectively, in the Idaho gap analysis (Scott et al. 1993). The accuracy of prediction tends to vary by taxonomic class, with predictions for birds generally being most reliable (Scott et al. 1993; Edwards et al. 1996). Distributions of rare, endemic, and patchily distributed species are least successfully predicted by habitat relationships models (Smith and Catanzaro 1996), which underscores the need to survey these species individually through a "fine filter" approach, rather than relying on modeling (Noss and Cooperrider 1994).

A more refined prediction of species occurrence can be made using knowledge of what biologists call "special habitat features" or "special habitat elements." These include habitat features other than vegetation type and seral stage that are required by many species. Snags, downed logs, springs, talus slopes, caves, and cliffs are examples of such features. Because these features usually cannot be surveyed remotely (i.e., from satellite or airplane), their measurement requires ground surveys, which are more costly. When predictions of vertebrate species distributions are made for large areas, as in gap analysis projects, the probability is relatively high that essential microhabitats are present; for smaller areas, predictions are less robust (Csuti 1996). Combining line (e.g., streams) and point data (e.g., locations of caves or other special habitats) with maps of vegetation can result in more accurate predictions for many species associated with these features over large areas (Edwards et al. 1995).

The California WHR system uses three basic habitat parameters—vegetation type, seral stage, and presence of special habitat elements—to predict presence or absence of vertebrate species (Airola 1988; Mayer and Laudenslayer 1988). Garrison (1992) applied the California WHR system to a 2,400-ha, second-growth (80–100-year-old) redwood forest watershed, to show how trends in species richness could be predicted and explained from simple knowledge of habitat requirements of vertebrates. The WHR model was used to predict the effects of a mixture of even-age and uneven-age management applied to the study area. Predicted responses included declines of 70 percent of the amphibian species and increases of 88 percent of the reptile species, 64 percent of the bird species, and 61 percent of the mammal species. Overall vertebrate species richness was predicted to increase with logging.

Although no attempt was made to verify Garrison's (1992) predictions, we suspect that at least the general trend (increased species richness as more seral habitat is created) would prove true (see generalization no. 4 in preceding section). Nevertheless, assigning positive conservation value to an abstraction such as "species richness" is foolhardy without considering the individual species added or lost when changes are made in the distribution of seral stages on a landscape. When an older forest is converted to earlier seral stages, the species likely to be lost are those closely associated with old forest, such as the spotted owl and marbled murrelet; that is, those that are rare regionally or globally and for which intensive conservation efforts are being undertaken. On the other hand, the species added to the forest tend to be relatively common or weedy ones, and may include exotics, such as wild pig. Therefore, all species cannot be treated as equal in conservation planning (Diamond 1976; Noss 1983; Noss and Cooperrider 1994).

Among other limitations, the California WHR does not consider the area requirements of particular species, the need of many species for contiguity or connectivity among patches of suitable habitat, or the need for spatial juxtaposition of different habitat types or habitat elements. The WHR assumes that one hectare of habitat is equivalent to 1,000 or 10,000 ha. And 100 ha are assumed to be of equivalent value in this system whether they are distributed in 10-ha patches or in one 100-ha block. This deficiency in the system has been exploited by unscrupulous companies and biologists in environmental assessments, such as timber harvest plan analyses, to mask the consequences of logging activities. One could harvest 999 out of 1,000 ha of old-growth redwood forest and show (on paper) no predicted decline in species richness or diversity. The authors and developers of the California WHR system have been forthright about this deficiency and have warned against such misuse (Airola 1988).

Another limitation of the WHR system is that many of the habitat relationships described for vertebrates of the redwood forest are nothing more than educated guesses. In searching the literature, we found few quantitative studies that would support these relationships. Despite these problems, the WHR system is widely used; therefore, it is essential that its foundations be well understood and, with future studies, validated.

Landscape Context

Wildlife biologists have known for a long time that spatial arrangement of the components in the landscape are critical determinants of the capability of the habitat to support wildlife species. More than sixty years ago Aldo Leopold wrote:

> The maximum population of any given piece of land depends, therefore, not only on its environmental types or composition, but also on the *interspersion* of these types in relation to the cruising radius of the species. Composition and interspersion are thus the two principle determinants of potential abundance. (1933:128–129)

Despite this knowledge, conventional wildlife habitat relationships models ignore landscape patterns, patch sizes, and other spatial phenomena critical to the persistence—if not the immediate presence—of many species. The models are incomplete in focusing on habitat *content* but not *context* (Noss and Harris 1986). In recent years, biologists and landscape ecologists have developed sophisticated ways of describing and organizing such spatial information. And with the advent of computer-based geographic information systems (GIS), analysis of such information has become easier. We describe here three of the most important spatial determinants of habitat suitability: patch size, juxtaposition, and connectivity. For descriptions of other spatial measurements, patch shape and porosity, for example, we refer readers to books on landscape ecology such as Forman and Godron (1986), Naveh and Lieberman (1990), Turner and Gardner (1991), Forman (1995), and Hansson et al. (1995). Chapter 7 of this book discusses application of principles of landscape ecology to conservation evaluation and landscape design.

PATCH SIZE. The size of a patch of habitat is important because a minimal area is necessary to support an individual animal or a breeding pair; a still larger area (either in one patch or several patches linked by dispersal; see below) is required to support a viable population—one with a specified probability of persisting over a given time period (Shaffer 1981; Soulé 1987). Thus, fifty 1-ha patches of old-growth redwood may not be adequate to support a single pair of a forest-interior bird species, whereas one 50-ha patch may support many breeding pairs. If animals are relegated to small patches of habitat that support only small populations, the probability of extinction may be high unless connectivity among patches is high (see below).

Fragmentation, which refers to the degree to which large blocks of habitat have been broken up into smaller, more isolated patches, is considered a leading threat to biodiversity (Wilcove et al. 1986; Noss and Cooperrider 1994; Noss and Csuti 1997). In portions of their range, redwood forests had a naturally patchy distribution; that is, they occurred in small, relatively isolated patches prior to the activities of modern humans. In other portions of the redwood region, redwood forests were relatively continuous—they constituted the landscape matrix. The spatial pattern of redwood forests today, especially in the northern section, is considerably more fragmented than the pattern before European settlement (Hurt and George, in prep.). Little research has been conducted on the effects of habitat fragmentation in the redwood region. Nevertheless, a recent study determined that the varied thrush, one of the most characteristic birds of the redwood forest, is fragmentation sensitive. Varied thrushes were present in 16 out of 17 fragments that were greater than 16 ha, but in only one of 21 fragments smaller than 16 ha (Hurt and George, in prep.). Presence of birds was also positively correlated with the size (dbh) of redwood trees in the fragments. The authors recommended maintaining redwood forest stands greater than 20 ha and older than 80 years of age to support breeding populations of varied thrushes. Similarly, a study of the red-backed vole in southwestern Oregon (Mills 1995) found this species sensitive to the area, isolation, and edge effects in forest remnants.

JUXTAPOSITION. Juxtaposition refers to the adjacency of one type of habitat to another type. For example, the adjacency or proximity of redwood habitat to riparian areas is critical for species that require both types of habitat to meet such life history needs as feeding, nesting, or thermal cover. If the required habitat patches are too far apart, then their utility to the species may be reduced or nonexistent.

Little is known of the juxtaposition requirements of redwoods fauna. Wide-ranging generalists such as black bears are known to require a mosaic of habitats to meet their life-history needs. Bats roosting in redwood hollows deep in the forest generally forage over watercourses and other habitat edges (Zielinski and Gellman 1999). Amphibians with terrestrial adult stages (with the exception of the fully terrestrial plethodontid salamanders, *Batrachoseps, Aneides,* and *Plethodon*) require adequate juxtaposition of aquatic breeding areas and terrestrial habitats.

CONNECTIVITY. Connectivity—specifically functional connectivity—refers to the potential for successful dispersal, population interchange, or daily or seasonal movement of organisms among two or more patches of habitat (Noss and Cooperrider 1994). Because of differences among species in life histories and movement capabilities, connectivity is best described relative to a particular species or biological phenomena. Connectivity may be present in the form of corridors (strips of a given type of habitat that connect larger patches of the same or similar habitat [Beier and Noss 1998]), but that is not the only way functional connectivity may be achieved. Some animals disperse freely through a landscape, as long as suitable resting, feeding, and predator avoidance habitat features are available. For example, connectivity among spotted owl management areas was defined in terms of a forest matrix that included a minimal canopy coverage of trees of a given size class (Thomas et al. 1990). Within the home ranges of owls, however, the total amount of suitable (old-growth) habitat is more important than the connections among patches of old growth. For less vagile organisms, even narrow (i.e., to us) barriers can prevent movement.

Connectivity is a necessary consideration in conservation planning because it potentially allows for persistence of species—for example, as metapopulations or systems of populations (Harrison 1994)—in a network of habitat areas, no single one of which is large enough to support a viable population. With functional connectivity, a whole can be greater than the sum of its parts (Noss and Harris 1986). Such considerations are crucial for species, such as the forest carnivores discussed later in this chapter, with demanding area requirements. Habitat fragmentation that poses barriers to movement can have severe effects on animals that need to travel from one kind of habitat to another daily or seasonally (see comments on juxtaposition, above). Roads, which cause a variety of ecological problems, often serve as barriers for species that hesitate to cross openings or that suffer high mortality from vehicles (Noss and Cooperrider 1994; Trombulak and Frissell 2000).

Special Habitat Characteristics and Features

Although relatively few vertebrates are endemic to the redwood region (table 5.2) and none strictly to the redwood forest, the redwood forest is a unique ecosystem that, nonetheless, shows commonalities with other portions of the Pacific coastal rain forest that stretches from the redwood region north to southeastern Alaska. Throughout this vast region, the life histories of most native vertebrates are closely tied to water and to the complex structure of natural forests (Bunnell and Chan-McLeod 1997). Most redwood forests are coastal and within the fog belt (see chap. 3 and 4). As with the rest of the coastal rain forest, the link between redwoods and coastal and estuarine habitat is critical to many species. For example, virtually all redwood forests have (or once had) streams with runs of anadromous salmonids. These runs supplied a pulse of food and nutrients for many terrestrial vertebrates, such as black bears, grizzly bears (now extirpated from the region), and bald eagles. The consequences of losing the fish runs—or having them greatly diminished—are poorly understood, but may include influences on forest productivity throughout the food web. Similarly, the connection between riparian areas and upland redwoods is critical; the richness and productivity of riparian forests, as well as their structural complexity and abundance of snags and downed wood under natural conditions, are crucial to the persistence of many populations (Bunnell and Chan-McLeod 1997).

The dependence of many terrestrial vertebrates on features of the natural old forest, such as snags, dead and downed woody material, and tree cavities, is well documented in other forests of the Pacific Northwest (Norse 1990; Ruggiero et al. 1991). What may be less apparent is that many terrestrial vertebrates surviving in managed (i.e., "cutover") forest stands may be there largely because of the legacies remaining from unmanaged forests. In British Columbia, for example, black bear dens could be found in second-growth forests, but primarily within structures (live trees, snags, or logs larger than 1 m in diameter) that were left over from the previous old-growth forest (Bunnell and Chan-McLeod 1997). Spotted owls may nest in second-growth redwoods only because of the structural legacies left from previous, old-growth stands (Noon and Murphy 1997). These observations leave us to wonder what will happen to native wildlife over subsequent, short rotations as the structural legacies from the natural forest gradually return to the soil.

Endangered and Imperiled Species

Forty-two vertebrate species of the redwood forest are considered by state and federal agencies endangered, threatened, or otherwise of concern (tables 5.2 and 5.3). Four invertebrate species are currently listed by the federal government as endangered or threatened; twelve others are federal species of concern (former and perhaps future candidates for listing) (table 5.4). None of these species is restricted to redwood forest. Our lack of knowledge, especially about forest

Table 5.3. Imperiled Terrestrial or Amphibious Vertebrates Found Within, But Not Endemic or Near-Endemic to, the Redwood Region.

Species	Category
AMPHIBIANS	
Southern torrent salamander (*Rhyacotriton variegatus*)	G4/S2S3, FSC, CSC, CP
Tailed frog (*Ascaphus truei*)	G3G4/S2S3, FSC, CSC, CP
Northern red-legged frog (*Rana aurora aurora*)	G4T2?/S2?, FSC, CSC, CP
California red-legged frog (*Rana aurora draytonii*)	G4T2T3/S2S3, FT, CSC, CP
Foothill yellow-legged frog (*Rana boylii*)	G3/S2S3, FSC, CSC, CP
REPTILES	
Western pond turtle (*Clemmys marmorata*)	G4S3, FSC, CSC, CP
BIRDS	
Aleutian Canada goose (*Branta canadensis leucopareia*)	wintering: G5T2/S2, FT
Tufted puffin (*Fratercula cirrhata*)	nesting: G5/S2, CSC
Osprey (*Pandion haliaetus*)	G5/S3, CSC
White-tailed kite (*Elanus leucurus*)	G5/S3, CP
Bald eagle (*Haliaeetus leucocephalus*)	G4/S2, FT, CE, CP
Northern harrier (*Circus cyaneus*)	G5/S3, CSC
Sharp-shinned hawk (*Accipiter striatus*)	G4/S3, CSC
Cooper's hawk (*Accipiter cooperi*)	G4/S3, CSC
Northern goshawk (*Accipiter gentilis*)	G4/S3, FSC, CSC
Golden eagle (*Aquila chrysaetos*)	G4/S3, CSC, CP
Merlin (*Falco columbarius*)	G5/S3, CSC
Peregrine falcon (*Falco peregrinus anatum*)	G3T2/S2, FE, CE, CP
Prairie falcon (*Falco mexicanus*)	G5/S3, CSC
Ruffed grouse (*Bonasa umbellus*)	G5/S4, CSC
Marbled murrelet (*Brachyramphus marmoratus*)	G3/S1, FT, CE, CSC
Northern spotted owl (*Strix occidentalis caurina*)	G3T2T3?S2S3, FT
Black swift (*Cypseloides niger*)	G4/S2, CSC
Purple martin (*Progne subis*)	G5/S3, CSC
Black-capped chickadee (*Parus atricapillus*)	G5/S3, CSC
Yellow warbler (*Dendroica petechia brewsteri*)	G5T2/S2, CSC
San Pablo song sparrow (*Melospiza melodia samuelis*)	G5T2??S2?, FSC, CSC
Alameda song sparrow (*Melospiza melodia pusillula*)	G5T2?/S2?, FSC, CSC
MAMMALS	
Salt-marsh wandering shrew (*Sorex vagrans halicoetes*)	G5T1/S1, FSC, CSC
Townsend's big-eared bat (*Corynorhinus townsendii townsendii*)	G5T3T4/S2S3, FSC, CSC
San Francisco dusky-footed woodrat (*Neotoma fuscipes annectens*)	G5T2T3/S2S3, FSC, CSC
Monterey dusky-footed woodrat (*Neotoma fuscipes luciana*)	G5T3?/S3?, FSC, CSC
White-footed vole (*Arborimus albipes*)	G4/S2S3, FSC, CSC

Note: Codes as in Table 5.2.

Table 5.4. Imperiled Invertebrates of the Redwood Region. Species listed are ranked G1, G2, S1, or S2 by the California Natural Diversity Data Base. None of the species is known to be restricted to redwood forest (most are narrow endemics, and many are restricted to caves or other unusual environments).

Species	Status
GASTROPODS	
Mimic tryonia (California brackish-water snail) (*Tryonia imitator*)	G2G3/S2S3, FSC
Pomo bronze shoulderband (snail) (*Helminthoglypta arrosa pomoensis*)	G2T1/S1, FSC
Rocky coast Pacific sideband (snail) (*Monadenia fidelis pronotis*)	G1/S1, FSC
ARACHNIDS	
Edgewood blind harvestman (*Calicina minor*)	G1/S1, FSC
Empire Cave pseudoscorpion (*Fissilicreagris imperialis*)	G1/S1, FSC
Dolloff cave spider (*Meta dolloff*)	G1/S1, FSC
Tiburon micro-blind harvestman (*Microcina tiburona*)	G1/S1, FSC
CRUSTACEANS	
Mackenzie's cave amphipod (*Stygobromus mackenziei*)	G1/S1, FSC
Tomales isopod (*Caecidotea tomalensis*)	G2/S2, FSC
California freshwater shrimp (*Syncaris pacifica*)	G1/S1, FE, CE
INSECTS	
Globose dune beetle (*Coelus globosus*)	G1/S1, FSC
Mount Hermon (barbate) June beetle (*Polyphylla barbata*)	G1/S1, FE
Ricksecker's water scavenger beetle (*Hydrochara rickseckeri*)	G1G2/S1S2, FSC
Fort Dick limnephilus caddisfly (*Limnephilus atercus*)	G1G3/S1, FSC
Lotis blue butterfly (*Lycaeides argyrognomon lotis*)	G5T1/S1, FE
Oregon silverspot butterfly (*Speyeria zerene hippolyta*)	G5T1/S1, FT

Note: Codes as in Table 5.2.

invertebrates, suggests that probably many other species are very rare or imperiled. Some imperiled species, listed or not, have presumably always been rare (i.e., such is the case with many endemics), whereas others have become rare because of human activities and face a high probability of extinction if those activities continue unabated. The case studies in the following sections provide examples of imperiled species and illuminate the forces that threaten them.

We cannot do justice, in one chapter, to the biology of the redwoods fauna. We noted earlier that this fauna has been poorly studied in comparison with many other major vegetation types. Yet what is known is fascinating. We offer, below, three case studies that give glimpses of the native fauna of redwoods that merit concern because of the activities of humans. The case studies are of (1) invertebrates, a group much more diverse than vertebrates but about which little is known for redwoods and other forests; (2) forest carnivores, a group that includes highly imperiled forms and is receiving increased research attention; and (3) the marbled murrelet, a federally threatened species that finds optimal habitat in old-growth redwoods. Together, these case studies provide strong lessons for conservation.

Invertebrates of the Redwoods and Other Northwest Forests

Invertebrate communities in redwood forests resemble those found throughout the temperate rain forests of the Pacific Northwest. Because information specific to redwood invertebrates is scarce, this similarity allows us to broadly characterize the redwood fauna using examples from other Pacific Northwest rain forests. Arthropods are some of the most diverse and best studied invertebrates of these habitats and will be emphasized in the following discussion.

An Extraordinary Diversity

As in most other terrestrial communities (Wilson 1992), invertebrates constitute the vast majority of animal species in redwood forests. An estimated 8,000 arthropod species comprise around 85 percent of all species of vertebrates, vascular plants, and arthropods in a relatively undisturbed Oregon Cascade forest (Asquith et al. 1990; Lattin 1990). Species densities can be amazingly high for invertebrates as well, with 200 to 250 species of invertebrates per square meter in undisturbed forest soils (Moldenke 1990). Arthropod abundance also can be high—more than 200,000 individual oribatid mites have been estimated for one square meter of forest floor, representing 75 species (Moldenke 1990). Temperate rain forests of the Pacific Northwest are also global and continental foci of diversity for several taxa. For example, the diversity and abundance of mites in old-growth forests here is thought to rival or exceed their diversities in many other terrestrial ecosystems, including tropical forests (Lattin 1990; Moldenke 1990). More species of slugs (24) live in these forests than any other in North America (Frest and Johannes 1993, 1996).

Comparing Temperate and Tropical Rain-Forest Faunas

The structure of arthropod communities in temperate rain forests differs from that observed in tropical rain forests in several ways. An abundance of mites, millipedes, isopods, and earthworms fills the role of major detritivores and decomposers in temperate rain forests, a function dominated by fungi and bacteria in tropical rain forests. For example, a Pacific Northwest cyanide-producing millipede (*Harpaphe haydeniana*) and sowbug (*Ligidium gracile*) are believed to be responsible for breaking down much of the litter and rotting wood into humus (Moldenke 1990; Lattin and Moldenke 1992).

Ants are exceptionally abundant in tropical forests; together with termites they constitute roughly a third of the dry-weight animal biomass, and are dominant consumers and predators (Wilson 1987). The relatively low diversity and abundance of ants in temperate rain forests is likely the result of the unfavorable cool and moist conditions of the forest floor. The scarcity of ants may contribute to the prevalence of land snails and slugs in temperate rain forests relative to tropical rain forests. Butterflies and native bees are also scarce in temperate old growth, leaving the role of pollinating understory plants to flies, moths, and beetles. Despite these differences, tropical and temperate rain forests share the vast majority of invertebrate higher taxa, and the complex old-growth forests of each have relatively similar structures and ratios of different functional guilds.

The Redwood Invertebrate Fauna

Distinct invertebrate assemblages are associated with three primary habitats in Pacific Northwest old-growth forests: the forest floor and understory, the forest canopy, and riparian habitats. The forest-floor invertebrate fauna of old growth harbors a large proportion of sedentary species (wingless and flightless forms) with narrow tolerance ranges to temperature and moisture conditions (Moldenke and Lattin 1990a; Lattin and Moldenke 1992; Frest and Johannes 1993). Such characteristics are seen in numerous species of oribatid mites, harvestman, millipedes, centipedes, springtails, beetles, flies, wasps, spiders, crickets, land molluscs, and isopods. Some taxa, such as millipedes, lack a waxy cuticle, making them sensitive to desiccation stress and restricting them to moist microhabitats. On an even finer scale, some soil arthropods, such as oribatid mites, are predictably associated with different combinations of soil temperature, moisture, structure, fungal abundance, limiting nutrients, and even the litter of different tree species (McIver et al. 1990; Moldenke 1990). Because of their specialized requirements, most forest-floor species appear to be poor at dispersing among old-growth fragments, and populations and locally endemic species are prone to extinction. Many old-growth associated species of millipedes and harvestmen known only from single patches of old growth, including some redwood fragments, have not been collected again since the type localities were logged (Olson 1992).

Knowledge of canopy invertebrates is rudimentary, but the few studies conducted in this challenging research environment reveal a highly diverse and distinctive fauna (Voeghtlin 1982; Schowalter 1989; Winchester 1993, 1996, 1997; Winchester and Ring 1996a,b). Most higher taxa found on the forest floor also are observed in the canopy on foliage, bark, and branches, or living in the detritus, moss, and root mats covering large branches in undisturbed temperate rain forests. These assemblages of canopy species, however, are quite distinct from forest-floor communities and those of younger forests, indicating a specialized arboreal old-growth fauna (Winchester 1996; Winchester and Ring 1996a,b). Canopy arthropod faunas are dominated by phytophagous and predator/parasitoid guilds (Schowalter and Crossley 1987; Schowalter 1989; Winchester and Ring 1996b), a pattern typical of functionally diverse and complex ecosystems (Winchester and Ring 1996b). Moreover, the amount and kinds of plant material lost to herbivory and the significant predator/parasitoid presence suggest that epizootic outbreaks are rare in late-seral forests (Winchester and Ring 1996b). Canopy invertebrates in moss mats are sufficiently numerous to support populations of clouded salamanders in redwood forests (see earlier discussion). Mating stoneflies and "pulses" of adult caddisflies observed in the canopies of old-growth Sitka spruce on Vancouver Island (Winchester 1993) also suggests a key role of canopy habitats in the life cycle of invertebrates more commonly linked to forest-floor and riparian habitats.

Although some canopy invertebrates are vagile dispersers with relatively widespread ranges, increasing evidence suggests that a sizable component of the canopy biota consists of species restricted to specialized microhabitats, such as moss mats. If such canopy species have limited geographic ranges, as observed in many forest-floor invertebrates (Olson 1992), then the loss and fragmentation of old-growth forests will lead to increasing extinction—before species are even discovered, in some cases (see Winchester and Ring 1996a). Given the diversity, specialization on old-growth environments, and limited ranges observed in many forest-floor and canopy invertebrates, it is highly likely that a wave of invertebrate extinctions already has accompanied the catastrophic loss of coastal old-growth forests.

Invertebrates associated with riparian and aquatic habitats in old-growth forests are poorly documented, but some species of stoneflies, caddisflies, and ground beetles are known to be restricted to these habitats. Some species are very localized, such as some ground beetles (e.g., *Nebria* spp.) whose only known localities are small stretches of streams (D. H. Kavanaugh, California Academy of Sciences, pers. comm.). Destruction of old-growth forests is known to severely affect streams and their biota (see Frest and Johannes 1996 and chap. 6, this volume).

The patterns of habitat specialization, low vagility, and restricted distribution displayed by a large percentage of the canopy, forest-floor, and aquatic invertebrate fauna are features often associated with forests that have enjoyed relatively stable and benign conditions for long periods of time, as have the redwoods of the Northern California Coastal Forest ecoregion (Lattin and Moldenke 1992; Ricketts et al. 1999). The ancient heritage of redwood forests is reflected in the presence of such primitive invertebrates as the silverfish, *Tricholepidion gertschi,* a living fossil that closely resembles the hypothetical ancestor of higher insects (Powell and Hogue 1979). Other relic species include the ant *Amblyopone oregonense*, the California swollen-footed pointed-wing moth *(Paraphymatopus californicus)*, which resembles the ancestor of moths and butterflies, and the colonial wood-feeding roach *(Cryptocercus punctulatus)*, a link between the roaches and termites (Lattin and Moldenke 1992). Many other distinctive taxa occur throughout the rain forests of the Pacific Northwest, including the taxonomically unique (e.g., monotypic genera) long-jawed nimble millipede searcher *(Promecognathus laevissimus)*, Matthew's ground beetle *(Zacotus matthewsi)*, the pitch-colored bark-dwelling carabid beetle *(Psydrus piceus)*, and Capizzi's blind rove beetle *(Fenderia capizzii)* (Lattin and Moldenke 1992).

Several higher taxa of invertebrates in Pacific Northwest rain forests have extraordinary disjunct ranges, with their closest relatives halfway around the world, a distribution reflecting the great age and relative stability of these rain-forest ecosystems (see chap. 2). Examples include the aquatic large west American unique-headed bug *(Boreostolus americanus)* with its nearest relative in Tierra del Fuego, the flightless moss lacebug *(Acalypta saundersi)* with similar species in the Mexican Highlands, the Ozarks, and the Appalachians (Lattin and Moldenke 1992), as well as a suite of longhorn (Cerambycidae) beetle genera found only in the Pacific Northwest and East Asia (Gorton-Linsley 1963).

Ecological Roles of Invertebrates

Forest invertebrates and their activities drive many key ecological processes, including control of decomposition processes and nutrient cycles, checking epizootic outbreaks, catalyzing natural disturbance and successional processes, and regulating growth and reproductive success of fungi, plants, and vertebrates (Schowalter and Crossley 1987; Schowalter 1989; Snyder 1992; Hoekstra et al. 1995; DellaSala and Olson 1996). In redwood forests, fungi and insects are implicated in the low viability (roughly 60 percent mortality) of redwood seed crops—for example, from the action of seed predators, such as the cone moth *(Commophila fuscodorsana)* and the roundheaded borer *(Phymatodes nitidus)*. Loss of redwood seedlings also occurs from herbivory of the gray millipede and banana slug *(Ariolomax columbianus)*. Herbivory and other forms of damage to adult trees are caused by mites and insects in diverse groups that include termites, true bugs, beetles, moths, and ants and wasps (Davidson 1970, 1971; Snyder 1992). Nevertheless, redwood trees are noted for their remarkable resistance to attack by insects (see chap. 4). This resistance is reflected in the small number of insects that feed on redwoods. Only four bark beetles and two species of miners attack redwoods, compared to 30 and 28 species, respectively, on Douglas-fir (Snyder 1992). Of the 54 insects recorded by Lauck (1964) to attack redwoods, few are believed to confine their activity to redwoods and none are capable of killing mature trees by themselves (Fritz 1957; Snyder 1992). Only the black-horned borer *(Callidium sempervirens)*, another borer *(C. pallidum)*, and the large cedar borer *(Semanotus ligneus sequoiae)* have been recorded exclusively from coast redwoods (Snyder 1992). Although few invertebrates are known to feed directly on redwoods, many species are associated with the moss, fern, lichen, and other epiphyte mats that cover the branches of mature redwoods (see chap. 3).

The ecological interactions of temperate rain-forest invertebrates can be complex, with many linkages to other processes and species. For example, the ferocious folding-trap-door spider *(Antrodiaetus pugnax)* occurs at high densities in old-growth forests and is thought to regulate populations of numerous invertebrate prey species (Coyle 1971). The handsome iridescent fly *(Eulonchus tristis)* is one of the most important pollinators of understory plants (i.e., native bees avoid shade), but its reproductive success is dependent on the availability of *Antrodiaetus* as hosts for its young (Lattin and Moldenke 1992).

Conservation of Invertebrates

The biota of late-seral redwood forests is sensitive to anthropogenic disturbances such as logging. The loss of large trees over extensive areas eliminates habitat for canopy invertebrates and creates desiccation stress, higher temperatures, and disturbed soil and litter habitats for forest-floor species (Olson 1992; Hoekstra et al. 1995). Selective logging of redwood forests can significantly alter guild structure, abundance, and diversity of invertebrates for long periods of time (Hoekstra et al. 1995)—predisturbance species composition may be absent even fifteen years after logging. Heavy logging of redwoods, and old-growth forests in general, causes the loss of large trees and complex microhabitats, disturbance of soil and litter habitats, loss of coarse woody debris, and substantial alteration of temperature and moisture regimes; it can also promote the invasion of alien species (Olson 1992; Frest and Johannes 1993). All of these trends seriously affect the native invertebrate biota associated with old-growth redwood forests. Flightless and mesophilic (moisture-loving) species are particularly sensitive to the loss of forest canopies and fragmentation of forests (Main 1987; Moldenke and Lattin 1990b; Olson 1992; Lattin 1993; Winchester and Ring 1996a; Winchester 1997).

The diversity and abundance of old-growth invertebrates may be irreversibly changed if logging is extensive and source populations are eliminated across the landscape (Vlug and Borden 1973; Schowalter 1990; Frest and Johannes 1993; Winchester and Ring 1996a; Winchester 1997). If many invertebrate species are restricted to local areas—and strong evidence suggests this—then the loss of more than 95 percent of the original redwood forest (U.S. Fish and Wildlife Service 1997) has almost certainly led to numerous extinctions of old-growth redwood invertebrates. A better understanding of patterns of invertebrate beta diversity (species turnover along environmental gradients), microhabitat specificity, and local endemism is urgently needed for conservation planning (see Olson 1992; Samways 1994; Winchester 1993, 1997; Winchester and Ring 1996a).

Forest Carnivores of the Redwoods Region

Forest carnivores (forest-dwelling mammals of the order Carnivora) merit conservation concern for their functional importance to ecosystems and for what they indicate about ecosystem condition. Carnivores affect the behavior and demography of their prey, cycle nutrients by scavenging carrion, complete or interrupt pathogen and parasite life cycles, and affect plant distribution—and likely landscape pattern—through dispersal and predation of seeds (Buskirk in press). Mammalian carnivores affect human economies and cultures positively (e.g., tourism generated by the spiritual and aesthetic value of observing them, harvest of some commercially important species) and negatively (e.g., depredations on livestock and fish at hatcheries, transmission of rabies). The population status of carnivores is a measure of ecosystem condition and integrity (Noss et al. 1996); maintaining and restoring their diversity, natural distributions, and abundances have ecological, spiritual, and economic value.

Historical Distributions and Abundances

An obvious yardstick by which to evaluate the status of mammalian carnivores in the redwood region is to compare their current diversity, distribution, and abundances to what they were before intensive human activities (i.e., settlement by Europeans). Unfortunately, little is known about the wildlife of the region prior to the twentieth century. The traditional uses of furbearing mammals by Native Americans—some of which are still practiced today—and the accounts by anthropologists of the customs of tribes of the region (e.g. Scott and Kroeber 1942), provide some insight on the diversity and abundance of mammalian carnivores. Most notable are the species that were either a potential threat to human life (i.e., grizzly bears) or which had important spiritual or cultural value. Grizzly bears appear to have been common throughout the region, especially along the mouths of the major rivers. Early historical information on smaller carnivores is reflected in the ceremonial use of their skins in regalia. The Hupa (Hoopa) tribe has existed on the eastern margin of the redwood region since before recorded history; their regalia include the skins of bobcat, fox, fisher, and ringtail (B. Colegrove, pers. comm.). Some species (e.g., marten and wolverine) are conspicuous by their absence in regalia, which may signify either they did not occur in the region or were extremely rare. Early European explorers and settlers left little recorded history of the wildlife of the redwood region. The Spanish, in the late 1700s, and then the Russians and Americans in the 1800s, focused their wildlife harvesting activities just offshore, on the California sea otter, whose pelt exceeded the value of any terrestrial furbearer.

The earliest scientific account of the ecology, distribution, and status of mammalian carnivores in California is a summary by Joseph Grinnell, Joseph Dixon, and Jean Lindsdale of trapping records, interviews, and field observations during the first decades of the twentieth century (Grinnell et al. 1937). They reported eighteen species of terrestrial mammalian carnivores as suspected to have occurred in the redwood region: two ursids, two procyonids, ten mustelids, two canids, and two felids (table 5.5). They ranged in size from the short-tailed weasel (ermine), at less than 1 kg, to the 650-kg grizzly bear, and can be categorized into

Table 5.5. Mammalian Carnivores That Historically Occurred in the Redwood Region, Their Typical Habitat in the Region, Their Status in the Region in the 1920s, and Their Current Status.

Species	Habitat	Status in 1920s	Status Today
Grizzly bear (*Ursus arctos*)	Generalist; grasslands, bays, forests, and mouths of major rivers	Extirpated but formerly common	Extirpated
Black bear (*Ursus americanus*)	Forests at all elevations	"Plentiful"	Regulated game species; limited season and bag limit; perhaps more abundant since extirpation of grizzly; commonly detected at track plate stations
Raccoon (*Procyon lotor*)	Lowlands and streamcourses in foothills and lower margins of fir forests; semi-aquatic	Moderate decline due to wetland drainage	Regulated fur-bearing game species; still common along streamcourses and near human habitation
Ringtail (*Bassariscus astutus*)	Open, deciduous forests on ridges and near streamcourses	'quite plentiful, especially in northwestern California'	Protected from trapping or hunting; more common at track plate stations in interior forests than in redwoods along the coast
Long-tailed weasel (*Mustela frenata*)	Habitat generalist; forest/grassland edges	Not specified	Unprotected nongame species; no closed season and no limit; uncommonly detected at track plate stations

Species	Habitat	Status in 1920s	Status Today
Short-tailed weasel (*Mustela erminea*)	Grasslands/meadows, especially at higher elevations	Not specified limit; uncommonly detected at track plate stations	Unprotected non-game species; no closed season and no bag
Mink (*Mustela vison*)	Aquatic; marshes along bays and mouths of rivers and along mountain streams; found away from water more along north coast than elsewhere in California	Moderate decline due to wetland drainage	Regulated fur-bearing game species; limited season but no bag limit
Marten (*Martes americana*)	Mature redwood forests and on higher ridges of Douglas-fir	'Sparse occurrence, though previously fairly numerous'	Protected; California Species of Special Concern; two *individuals* verified within the redwood range in past 50 years
Fisher (*Martes pennanti*)	Mature mixed coniferous and deciduous forests at all elevations but most common in mountains	Is becoming more rare, 'probably less than 300 in entire state'	Protected; California Species of Special Concern; twice petitioned to be listed under federal Endangered Species Act; commonly detected at trackplate stations in the redwood zone but less so than in interior coast ranges
Wolverine (*Gulo gulo*)	Unconfirmed, high-elevation forests suspected	Extinct, "formerly suspected to occur in the mountainsof the northern coast region"	Protected; California Threatened Species; sporadic unconfirmed sightings
Otter (*Lutra canadensis*)	Aquatic; marshes along bays and mouths of rivers and in mountain streams	"Numbers are stable and the species is expected to persist"	Protected
Spotted skunk (*Spilogale gracilis*)	Drier forests and upland rocky or brushy unforested areas	"Population has maintained itself rather consistently"	Unprotected non-game species; no closed season and no bag limit; commonly detected at track plate stations
Striped skunk (*Mephitis mephitis*)	Brushy woodlands at low elevation, regions with sod-forming vegetation, beaches	"The most abundant of the fur-bearing mammals"	Unprotected non-game species; no closed season and no bag limit

(continues)

Table 5.5. (*continued*)

Species	Habitat	Status in 1920s	Status Today
Badger (*Taxidae taxus*)	Open, unforested regions and forests interspersed with grasslands	"Numbers but little reduced from those of former times"	Regulated fur-bearing game species; limited season but no bag limit
Gray fox (*Urocyon cinereoargenteus*)	Brushy woodlands and forest from sea level to high-elevation forests.	"Trapping has caused relatively little change in numbers"	Regulated fur-bearing game species; limited season but no bag limit
Coyote (*Canis latrans*)	Brushy woodlands and forest from sea level to high-elevation forests; openings caused by timber harvest	Recent immigrant to north coast region; numbers increasing	Unprotected nongame species; no closed season and no bag limit
Wolf (*Canis lupus*)	Unknown	Unconfirmed in redwood region.	Protected
Mountain lion (*Puma concolor*)	Habitat generalist; forests, woodlands and brush at all elevations	"Breeding stock has not been greatly reduced"	Protected
Bobcat (*Felis rufus*)	Habitat generalist; forests and brushlands with rocky ground, river bottoms	"Wide distribution ... would be last fur animal to become extinct in California"	Regulated fur-bearing game species; limited season and bag limit; relatively commonly detected at track plate stations

Source: Grinnell et al. (1937).

six groups by habitat: aquatic or semiaquatic species (otter, mink), riparian associates (raccoon, ringtail), habitat generalists (grizzly bear, black bear, coyote, mountain lion), forest generalists (gray fox, spotted skunk, bobcat, long-tailed weasel), grassland/meadow associates (striped skunk, short-tailed weasel, badger), and mature forest associates (marten, fisher, wolverine).

Only one species of mammalian carnivore, the American marten, includes a recognized subspecies that is considered endemic to the humid redwood belt. The Humboldt marten *(M. a. humboldtensis)* is distinguished from the other California subspecies, *M. a. sierrae*, by its darker color, reduced size of orange throat patch, and various cranial skeletal features (Grinnell and Dixon 1926; Grinnell et al. 1937). Genetic analysis to establish the genetic uniqueness of this subspecies is under way (Bayliss and Zielinski, in prep.).

Although the classic work by Grinnell et al. is the most extensive summary of the ecology of mammalian carnivores of California, by the time it was undertaken, the trapping and hunting of terrestrial furbearers had already begun to decimate their populations: "Since the State was first occupied by white man, the major population trends of fur animals has been downward" (Grinnell et al. 1937:24). The grizzly was extirpated from the redwood region by the mid-1800s, the wolverine was absent and suspected to have been extirpated, and the marten and fisher were undergoing significant population declines in the early twentieth century. The decline of some species apparently occurred so soon after European colonization that it is unclear whether they ever occurred in the region. No documentation exists of the historic occurrence of wolverine or wolf in the redwood region. The wolverine was suspected to have occurred in the "mountains of the northern coast region" (Stephens 1906, cited in Grinnell et al. 1937), but this could not be verified. And of the wolf, Grinnell et al. stated, "If the northwest timber wolf ever occurred in California, which is not unlikely, it probably was restricted to the northwest coastal strip of high humidity and heavy timber." Thus, historical information provides little basis for interpreting the current apparent absence of wolverines and wolves in northwestern California.

Early Conservation Efforts

By the end of the nineteenth century, the reduction in numbers of fur-bearing mammals had become so great that the "attention of thoughtful people was directed to it" (Grinnell et al. 1937). Laws were enacted in 1913 to protect the sea otter and in 1917 to set seasonal and bag limits on the take of several species of terrestrial carnivores. By the 1920s, it was clear that species with high market value—the marten, fisher, and wolverine—would require special conservation efforts (Dixon 1924), a plea that was reinforced in the 1940s (Twining and Hensley 1947). The marten was protected from trapping in 1946 in northwestern California, and the fisher season was closed statewide in the same year.

The protection of vulnerable species from commercial trapping in the mid-1900s was a pivotal conservation measure. Nevertheless, government trappers continued to trap and poison carnivores in the interest of protecting livestock. In four counties in California (including Humboldt Co.), from 1920 to 1930, government predator control agents killed an estimated 40,000 carnivores, rare forest carnivores such as fisher and marten no doubt among them.

Although targeted trapping of forest carnivores was now regulated, the second major threat to forest-dwelling mammals—timber harvest—was intensifying in the redwood region in the mid-1900s (see chap. 2). Most vulnerable are the species associated with old forests—marten and fisher. These species are typically associated with mature, mesic forests, noted for their structural complexity (uneven aged with high densities of snags and logs), large-diameter trees for resting and denning, and closed canopies (Buskirk and Ruggiero 1994; Powell and Zielinski 1994). Clear-cutting has been the dominant silvicultural method for the harvest and regeneration of redwood (see chap. 8) and is the greatest current threat to the recovery and conservation of forest carnivores.

Current Status of Carnivores

The end of the twentieth century brings renewed interest in the status of carnivores. Although the threat of trapping—at least to those species most vulnerable—has abated, most of the original redwood forest to which carnivores were adapted has been forever changed. With adequate habitat and source populations, protection from trapping usually leads to the restoration of furbearer populations (Coulter 1960; Buskirk 1993). The quality of the habitat in the redwood zone, however, has apparently declined since the fisher and marten were protected in the 1940s. The offsetting effects of protection from trapping and increasing habitat loss can be evaluated by examining the current diversity, distribution, and relative abundance of carnivores.

Since the late 1980s, carnivore surveys using standard detection methods (Zielinski and Kucera 1995) have been conducted with increasing frequency. Most surveys have been conducted in the northern redwood zone; areas south of Humboldt County have received little effort, and regions south of San Francisco Bay virtually none. The most comprehensive surveys in redwoods have been those by Klug (1996) and Beyer and Golightly (1995), but the results of other surveys conducted by timber companies, the USDA Forest Service, and environmental consulting companies have produced similar results. Considering the entire redwood region, including both redwood-dominated areas and those with substantial coverage of other tree species, the most commonly detected species are the gray fox, spotted skunk, bear, and fisher. Next most frequently detected are the raccoon, bobcat, and ringtail, with detections of weasel (*Mustela erminea* and *M. frenata* cannot be distinguished by their tracks) and striped skunk the least frequent. Each species has characteristics that place it on a continuum from relatively easy to very difficult to detect using enclosed track plate methods. Several of the most frequently detected species appear to have a low threshold for entering and re-entering track plate enclosures. Other species, such as the coyote, which appears to avoid novel items associated with humans, and the bobcat, which like many cats does not regularly consume carrion, are less effectively detected using these methods. For this reason, and due to the varying densities and home range sizes of the species, the number of detections is not a reliable index of relative abundance. For those species that regularly visit baited track plates, however, detections are a reliable measure of presence.

Survey results are somewhat different considering only redwood-dominated stands. Fishers are less common in redwood-dominated stands than in mixed redwood–Douglas-fir or Douglas-fir dominated forests that occur farther inland (Beyer and Golightly 1995; Klug 1997). Ringtails also appear less common in the redwood-dominated stands. Black bear, gray fox, and spotted skunk are again the most commonly detected carnivores in redwood forests.

In reviewing surveys in the northern redwood zone over the past ten years, the most disturbing conclusion is the absence of the Humboldt marten. Since 1990, more than 1,000 track plate or camera stations have been deployed in more than 15,000 survey days without a single marten detection (Zielinski and Golightly 1996). This finding is especially noteworthy because martens are one of the species most readily attracted to and detected at track plate stations (Kucera et al. 1995; Zielinski et al. 1997).

Moreover, fisher researchers or commercial trappers seeking legal quarry have not trapped marten incidentally, nor has a single road-killed marten been reported to any state or federal resource agency in the redwood zone (Zielinski and Golightly 1996). Recently, marten have been detected at two locations near the eastern limit of the subspecies' original range, 20 km apart, during surveys conducted in Douglas-fir forests (Zielinski, unpubl. data.). Although these are the first martens verified within the historic range of the Humboldt subspecies, their genetic relationship to the martens that once occurred in the low-elevation redwood forests has yet to be established. Even considering these recent detections, the prognosis for the natural recovery of marten in the redwood forest is poor.

Consequences of Logging for Forest Carnivores

The absence of verified records of wolverine in the redwood region and the loss of marten as a common predator in redwood ecosystems means that little local information is available on their habitat requirements. For marten, we can surmise the effects of logging based on their responses elsewhere, but the lack of data from the Pacific states makes this difficult for wolverines. Hence, most of our discussion must focus on the fisher, the only species of forest carnivore that remains in sufficient numbers to be the subject of recent research.

Almost everywhere they occur in the western United States, and like martens, fishers are associated with extensive, structurally complex, late-seral forests (Rosenberg and Raphael 1988; Buskirk and Ruggiero 1994; Powell and Zielinski 1994). Patterns of habitat association in wide-ranging carnivores, like many other species, are complex because of the interaction of processes operating at multiple scales. For example, detections of fishers in second-growth redwood stands belie a simplistic characterization of the species as an old-growth associate (Harris 1984). Though younger stands may retain habitat value at the patch scale (see discussion on WHR in the preceding section), landscape and regional-scale changes in redwood ecosystems give reason for concern about the long-term viability of coastal fisher populations—and the recovery of marten—under current timber-harvesting regimes.

Stand-level changes in redwood habitat likely to affect forest carnivores include reduction in canopy closure, tree diameter, and snag and log abundance, as well as changes in floristic composition. High levels of overhead cover provide protection from predation, lower the energetic costs of traveling between foraging sites, and provide more favorable microclimatic conditions (Powell and Zielinski 1994; Buskirk and Ruggiero 1994; Thompson and Harestad 1994). Abundance or vulnerability of preferred prey species may be higher in areas with higher canopy closure (Buskirk and Powell 1994) or coarse woody debris. Clear-cutting, or any logging that reduces the density of large trees, affects the availability of denning and resting sites and structural components associated with prey vulnerability (Thompson 1986; Thompson and Colgan 1987; Thompson and Harestad 1994; Sturtevant and Bissonette 1997). Runoff from clear-cut sites also can severely degrade riparian areas, which not only are essential habitat for aquatic and semiaquatic carnivores but also provide critical foraging, travel, and resting habitats for marten and fisher (Powell and Zielinski 1994).

Changes in species composition associated with logging may include an increase in hardwood species (red alder and tanoak) and Douglas-fir. These floristic changes may have favored the expansion of fishers in the redwood region at the expense of martens, which are more closely associated with conifer than mixed forests. Marten have been shown elsewhere to suffer from competition with fishers, especially in regions lacking deep snow (Krohn et al. 1995). The lack of snow throughout most of the redwood zone, which if deep enough would inhibit fisher movement more than marten, may be an impediment to the recolonization of marten in regions where fishers are common.

Several structural features may enhance the resiliency of redwood forest to disturbance from logging. Regeneration by basal sprouting allows stands to recover canopy closure more quickly than in trees that regenerate from seed (see chap. 4). Historical logging methods often left behind coarse woody debris, snags, and residual "wolf" trees that provide a structural "legacy" that are favored rest sites for fishers (R. Klug, pers. comm.). Current harvesting methods are more intensive, however, and as harvested redwood stands enter short-term rotations they will increasingly lack the structural and floristic complexity they possessed before harvest (see chap. 8).

Changes in landscape patterns in the redwood region, such as the fragmentation described earlier, may limit the positive effects of stand-level resiliency. The current landscape mosaic in the redwood zone differs in several respects from the pre-settlement condition. Historically, redwood stands formed a mosaic with other forest types, such as oak woodland and closed-cone conifer forest, with fire playing a significant role (see chap. 3 and 4). The current landscape condition shows a lower abundance of older stands, smaller patch size, lower spatial and temporal variability in stand age-class distribution, and reduced natural fire frequency. These trends are likely to lead to reduced functional connectivity for species associated with older forests, as has been shown elsewhere for American marten (Hargis and Bissonette 1997).

Increased road access is often suggested as a negative collateral effect of logging primary forest. Roads provide access for trappers and are a source of vehicle-caused mortality. None of the rare forest carnivores is currently legally trapped, but they are captured incidentally in traps set for other species (Lewis and Zielinski 1996), and roadkills are not uncommon. Analyzing the effect of road density on abundance and distribution is difficult, however. For example, if productive forestlands are preferentially targeted for logging and associated road-building, a complex interaction of road density and habitat quality may be evident. Fisher detections were positively correlated with low-use road density in the eastern Klamath region, but this conclusion is difficult to interpret in light of the absence of a negative effect of high-use roads (Dark 1997). Perhaps because of the scarcity of primary forest in the redwood region, older second-growth forest with low to moderate road density probably is a critical factor as forest carnivore habitat.

Habitat Value of Redwoods for Forest Carnivores

Klug's (1996) extensive surveys for fishers in the redwood zone found positive correlations between fisher detection rates and such stand-level attributes as diameter and basal area of hardwoods, canopy closure, and volume of logs. Similarly, Carroll et al. (1999), using survey data from the adjacent inland forest types, found fisher detection rates were highest at sites with large hardwoods in landscapes of dense, mixed hardwood–conifer forests.

Lower detection rates for fishers in redwood forests when compared to the redwood–Douglas-fir transition zone led Klug (1996) to conclude that redwood habitat supported smaller populations. The generality of this conclusion may be limited, however, by the predominantly early-seral character of the commercial timberlands surveyed in Klug's study. The habitat model of Carroll et al. (1999) predicts greater variability of detection rates in the redwood forest, in part because of the high contrast in average tree age and diameter between protected areas and commercial timber lands. Beyer and Golightly (1995) report generally lower than expected detection rates in redwoods; they note, however, that a survey transect in Redwood National Park recorded high detection rates. Although floristic variation, especially the abundance of large hardwoods, appears to be a major factor influencing fisher distribution, the relative habitat value of late-seral redwood stands cannot be characterized without additional surveys.

Regional-scale variation in fisher abundance confounds efforts to unravel stand-level habitat associations. Biogeographic correlations between fisher distribution and such regional factors as elevation and distance to the ocean (Klug 1996) or precipitation and geographic variables (Carroll 1997) may be as strong as those with patch or landscape-level habitat attributes. The higher detection rates in Redwood National Park as compared to Humboldt State Redwoods

Park (Beyer and Golightly 1995) and the low rate of detections on commercial timberlands in the north and south of Klug's (1996) study area may reflect these regional-scale factors.

One hypothesis links the coarse-scale distribution patterns described above to regional fisher metapopulation structure (Carroll et al. 1999). The concept of a metapopulation of populations linked by occasional dispersal may be especially applicable to fishers inhabiting the fragmented habitat of the redwood region. We use the term *metapopulation* in a broad sense here, as real-world metapopulations often do not conform to the classic "island-island" model (Harrison 1994). Instead, "Central patches are united by dispersal into a single population, slightly more isolated ones undergo extinction and recolonization, and still more isolated patches are usually vacant" (Harrison and Taylor 1997). Limited evidence concerning the structure of the regional fisher metapopulation suggests that the lower Trinity River watershed may function as a central "mainland" population. This pattern would favor occupation by fishers of redwood forest lying to the west of this area.

As with the spotted owl, fisher metapopulations may show nonlinear responses to the size and spacing of habitat clusters. If clusters of suitable habitat become too small or isolated, the imbalance between immigration and emigration can limit long-term viability (McKelvey et al. 1993; Noon and McKelvey 1996). The decline in distribution of the fisher in the western United States may be due to such regional-level dynamics (Heinemeyer and Jones 1994). Existing land management planning processes are poorly adapted to decision-making across administrative boundaries. A comprehensive plan to ensure the viability of forest carnivores and maintain well-distributed populations would need to include measures to increase levels of canopy closure on managed lands and restore connectivity through development of landscape linkages between the redwood parks and stands farther inland. The survival of the fisher and marten, like that of other wide-ranging carnivores, may ultimately depend on multiownership cooperative management on a regional scale (Mladenoff et al. 1995).

Marbled Murrelets in Redwoods

The marbled murrelet is a seabird that nests up to 88 km inland in coastal, older-aged coniferous forests throughout most of its range from Alaska to central California. Murrelets fly at high speeds (up to 158 km/h), attend their breeding sites primarily during low light levels, and nest solitarily (Nelson and Hamer 1995a; Nelson 1997). The secretive aspects of their breeding biology have made their nests difficult to locate; the first murrelet nest was not discovered until 1974 (Big Basin State Park, California; Binford et al. 1975). Since then, more than 170 nests have been found, 18 of them in California (table 5.6).

Table 5.6. Characteristics of Marbled Murrelet Tree Nests in California.

Site	Tree species[a]	Diameter (dbh, cm)	Tree height (m)	Limb height (m)	Limb diameter at bole (cm)	Limb diameter from nest (cm)	Distance from trunk (cm)	Vertical cover (%)	Moss depth (mm)	Moss (%)	Outcome/stage[b]
Big Basin State Park 1974[c]	PSME	167.0	61.0	45.0	41.0	—	6.8	100	0.8	100	failed/chick fell out
Big Basin State Park 1989[d]	PSME	210.0	61.2	43.7	—	47.7	122.0	100	1.0	100	predation by CORA / egg
Big Basin State Park 1989[d]	PSME	196.0	76.2	38.5	—	36.3	61.0	100	1.0	100	predation by STJA / chick
Big Basin State Park 1991[e]	SESE	533.0	79.2	41.1	61.0	61.0	0	100	0	0	successful
Big Basin State Park 1992[e]	SESE	533.0	79.2	53.2	42.0	42.0	0	100	0	0	successful
Pacific Lumber 1992[f]	SESE	229.0	86.5	67.0	21.2	21.0	0	100	0	0	failed/chick died
Pacific Lumber 1992[f]	SESE	254.0	74.3	67.5	25.0	20.0	18.0	100	0	0	unknown
Pacific Lumber 1993[g]	PSME	183.0	70.0	37.0	37.0	30.0	0	5	3.8	100	inactive
Pacific Lumber 1993[g]	SESE	300.0	83.0	60.0	37.0	37.0	5.0	50	0	5	unknown
Jedediah Smith State Park 1993[h]	SESE	338.0	79.5	43.6	21.0	16.0	90.0	99	0	1	inactive
Prairie Creek State Park 1993[h]	TSHE	139.0	66.7	33.3	29.0	23.0	20.0	98	8.1	100	inactive
Big Basin State Park 1994[e]	SESE	533.0	79.2	41.1	61.0	61.0	0	100	0	0	predation/egg
Butano State Park 1995[i]											predation by CORA / egg

(continues)

Table 5.6. (*continued*)

Site	Tree species[a]	Diameter (dbh, cm)	Tree height (m)	Limb height (m)	Limb diameter at bole (cm)	Limb diameter from nest (cm)	Distance from trunk (cm)	Vertical cover (%)	Moss depth (mm)	Moss (%)	Outcome/ stage[b]
Big Basin State Park 1996[j]	SESE	533.0	79.2	53.2	42.0	42.0	0	100	0	0	predation/egg
Big Basin State Park 1996[j]	SESE	174.0	48.8	31.7	18.0	18.0	0	80	0	0	successful
Big Creek Lumber predation/chick 1997[k]											
Pescadero Creek County Park 1997[k]											predation by RSHA/egg unknown
Overall (mean + SE)		308.7 + 41.7	73.1 + 2.8	46.9 + 3.1	36.3 + 4.2	35.0 + 4.3	23.1 + 10.5	88.0 + 7.4	1.1 + 0.6	36.1 + 13.2	

[a] PSME = *Pseudotsuga menziesii*, SESE = *Sequoia sempervirens*, TSHE = *Tsuga heterophylla*
[b] CORA = Common Raven, STJA = Steller's Jay, RSHA = Red-shouldered Hawk
[c] Binford et al. 1975
[d] Singer et al. 1991
[e] Singer et al. 1995
[f] Kerns and Miller 1995
[g] S. Chinnici, pers. comm.
[h] T. Hamer, pers. comm.
[i] D. Suddjian, pers. comm.
[j] S. W. Singer, pers. comm.
[k] E. Burkett, pers. comm.

Murrelet nests in California, like other tree nests throughout their range (Hamer and Nelson 1995), occur in large, tall redwood, Douglas-fir and western hemlock trees (table 5.6). These nests are located above the 30 m height on large limbs with high vertical cover. In contrast to other regions, nest limbs in California generally have less moss because of drier conditions and higher daytime temperatures compared to farther north. Key components of nesting habitat in California include the percentage of old-growth canopy cover and tree species composition (>50 percent coast redwood; Miller and Ralph 1995). The amount of habitat along major drainages and at lower elevations is also useful in predicting murrelet occurrence.

Marbled murrelets in California are known to occur primarily in the coast redwood forests. Only a few observations (but no nests) have been recorded in Douglas-fir dominated forests. The distribution of redwood forests closely matches the inland extent of marine air influences and summer fog (Barbour and Major 1988; and see chap. 3 and 4). At sites farther inland, where the maritime influence is limited by rugged topography, Douglas-fir and tanoak forest types dominate. Although these forests generally have multiple tree layers, large trees, and potential nesting platforms (Jimerson et al. 1996), summer temperatures are higher, resulting in a lack of moss on tree limbs and hot, dry conditions that may be unsuitable for nesting by murrelets. The distance to the ocean is also greater, increasing energetic demands.

Murrelet nests are frequently unsuccessful; overall nest success has been only 37 percent (n = 60; Nelson and Hamer 1995b; S. K. Nelson and I. A. Manley, unpub. data). Of the twelve nests with known outcomes in California, 75 percent (n = 9) were unsuccessful (table 5.6). Of those that failed, seven (78 percent) failed from predation during the egg (n = 5) and chick (n = 2) stages. Known predators at these nests have included common ravens, Steller's jays, and red-shouldered hawks. Habitat fragmentation (smaller stand size, more edge habitat) and distance of nests from the tree trunk and stand edge (roads, clearcuts, younger forests) are factors that appear to reduce nest success (Nelson and Hamer 1995b; S. K. Nelson and I. A. Manley, unpub. data).

An estimated 60,000 or more marbled murrelets were found historically along the coast of California (Larson 1991); today, only about 6,000 birds (including breeders and nonbreeders) remain (Ralph and Miller 1995)—a 90 percent decline. Murrelet populations are projected to be declining at a rate of 4–7 percent per year (Beissinger 1995), primarily because of habitat loss and fragmentation from logging and development. This species was federally listed as threatened in Washington, Oregon, and California (U.S. Fish and Wildlife Service 1992) and state listed in California as endangered (California Fish and Game Commission 1992) in 1992.

Less than 4 percent of the original redwood forests, once totaling more than 770,000 ha, remain in California (U.S. Fish and Wildlife Service 1997).

Suitable marbled murrelet forest habitat currently occurs in three separate areas: (1) Humboldt, (2) Del Norte Counties in the north, and (3) San Mateo and Santa Cruz counties in central California; the northern and central areas are separated by a distance of about 480 km, as most of Mendocino County no longer contains nesting habitat. Most of the remaining redwood forests occur in state, county, and national parks; however, more than 36,000 ha apparently remain on private land, especially in Humboldt County. In California, critical murrelet habitat (i.e., critical for species survival) was designated primarily on federal land (64 percent; 193,150 ha). Nevertheless, 88,750 ha on state and county lands and 18,080 ha of private forests were also designated as critical because habitat protected in parks and national forests alone will not guarantee the long-term survival of the murrelet (U.S. Fish and Wildlife Service 1996). Most of the private critical habitat occurs on lands owned by the Pacific Lumber Company, which are currently the subject of a multispecies habitat conservation plan (HCP) under section 10(a) of the federal Endangered Species Act. Approved just before this book went to press, this HCP would allow some incidental take of murrelets and, many feel, does not replace the old-growth forest structure required by murrelets.

In the Marbled Murrelet Recovery Plan (U.S. Fish and Wildlife 1997), goals for recovery call for preventing the loss of any additional occupied habitat. Developing new habitat and improving the quality of existing habitat by minimizing human impacts in the parks is also recommended to ensure the long-term survival of this species. In California and elsewhere, however, many private timber companies are currently negotiating HCPs with the U.S. Fish and Wildlife Service that would allow them to log some occupied and critical habitat and cause "incidental take" of murrelets. These negotiations, under the auspices of section 10(a) of the federal Endangered Species Act, are likely to cause a continual decline in suitable murrelet habitat. The next fifty years have been identified as the most critical for murrelets, a time period in which, under the Northwest Forest Plan (U.S. Departments of Agriculture and Interior 1993), some habitat on federal lands will become suitable for nesting by murrelets. The current downward trend of available murrelet nesting habitat, however, along with pressures on their food resources from El Niño and overfishing, and adult mortality in oil spills and the gill-net fishery, place the continued survival of this seabird in jeopardy.

Conclusions

The fauna of the redwood forest and region, though poorly known compared to many forest regions, reflects the long evolutionary history of the redwood lineage and the relative stability of this forest over millennia. Many of the animals of the region are adapted to moist, intact forest distributed over large areas and to the structural characteristics of late-seral forest, ranging from the coarse woody debris on the forest floor to the complex canopies 100 m above the ground. Although they occurred in a mosaic with many other kinds of vegetation, old-growth redwood forests dominated the region for a long period of time before settlement by Europeans. The redwood fauna reflects this antiquity.

The drastic diminishment of old-growth redwood forests in the span of just several human generations threatens many forest species with extinction. The reduction in total area of old-growth redwoods is only part of the problem—just as threatening to many species is the continued simplification of the structure of these forests and their fragmentation into smaller, more isolated pieces. Undoubtedly, some species of invertebrates have been lost before they were ever described by science, and some vertebrates (i.e., several amphibians, Humboldt marten, fisher, northern spotted owl, and marbeled murrelet) are seriously imperiled. Although no species or subspecies is strictly endemic to the redwood forest, several are endemic to the region, and the long-term persistence of many more likely depends on conservation actions taken here.

Chapter 6

AQUATIC ECOSYSTEMS OF THE REDWOOD REGION

Hartwell H. Welsh Jr., Terry D. Roelofs, and Christopher A. Frissell

The primary purpose of this chapter is to describe aquatic ecosystems within the redwood region and discuss related management and conservation issues. Although scientists from many disciplines have conducted research in the redwood region, few comprehensive interdisciplinary studies exist (but see Ziemer 1998b) and no regionwide overview or synthesis of the aquatic systems in the redwoods has been published. Private ownership of most of the region has limited access and, therefore, scientific study in many areas. Fortunately, a large body of applicable science exists on riparian and aquatic systems and the relationships between geomorphological, hydrologic, and biotic processes in the Pacific Northwest, some inclusive of the redwood region (e.g., Meehan 1991; Spence et al. 1996; National Research Council 1996; Stouder et al. 1997).

The redwood forests are in the coast range ecoregion (Omernik and Gallant 1986), mainly in northern California. Coastal streams with basins of 60–80 km^2 (e.g., Redwood Creek) may be entirely within the redwood zone, but the larger

Author contributions: Welsh and Roelofs, material on stream ecosystem processes and forest practices; Welsh, vertebrate taxa besides fish; Roelofs, plants, invertebrates, and fish; Frissell, stream responses to catchmentwide processes.

rivers (e.g., Russian, Eel, Mad, Klamath, and Smith Rivers) flow through the redwood zone only in their lowermost sections, with headwaters well inland from the coast and the redwood belt. The influence of redwoods on stream communities, therefore, is more pronounced in the small coastal streams than in larger rivers. Smaller streams also are more directly linked to and influenced by the adjacent riparian plant community. For this reason, the primary focus of this chapter is on the relatively small streams that originate and enter the ocean within the coastal redwood zone, although we do reference relevant research from the larger riverine systems that transect the redwood region.

Our approach here is, first, to discuss in some detail the structures and processes, and the spatial and temporal dynamics, that constitute a healthy riparian/aquatic ecosystem in the redwood region. This perspective is critical to understanding the ecological and evolutionary context within which the aquatic and riparian species of the redwoods have evolved over the millennia. This background also provides the appropriate frame of reference for discussing recent anthropogenic changes resulting from timber harvesting, road building, and other activities.

Stream Ecosystem Processes in Pristine Watersheds

Although few watersheds in the redwood region are pristine today, some understanding of how unaltered watersheds function is necessary to guide conservation and restoration programs. Here, we discuss the roles of riparian vegetation and large woody debris in streams, the river continuum concept, stream responses to catchmentwide processes, and the role of disturbance in stream ecosystems.

The Role of Riparian Vegetation and Large Woody Debris

Riparian plants play a dominant role in stream ecosystems. Terrestrial plants provide shade, regulate microclimates, and contribute both large wood pieces that add habitat complexity and small organic materials that serve as food for aquatic organisms. They also stabilize stream banks, control sediment inputs from surface erosion, and regulate nutrient and pollutant inputs (fig. 6.1; Gregory et al. 1991; Naiman et al. 1992; Spence et al. 1996; Naiman and Decamps 1997; Naiman et al. 1998). Redwood trees, by their size, age, and resistance to fire, floods, and decomposition, play a central role in shaping the physical and chemical conditions within the aquatic zone and thereby strongly affect the aquatic community. Among the world's tallest trees, redwoods cast long shadows and shade even wide streams. Large redwoods entering streams as a result of blowdown, bank undercutting, or mass wasting can remain there for centuries, functioning as stream features that rival bedrock sills or outcrops in regulating channel processes. Historical harvesting of redwoods throughout

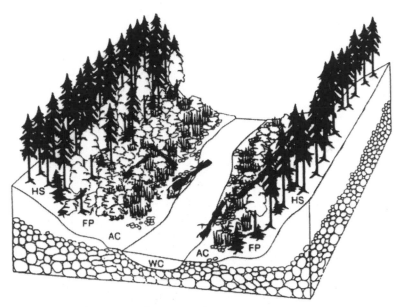

Figure 6.1. Typical patterns of riparian plant communities associated with different geomorphic surfaces of river valleys in the Pacific Northwest. Scattered patches of grasses and herbs occur on exposed portions of the active channel (AC), but little terrestrial vegetation is found within the low-flow wetted channel (WC). Floodplains (FP) include mosaics of herbs, shrubs, and deciduous trees. Conifers are scattered along floodplains and dominate older surfaces. The overstory species in riparian forests on lower hillslopes (HS) consist primarily of conifers. (From Gregory et al. 1991.)

most of the region, however, with much of the early harvest taking place along streams, has significantly altered stream ecosystems. Current harvesting is often still focused around streams because trees grow larger and more quickly there.

Stream channels in pristine watersheds exhibit a complex array of hydraulic conditions (pools, riffles, alcoves, side channels, single and multiple channel sections), substrate sizes, and accumulations of wood and other organic matter. Woody debris is one of five watershed elements that create and modify stream habitats, the others being water, sediments, nutrients, and heat (table 6.1; Naiman et al. 1992). Large woody debris consists of large logs that fall into stream channels, either from natural tree death, windthrow, or bank failure, and then play a central role in structuring stream habitats (table 6.1; Sedell et al. 1988). In the redwood region, large redwood logs can have a dominant influence on stream processes and channel complexity (Keller et al. 1995). In headwater streams flowing through old-growth redwoods, nearly all pools are either formed directly or influenced significantly by large organic debris (i.e., logs) (Keller et al. 1995). Welsh and Ollivier (unpub. data) reported an average

Table 6.1. Riparian Forests: Important Ecological Elements and Their Functions for Aquatic and Riparian Vertebrate and Invertebrate Communities.

Location and Function	Element						
	Large Woody Debris	Organic Litter	Roots	Substrate	Overstory Vegetation	Understory Vegetation	Soil
STREAM CHANNEL							
Provides breeding, feeding, and shelter habitat for many species	X	X	—	X	X	X	—
Controls aquatic habitat dynamics	X	—	—	—	—	—	—
Contributes to formation of islands and floodplains	X	—	—	—	—	—	—
Sort, routes, and retains substrates	X	—	—	—	—	—	—
Filters and stores sediments and organic material	X	—	—	—	—	—	—
Increases primary and secondary production	—	X	—	—	—	—	—
Self sorts and provides habitat heterogeneity	—	—	—	X	—	—	—
STREAM BANKS							
Provides breeding, feeding, and shelter habitat for many species	X	—	X	X	X	X	X
Stabilizes stream banks	X	—	X	—	—	—	—
Provides coarse, woody debris to channel	X	—	—	—	—	—	—
Provides substrate to channel	—	—	—	X	—	—	—
FLOODPLAIN							
Provides breeding, feeding, and shelter habitat for many species	X	X	—	X	X	X	X

Location and Function	Large Woody Debris	Organic Litter	Roots	Substrate	Overstory Vegetation	Understory Vegetation	Soil
					Element		
Regulates air temperature, humidity, and solar penetration	—	—	—	—	X	X	—
Provides suitable microclimatic conditions	X	X	—	X	—	—	—
Provides woody debris to channel	X	—	—	—	X	—	—
Provides organic litter to channel	—	—	—	—	X	X	—
Retards movement of coarse woody debris	—	—	—	—	X	X	—
Filters sediment	X	X	—	X	X	X	—
Dissipates water energy	X	X	—	X	X	X	—
ADJACENT HILLSLOPE							
Provides breeding, feeding, and shelter for many species	—	—	—	—	X	X	—
Regulates air temperature, humidity, and solar penetration	—	—	—	—	X	X	—
Provides coarse, woody debris to floodplain and channel	X	—	—	—	X	—	—
Provides organic litter to floodplain and channel	—	X	—	—	X	X	—
Protects riparian forest from effects of wind	—	—	—	—	X	X	—
Provides recruitment source to floodplain and channel	—	—	—	X	—	—	—
Filters sediment	X	X	—	—	X	X	—
Dissipates water energy	X	X	—	—	X	X	X

of 55 logs > 50 cm dbh/km (S. D. 11) for 10 streams in an old-growth red-wood forest in Prairie Creek Redwoods State Park (Humboldt Co.); 99 percent were conifer logs, primarily redwood. Keller et al. (1995) aged 30 individual logs in the Prairie Creek Basin and found about half of them to have been in the stream for more than 100 years, some in excess of 200 years. In coastal Oregon streams, 70 percent of pools > 1 m³ in volume were formed by wood (Andrus 1988). Andrus also noted that conifers are markedly more decay-resistant than hardwoods, and that redwood and cedar logs outlast Douglas-fir and hemlock logs in streams. In a natural system with intact riparian forests, a large proportion of these logs would enter streams from the highest channels in the stream network (i.e., the first-order channels; Strahler [1957]) during large storm events (Sedell et al. 1988). Because they provide large woody debris and a variety of sediment types, headwater or first-order stream channels strongly influence the type and quality of downstream fish habitat (Sedell et al. 1988). Stated succinctly, "Reaches that are themselves inhospitable to salmonids may contribute to the maintenance of salmonid populations downstream" (G. Reeves in Reid 1998).

The River Continuum Concept

The river continuum concept (fig. 6.2; Vannote et al. 1980; Minshall et al. 1983, 1985, 1992) describes the changes natural stream communities undergo from headwater areas to lower elevations in response to riparian and fluvial processes. Relative contributions of energy from riparian versus aquatic primary producers (plants), and the size and type of organic material transported from upstream areas, change predictably through the progression from headwaters to the mouth in natural stream communities. Small headwater streams have steep gradients, confined channels, and cool temperatures when adequately shaded by riparian vegetation. Consequently, they obtain most of their organic material in the form of leaves, needles, branches, and other plant parts from the riparian zone rather than from primary producers within the stream. Farther down the river continuum, stream channels are less confined and have more extensive floodplains. Instream primary production increases in response to greater light penetration. Daily temperature fluctuations in these lower stream sections also are more pronounced because of increased exposure to solar warming (Beschta et al. 1987). Historically, the presence of old-growth redwoods and a low level of forest disturbance in the riparian zone of small and intermediate-size streams resulted in extensively shaded streams with limited temperature gains and primary production (table 6.1). Large woody debris input also changes along the stream continuum, with smaller channels contributing more large woody debris to the stream network than larger channels (Sedell et al. 1988).

Biological communities in streams are highly complex and dynamic entities comprising hundreds of plant and animal species structured and organized in

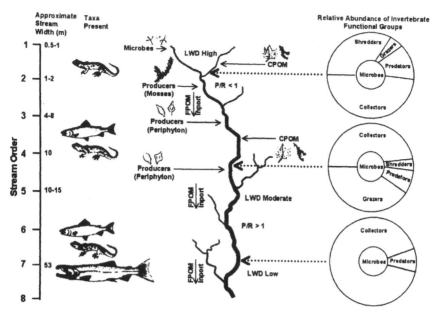

Figure 6.2. The river continuum concept (adapted from Vannote et al. 1980). An example showing stream width, dominant predators, producer groups, production (P) to respiration (R) ratios, importance of wood, and proportions of invertebrate functional groups. CPOM = coarse particulate organic matter, FPOM = fine particulate organic matter. (Adapted from Sedell et al. 1988.)

response to physical, chemical, and biological interactions. Interactions among the major abiotic factors (solar energy, climate, geology, geomorphology, and hydrology) provide the physical environment (Bencala 1984; Boulton and Lake 1991) that influences the species composition of the community. Biological interactions between organisms include food-web linkages (e.g., herbivory, predator-prey), competition, mutualism, and disease-host or parasite-host relationships (Spence et al. 1996). Energy flow within stream communities is often represented by the feeding relationships of these functional groups (table 6.2; fig. 6.2). The base of the food web is provided by plants, either from the riparian zone (allochthonous material) or instream primary producers (autochthonous material), such as algae, mosses, and rooted aquatic vascular plants.

The river continuum concept (Vannote et al. 1980) predicts a systematic change in consumer functional groups from headwater areas downstream (table 6.2; fig. 6.2). Small streams are dominated by invertebrates (shredders and collectors) that process riparian litter and its residues (Cummins and Klug 1979). Intermediate-size streams have more scrapers in response to increased periphyton growth from greater light availability. In the extensively shaded, medium-size streams of the coastal redwood zone, however, community structure tends

Table 6.2. Influences of Timber Harvest on Physical Characteristics of Stream Environments, Potential Changes in Habitat Quality, and Resultant Consequences for Salmonid Growth and Survival.

Forest practice	Potential change in physical stream environment	Potential change in quality of salmonid habitat	Potential consequences for salmonid growth and survival
Timber harvest from streamside areas	Increased incident solar radiation	Increased stream temperature; higher light levels; increased autotrophic production	Reduced growth efficiency; increased susceptibility to disease; increased food production; changes in growth rate and age at smolting
	Decreased supply of large woody debris	Reduced cover; loss of pool habitat; reduced protection from peak flows; reduced storage of gravel and organic matter; loss of hydraulic complexity	Increased vulnerability to predation; lower winter survival; reduced carrying capacity; less spawning gravel; reduced food production; loss of species diversity
	Addition of logging slash (needles, bark, branches)	Short-term increase in dissolved oxygen demand; increased amount of fine particulate organic matter; increased cover	Reduced spawning success; short-term increase in food production; increased survival of juveniles
	Erosion of streambanks	Loss of cover along edge of channel; increased stream width; reduced depth	Increased vulnerability to predation; increased carrying capacity for age-0 fish, but reduced carrying capacity for age-1 and older fish
		Increased fine sediment in spawning gravels and food production areas	Reduced spawning success; reduced food supply
Timber harvest from hillslopes; forest roads	Altered streamflow regime	Short-term increase in streamflows during summer	Short-term increase in survival
	Accelerated surface erosion and mass wasting	Increased severity of some peak flow events	Embryo mortality caused by bedload movement
	Increased fine sediment in stream gravels	Reduced spawning success; reduced food abundance; loss of winter hiding space	
		Increased supply of coarse sediment	Increased or decreased rearing capacity

Forest practice	Potential change in physical stream environment	Potential change in quality of salmonid habitat	Potential consequences for salmonid growth and survival
		Increased frequency of debris torrents; loss of instream cover in the torrent track; improved cover in some debris jams	Blockage to migrations; reduced survival in the torrent track; improved winter habitat in some torrent deposits
	Increased nutrient runoff	Elevated nutrient levels in streams	Increased food production
	Increased number of road crossings	Physical obstructions in stream channel; input of fine sediment from road surfaces	Restriction of upstream movement; reduced feeding efficiency
Scarification and slash burning (Preparation of soil for reforestation)	Increased nutrient runoff	Short-term elevation of nutrient levels of streams	Temporary increase in food production
	Inputs of fine in organic and organic matter	Increased fine sediment in spawning gravels and food production areas; short-term increase in dissolved oxygen demand	Reduced spawning success

Source: From Hicks et al. (1991).

toward that found at the upper end of most forest stream continua. Farther downstream, collectors (filter feeders) predominate; these species use the fine particulate organics that upstream communities produce and fail to retain. The recycling of nutrients and ingestion of fine particulate matter from upstream areas is referred to as *nutrient spiraling* (Newbold et al. 1982; Mulholland 1992). Predators occur all along the continuum, eating each other and other consumers. Besides invertebrate predators, most aquatic and semiaquatic vertebrates are also predators (table 6.2, fig. 6.2), although some fish (e.g., California roach and Sacramento sucker) eat significant amounts of plant material (Moyle 1976). The carcasses of spawned-out anadromous fish, before they were depleted, were a major component of the nutrient cycle in stream ecosystems, providing inputs from the marine environment (Cederholm et al. 1989; Bilby et al. 1996). Assigning a species to a particular functional group or trophic level is not always possible because many species exhibit ontogenic shifts in feeding habits. For example, many aquatic insects in early life-history stages are collectors, later becoming shredders or predators (Merritt and Cummins 1978).

An important, but little studied or understood component of the stream environment is the hyporheic zone, the area of flow below the streambed (fig. 6.3; Stanford and Ward 1988; Grimm and Fisher 1984; Triska et al. 1989a; Ward 1994; Duff and Triska 1990). This zone may extend vertically to depths of several meters and laterally from tens to hundreds of meters depending on channel confinement and floodplain geology (Stanford and Ward 1988). The stream bottom, although sometimes viewed as a boundary for the biological community, is the transition zone between the more rapidly flowing water of the active channel and the slower moving water of the hyporheic zone. In some streams, this subsurface flow can comprise 30–60 percent of the streamflow (Spence et al. 1996) and, in seasonally intermittent streams, all of the flow. Physical, chemical, and biological interactions within the hyporheic zone affect

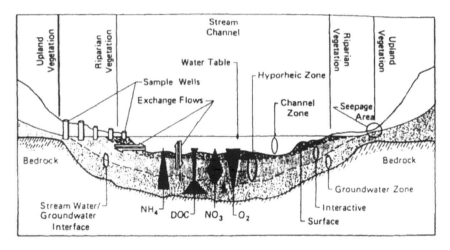

Figure 6.3. Conceptual model of the groundwater–surface water linkage at Little Lost Man Creek. Waters are divided into three zones: a channel zone containing surface water, a hyporheic zone, and a groundwater zone. The hyporheic zone has been divided into a surface hyporheos with virtually identical chemistry to channel waters, and containing > 98 percent advected channel water, and an interactive hyporheos characterized by physical-chemical gradients (e.g., NH_4, O_2, DOC, NO_3, O_2, temperature). The interactive zone consists of less than 98% but more than 10% advected channel water. The overall solute concentration is theoretically represented by the shape of the concentration profile. Transport of solutes across the groundwater–stream water interface (the stream boundary) is a function of the hydologic head of the adjacent landscape or response to a concentration gradient, whereas transport from the channel zone to the hyporheic zone is dominated by advective processes. (From Triska et al. 1989b.)

nutrient cycling, dissolved oxygen, microbial processes, and stream temperature. A rich biological community lacking primary production by photosynthesis inhabits this subsurface habitat. Although dominated by meiofauna (bacteria, fungi, protozoans) and invertebrates, some vertebrates use the hyporheic zone for reproduction (e.g., lampreys, sculpins, and salmonids), shelter from high streamflows (e.g., salmonids and Pacific giant salamanders), or feeding (e.g., Pacific giant salamanders). Triska et al. (1989a) and Duff and Triska (1990) studied nitrogen dynamics and flow in the hyporheic zone of Little Lost Man Creek (tributary to Prairie Creek in Humboldt Co.), where approximately 92 percent of the streamside vegetation is old-growth redwoods. Dissolved oxygen levels in the hyporheic zone immediately adjacent to the stream were 100 percent saturated, but that saturation dropped to as low as 7 percent 11 m inland from the stream (fig. 6.3). Observed denitrification in the hyporheic zone demonstrated that microbial activity changed the water quality as it passed through the zone.

Stream Responses to Catchmentwide Processes

Riparian vegetation mediates the delivery and flow of material in stream channels, but in the redwood region stream conditions are also strongly controlled by dynamic, catchment-scale physical processes. A geologic terrain dominated by tectonically sheared and uplifted rocks overlain by unconsolidated sediments and locally deep colluvium, exposed to a climate of wet winters with periods of intense rainfall and runoff, produces a region known for some of the highest recorded rates of natural erosion and sediment transport (Janda et al. 1975; Kelsey 1980; Mattole Sensitive Watershed Group 1996). Large debris slides, extensive and deep-seated earthflows, and long-runout debris flows and debris-charged floods are frequent and widespread in the redwood region; their effects are often of sufficient magnitude and persistence that riparian vegetation along streams in the redwoods region, rather than resisting them, is modified and strongly shaped by them (Lisle 1989). This phenomenon is exemplified by the loss of old-growth redwood stands along Redwood Creek from channel expansion and accelerated lateral erosion caused by increased sediment loads from upstream and upslope sources (Janda et al. 1975). Even centuries-old redwood trees cannot resist the physical forces that are magnified by recent human alteration of erosion and sedimentation regimes in the catchment of Redwood National Park.

Upland or catchmentwide land use exacerbates naturally high erosion and sedimentation rates in many ways. Roads built to access timber alter flow and sediment patterns by extending drainage networks and providing direct connectivity of sediment delivery from eroding road surfaces to channels at stream crossings (Wemple and Jones 1996; Ziemer and Lisle 1998). Plugging or overflow of road culverts causes flow diversion and gully erosion, which deliver large quantities of sediment to downstream channels (Hagans and Weaver 1987; Weaver et al. 1995). Removal of forest cover from steep or marginally stable hillslopes can trigger or accelerate mass failure through several mechanisms that affect soil water and root strength (Kelsey 1980; Sidle et al. 1985). The incidence or size of landslides typically increases in logged catchments, and some landslides propagate into debris floods or sediment-charged debris floods that reshape channel morphology and riparian vegetation for many kilometers downstream of the site of initiation (Kelsey 1980; Frissell et al. 1997; Pacific Watershed Associates 1998).

These changes in erosion and sediment regimes pose substantial and long-standing consequences for stream channels and floodplains, as well as their biota. Masses of deposited sediment can be flushed from steep tributary streams within years or decades (Madej 1987), but the delivery, residence, and transport of the sediment reconfigure and destabilize headwater stream habitats. The sediments are exported downstream to low-gradient reaches that were of greatest historical importance for production of salmon and other aquatic biota (Hagans and Weaver 1987; Frissell 1992; Frissell et al. 1997). Sediments generated as a result of natural or human disturbances in the landscape, once deposited in alluvial channels and floodplains, may have long residence times and thus influence channel conditions for many decades, in some cases probably more than a century (Madej 1987; Madej and Ozaki 1996). Alluvial stream channels in the redwood region and adjacent coastal regions respond by widening, shallowing, and losing pools and summer surface flow (Lisle 1982; Frissell 1992). Floodplain surfaces, channel beds, and large woody debris structures are destabilized (Lisle 1989; Madej and Ozaki 1996), leading to increased incidence of scour and fill of sufficient magnitude to kill the eggs and fry of salmon (Frissell et al. 1997), as well as other biota that seek refuge in bed interstices (Welsh and Ollivier 1998) or off-channel areas during high flows.

Comparative chronosequential studies of historical air photos, maps, and other documentary sources indicate that many streams in the redwood region, with their catchments extensively disturbed by logging, agriculture, roads, and other land uses, have experienced dramatic physical changes during the past three to seven decades, consistent with the described pattern of stream response to increased sediment loading (Madej 1987; Madej and Ozaki 1996). Partial physical recovery has been observed in many streams in the years following large storm events, but for many other streams, full recovery of natural sediment-transport regimes and channel morphology and behavior will likely take decades or centuries.

Few catchments in the redwood region have escaped extensive human disturbance and the accompanying transformation of fluvial dynamics over the past forty to one hundred years; most of the relatively pristine basins are small. These catchments should be considered landscape refugia, isolates where a semblance of natural-historical erosion and sedimentation regimes remain locally intact, and which consequently are likely to be associated with high values for salmon and other aquatic biota that have declined precipitously elsewhere in the region (Frissell 1992; Frissell and Bayles 1996; Mattole Sensitive Watershed Group 1996).

The Role of Disturbance

The role of disturbance in structuring stream communities has been a primary focus of aquatic ecology for the past decade (Resh et al. 1988; Townsend 1989; Stanford and Ward 1992; Reeves et al. 1995; Spence et al. 1996; Wootton et al. 1996). Some changes are cyclic, such as seasonal variation in solar radiation, temperature, discharge, and leaf-fall. Less predictable but longer-lasting disturbance events, such as major floods, fires, mass-wasting events, and extreme winds that lead to windfall in the riparian zone, can influence the physical environment, and thus the biological community, for decades to centuries. Natural disturbances (i.e., floods, fires, mass wasting caused by previous events and earthquakes) and human activities (i.e., timber harvest, agriculture, mining, urban development, dams, and bank channelization) alike change the riparian communities associated with stream systems (Gregory et al. 1989; Schlosser 1991; Sedell et al. 1997; Frissell et al. 1997).

In the absence of human activities, based on the resistance-resilience model of ecosystem stability (Waide 1995), the late-seral redwood ecosystem is one of the most stable on the planet, achieving a self-perpetuating steady state (Bormann and Likens 1979), with a low incidence of severe fire (Veirs 1982). In such a stable ecosystem, disturbances tend to be localized within subdrainages (e.g., forest gap dynamics and landslides), their effects very limited in scope. Logging introduces a new disturbance regime. In coastal Oregon, the species diversity of salmonid fish assemblages was reduced in watersheds with more than 25 percent of the old-growth forest harvested (Reeves et al. 1995). Though large, natural disturbance events tend to be infrequent and cause large but localized changes to stream systems (pulse or stochastic events), human-caused changes are often more frequent and affect larger regions of the landscape (press or deterministic events) (Yount and Niemi 1990). In the redwood region, stream communities are shaped both by past natural disturbances and by a pervasive legacy of past and present human activities. Timber harvesting, as currently practiced in the region, represents an extreme disturbance to natural ecosystem processes—it is occurring on a scale and at a rate well beyond any natural disturbance the landscape has experienced in recent history (see Reeves et al. 1995; chap. 2, this volume). As indicated by the extensive loss of native biota, especially aquatic and riparian forms such as the many salmonid stocks in decline in northwestern California, the disturbance may already be irreparable for many species (Stouder et al. 1997).

The Aquatic Biota

In this section, we examine the aquatic and riparian life-forms of the redwood region and discuss their natural histories and roles in the ecosystem. We devote most attention to fish and amphibians, but also review briefly what is known about other taxa. (Appendix 6.1 lists plant and vertebrate species associated with aquatic ecosystems of the redwood region.)

Bacteria, Fungi, and Metazoans

The bacteria, fungi, and metazoans (protozoa, rotifers, copepods, nematodes, etc.) are extremely important but little-studied organisms of streams in the redwood region. These microscopic life-forms live in the hyporheic zone and the stream substrate, constituting part of the biofilm covering the streambed (Ward 1994). They fix dissolved organic carbon, break down detritus (fine and coarse particulate organic materials), and form a vital series of links in the aquatic food web (fig. 6.4) (Dahm 1981; Allan 1995).

The metazoans are microconsumers, which, in turn, are eaten by aquatic

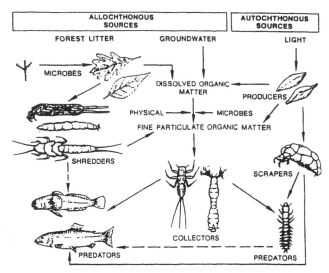

Figure 6.4. Energy sources for energy-flow pathways in, and the trophic structure of, woodland stream ecosystems. Deciduous leaves, photosynthesis by diatoms, and dissolved organic matter in groundwater are major energy sources. Genera that typify consumer functional groups are the shredders *Pteronacys* (above), *Tipula*, and *Pycnopsyche*; the collectors *Stenonema* (middle) and *Simulium*; the scraper *Glossoma*; and the predators *Nigronia* (lower right), *Cottus*, and *Salmo* (left). Litter microbes are characterized by hyphomycetes fungi. The dashed arrow indicates infrequent exchange. Organisms are not drawn to scale. (From Murphy and Meehan 1991 as adapted from Cummins 1974)

insects and other macroconsumers. The hypomycete fungi, which break down leaves and other riparian litter, are important energy sources for the shredders, which convert coarse particulate material to fine particulate material (fig. 6.4). The studies by Triska et al. (1989a,b) and Duff and Triska (1990) are among the few published accounts of bacterial activities in streams of the redwood region. Triska et al. (1989b) state that "further understanding of hyporheic metabolism and nutrient dynamics is required if we are to holistically address the biological continuum of river networks." We are not aware of any studies involving production or food web dynamics of bacteria, fungi, or metazoans in aquatic habitats of the redwood region.

Plants

Algae, mosses, and vascular plants constitute the aquatic primary producers in streams, providing autochthonous organic material. These plants require dissolved nutrients and a suitable substrate to fix solar energy. Iwatsubo et al. (1975, 1976) provided species lists of plants identified in Redwood National Park and reported that the periphyton community (food source for scrapers or grazers) is dominated by the diatoms *Achnanthes lanceolata, Diatoma vulgare, Gomphonema angustatum,* and *Melosira varians.* Power (1991, 1992) studied algal dynamics *(Cladophora glomerata)* in the South Fork Eel River on the eastern edge of the central redwood section. Mosses, which are major primary producers in headwater streams, and vascular plants (e.g., *Potomogeton* and *Rupia*) are locally abundant in larger streams and estuaries but are little studied in the redwood region and clearly warrant more research.

Macroinvertebrates

Iwatsubo et al. (1976), Iwatsubo and Averett (1981), and Harrington (1983) provided species lists of macroinvertebrates in Redwood National Park. Harrington (1983) found seven orders of aquatic insects, including mayflies, stoneflies, caddisflies, beetles, dragonflies and damselflies, and true flies. No endemic aquatic macroinvertebrates have yet been described for the redwood region, but this may be due to a lack of systematic research efforts. Much more research is warranted on these important aquatic species, particularly interdisciplinary studies that include chemical, physical, and biological interactions.

Fish

Compared with other regions of North America, the Pacific Coast has low species diversity of fish (Moyle 1976; Moyle and Herbold 1987; Reeves et al. 1998). The redwood region includes portions of two of the six native fish provinces within California (the Klamath and the Sacramento–San Joaquin) (Moyle 1976), but none of the endemic species of these two provinces is confined to the redwood region. The redwood region supports a few endemic terrestrial vertebrates (chap. 5), but no endemic fish or other aquatic vertebrates. Most of the fish species native to the redwood region are anadromous (e.g., salmon, sturgeon, and lampreys), euryhaline (able to move between marine and fresh waters, e.g., sticklebacks and sculpins), or are descended from marine ancestry (e.g., three species of stream-dwelling sculpins). The only naturally occurring (nonintroduced) representatives of truly freshwater fish species (those not capable of colonizing via the marine environment) in the redwood region are two sucker and five minnow species (table 6.3), all derived from the Sacramento–San Joaquin River system (Moyle 1976).

Moyle (1976) describes three fish zones in streams within the California coastal region: (1) a resident trout zone (rainbow trout in streams south of the Eel River and either rainbow or cutthroat trout in streams from the Eel River north) located in upper stream sections above barriers to anadromous fish; (2) an anadromous fish zone (lampreys, salmonids, sculpins, suckers, and minnows); and (3) an intertidal zone (euryhaline species listed in table 6.3) or estuary that may extend inland from the ocean for several kilometers (e.g., Navarro, Eel, Mad, Klamath, and Smith Rivers), or much less than a km ending at the first rocky riffle (e.g., Wilson and Mill Creeks in Humboldt Co.). The presence or absence of fish is a primary determinant in classifying streams under the California Forest Practice Act (discussed later in this chapter and in chap. 8).

Anadromous salmonids in the redwood region (i.e., Chinook and coho salmon, and rainbow [steelhead] and cutthroat trout) have somewhat similar life-history patterns and habitat requirements; each has been considered for listing under the Endangered Species Act (ESA) (NMFS 1998). These four species coexist in some streams by spawning and rearing in different areas or at different times. Chinook salmon typically spawn lower in a basin than coho salmon, with steelhead and cutthroat trout spawning progressively farther upstream. Juvenile salmonids (known by several names, including fry, fingerlings, parr, yearlings, young-of-year [YOY], etc.) grow in streams for several months to several years, depending on intrinsic (genetic makeup) and environmental conditions. At the end of their stream-rearing stage, the juveniles undergo a physiological change, the parr-smolt transformation, which adapts them for marine conditions.

Chinook salmon, the largest of the Pacific salmon, once populated large coastal streams as far south as the Ventura River (Moyle 1976). Their present range in coastal streams extends south only to the Mattole River in Humboldt County (G. Bryant, pers. comm.). Four different Chinook runs or life-history patterns occur in California (Yoshiyama et al. 1998), the late-fall and winter runs occurring only in the Central Valley and the latter listed as endangered under the ESA. Only fall Chinook both spawn and rear in the redwood region. Spring Chinook salmon of the Klamath River system migrate inland of the redwood zone to spawn in the Salmon and Trinity Rivers. The Southern Oregon/

Table 6.3. Consumer Functional Groups of Stream Animals.

Functional Group	Subdivision Based on Feeding Mechanisms or Dominant Food	Known Redwood Region Organisms
Shredders	Detritivores: decaying vascular plant tissue	Trichoptera (*Hydatophylax*, Lepidostomatidae)
		Plecoptera (Peltoperlidae, Pteronarcidae)
		Diptera (*Holorusia*)
	Herbivores: living vascular plant tissue	Trichoptera (*Phryganea*, *Leptocerus*)
		Lepidoptera (Pyralidae)
		Coleoptera (Chrysomelidae)
		Diptera (*Polypedilum, Lemnaphila*)
Scrapers	Rock substrate	Ephemeroptera (Heptageniidae, Baetidae, Ephemerellidae)
		Trichoptera (Glossosomatidae, *Neophylax*)
		Lepidoptera (*Parargyractis*)
		Coleoptera (Psephenidae)
		Diptera (Thaumaleidae, Deuteraphlebiidae)
	Wood substrate	Ephemeroptera (Caenidae, Leptophlebiidae)
		Trichoptera (*Heteroplectron*)
		Snails (*Juga*)
Collectors	Filter feeders	Ephemeroptera (*Isonychia*)
		Trichoptera (Hydropsychidae, Brachycentridae)
		Diptera (Simuliidae, *Rheotanytarsus*, Culicidae)
	Deposit feeders	Ephemeroptera (Ephemeridae, *Baetis, Paraleptophlebia*)
		Diptera (Chironomini, Psychodidae)
Predators	Swallowers of whole animals	Odonata
		Plecoptera (Perlidae)
		Megaloptera
		Trichoptera (Rhyacophilidae)
		Coleoptera (Amphizoidae)
		Diptera (Tanypodinae, Empididae)
		Fish (Salmonids)
		Amphibians (frogs, toads, salamanders)
		Reptiles (lizards, snakes, turtles)
		Birds (kingfishers, mergansers)
		Mammals (otters, bears)
	Piercers of tissue fluids	Hemiptera (Belastomatidae, Notonectidae)
		Coleoptera (Dytiscidae)
		Diptera (Tabanidae)

Source: Modified from Cummins (1973); Merritt and Cummins (1978).

Northern California Chinook salmon Evolutionarily Significant Unit (ESU) was proposed for listing as threatened in March 1998 (NMFS 1998).

Fall Chinook salmon in coastal California typically return to their natal streams during early fall after two to four years in the ocean and spawn between November and February. Female salmon construct their nests (redds) in the stream substrate of mainstem rivers and larger tributaries, where the eggs are buried after fertilization by males. Egg numbers can range from 2,000 to 14,000, depending on the size of the female (Moyle 1976). Fry emerge from the gravel in spring and begin feeding on drifting aquatic and terrestrial insects. They usually complete their stream growth and migrate to the ocean within three to six months at a total length of 70–90 mm (Healey 1991).

Coho salmon historically may have occurred as far south as the Big Sur River, but the southernmost naturally spawning populations now are in Scott and Waddell Creeks (Brown et al. 1994). All coho salmon populations in the coastal redwood region currently are listed as threatened (NMFS 1998). The life history of coho salmon in California was detailed in a classic study carried out primarily on Waddell Creek by Shapovalov and Taft (1954). Sandercock (1991) provided a detailed review of coho salmon biology throughout the range of the species. Unlike Chinook salmon, with several genetically distinct life history patterns, all coho salmon populations in California have similar life cycles. They typically spend fifteen to eighteen months in streams before smolting and entering the ocean. Most adult coho return to natal streams about eighteen months later, at age three, for spawning in late fall or early winter. Coho generally spawn in smaller streams than Chinook salmon, but in some streams (e.g., Prairie Creek) both species may spawn in the same area, Chinooks spawning earlier than coho. Female coho usually construct redds at the outlet of a pool (upstream end of a riffle) in small- to medium-size gravel. Depending on the size of the female, 1,000 to 5,000 eggs are deposited in the redd (Moyle 1976).

Coho salmon fry emerge from the gravel in early spring, move to the stream margins, and begin feeding on small invertebrates. As they grow they move into progressively deeper water, favoring pools with wood, shade, and other forms of cover. Juveniles feed primarily on aquatic invertebrates, but terrestrial prey (e.g., ants, beetles, spiders) may predominate in late summer to early fall (Sandercock 1991). In late fall to early winter, juvenile coho move into alcoves or side channels along the stream channel or congregate in deeper pools with wood to overwinter during times of high flows. Smolting and migration to the ocean typically take place in March through May, when the fish are 10–14 cm in total length.

Steelhead trout (anadromous rainbow trout) currently occur along the California coast as far south as Malibu Creek (NMFS 1998), well south of the redwood zone and the southernmost distribution of any anadromous salmonid. The three southernmost ESU populations of steelhead are currently listed as endangered (Southern California ESU) or threatened (South-central California Coast, and Central California Coast ESUs) (NMFS 1998). National Marine Fisheries Service (NMFS) (1998) determined that listing was not presently warranted either for the Northern California ESU or the Klamath Mountain Province ESU.

Steelhead in California exhibit one of two life histories, either summer-run or winter-run. These run types are based on the timing and state of gonadal development when adult steelhead reenter freshwater from the ocean and location and timing of spawning (reviewed by Roelofs 1983). Steelhead in the redwood region are winter-run populations, except for a small run of summer steelhead in Redwood Creek, typically fewer than fifty fish annually (David Anderson, pers. comm). Summer steelhead migrate upstream beyond the redwood zone in the Eel, Mad, and Klamath Rivers (Roelofs 1983).

Shapovalov and Taft (1954) noted that "unlike silver (coho) salmon, steelhead migrate to sea at various ages and over a long period within a season, spend varying amounts of time in the ocean, are capable of spawning more than once, sometimes spawn before their first journey to the sea, and may even remain in fresh water for their entire lives." Adult winter-run steelhead return to spawn in coastal streams beginning in late fall and continuing through winter and early spring (April). There may be some overlap in spawning areas used by coho salmon, but steelhead usually migrate farther upstream to construct redds and deposit eggs. The number of eggs produced ranges from around 1,000 to more than 10,000, depending on female size (Shapovalov and Taft 1954). Fry emerge in late spring through early summer and begin feeding on small invertebrates at the stream margin. Individual fry that survive and grow move into deeper, faster water, where they continue feeding on drifting prey.

The most common life-history pattern for coastal steelhead populations south of Alaska is two years of freshwater rearing and returning to spawn after two years in the ocean (Busby et al. 1994). In Waddell Creek between 1933 and 1941, 56 percent of the steelhead returning to spawn had reared one year in the ocean; 44 percent had spent two years (Shapovalov and Taft 1954). Adult steelhead can survive spawning, return to the ocean, and spawn a second or third time (Busby et al. 1994). In a sample of 3,888 adult steelhead, Shapovalov and Taft (1954) found that 83 percent were on their first spawning run, 15 percent on their second, and 2 percent on their third. Steelhead that survive to spawn a second or third time are most likely to be females (Busby et al. 1954).

Coastal cutthroat trout in the redwood region occur south to the Eel River and range inland 8–48 km, with the Smith River having the most substantial populations (Gerstung 1997). All coastal cutthroat trout populations in California, Oregon, and Washington are currently being evaluated for listing under the ESA (NMFS 1998) and are classified as a Species of Special Concern by the California Department of Fish and Game (Gerstung 1997). The life history of coastal cutthroat trout is complicated and poorly documented. Populations can be resident (existing upstream of migration barriers to other migratory fish), potamodromous (spawning in small tributaries but residing in larger streams as adults), or anadromous, although anadromy is "sporadic or uncertain in California populations" because adult fish may remain in estuaries instead of entering the ocean (Gerstung 1997).

Coastal cutthroat spawn during winter and early spring in small tributaries upstream of areas used by steelhead trout if accessible (Gerstung 1997). Fry emerge at a size of about 25 mm and move to the stream margins to feed on small invertebrates, primarily insects (Trotter 1997). When sympatric, juvenile cutthroat trout utilize stream habitats and food resources similar to those of juvenile coho and steelhead and may be limited by competitive interactions (Trotter 1997). Mitchell (1988) found that sympatric populations of steelhead and cutthroat greater than 10 cm long partitioned stream habitat by selective segregation, cutthroat rearing in deeper, slower water, steelhead in shallower, swifter water of runs, rapids, and the heads of pools. Age one and older cutthroat trout are opportunistic predators that feed on invertebrates and fish, including other salmonids (Trotter 1997). Most anadromous cutthroat trout smolt at ages two, three, or four, but can be as young as one or as old as six (Trotter 1997). Like steelhead, coastal cutthroat trout can spawn repeatedly in successive years, with up to five times reported (Trotter 1997).

Other fish in table 6.3 mentioned by Moyle et al. (1995) as species of special concern include green sturgeon, eulachon, longfin smelt, and tidewater goby (now listed as federally threatened). These authors also note that Pacific lamprey numbers are declining in California.

Herpetofauna

Streams, ponds, lakes, and associated riparian habitats of the coastal redwood forests of northern California are relatively rich in herpetofauna (amphibians and reptiles), supporting up to eighteen species of amphibians (one toad, four frogs, and thirteen salamanders) and eleven species of reptiles (four lizards, six snakes, and a turtle) (Stebbins 1985). With two exceptions—western pond turtle and Oregon aquatic garter snake—these reptiles are not aquatic but may frequent the riparian zone, where they often forage on the abundant invertebrates at the water's edge. The nonaquatic reptiles (western fence lizard, western skink, northern and southern alligator lizards, rubber boa, ringneck snake, sharptail snake, and the western terrestrial, northwestern, and common garter snakes) are discussed in chapter 5.

The seven terrestrial, direct-developing salamanders of the family Plethodontidae are discussed in chapter 5. The other ten amphibians of the redwood region can be placed into one of two main groups based on their reproductive behavior—those species that breed primarily in lentic habitats (ponds and lakes) and those that breed primarily in lotic (stream) habitats. Some of these species, however, will breed in both types of aquatic habitat. The species most closely associated with lentic habitats are the northwestern salamander, rough-skinned newt, Pacific tree frog, red-legged frog, and the western toad (Stebbins 1985). The rough-skinned newt, red-legged frog, and western toad also breed in slow-moving streams. Those species most closely associated with streams for breeding are the giant salamanders, with two species in the redwoods, the Pacific giant salamander and the California giant salamander (Good 1989), the southern torrent salamander (Good and Wake 1992), the red-bellied newt, the tailed frog, and the foothill yellow-legged frog (Stebbins 1985).

Pond- and lake-breeding species show site fidelity from generation to generation, with most individuals returning to the same body of water to reproduce (e.g., Semlitsch et al. 1996). Adult red-legged frogs, northwestern salamanders, and Pacific tree frogs usually migrate from upland foraging and resting habitats to breeding sites in late winter or early spring (Stebbins 1985). The rough-skinned newt migrates in early to late spring and the western toad from late spring to early summer (Stebbins 1985). These migration events are not absolutely fixed in time but are triggered by changes in climatic conditions in the vicinity of breeding sites in the appropriate season (e.g., Packer 1960; Semlitsch et al. 1993; H. Welsh, unpub. data).

Adult red-legged frogs, northwestern salamanders, and rough-skinned newts—all forest-dwellers—migrate to slow-moving streams, ponds, or lakes to seek mates, amplex, and deposit eggs. The western toad and the Pacific tree frog are the least specialized, occurring across a wide range of habitat types, from forest to meadow, and breeding in a variety of standing waters. Western toads select larger bodies of water than tree frogs and often take advantage of side pools along rivers, such as the Eel and the Trinity, during the waning flows of late spring. Thousands of newly metamorphosed toadlets are often encountered along riverbanks adjacent to these backwaters in late summer. Pacific tree frogs will breed in cattle tanks, wallows, roadside ditches, and large puddles, as well as in larger bodies of standing water (Stebbins 1985).

Lotic or stream-breeding amphibians make up a major component of the biomass in streams throughout the Pacific Northwest and can exceed fish in numbers and total biomass (e.g., Hawkins et al. 1983). Recent studies in the redwoods indicate that stream habitats in these forests support large numbers of amphibians (Welsh et al. 1997; Welsh and Ollivier 1998). Amphibians that breed in lotic habitats often show considerable specialization in their use of such waters. For example, the southern torrent salamander is a headwater specialist, breeding in springheads and other emergent water sources (Nussbaum 1969). It requires cold, clear, shallow water flowing over clean, coarse streambed substrates (Welsh and Lind 1996). In the interior parts of its range in California, it is closely associated with late-seral forests (Welsh 1990).

Pacific giant salamanders breed in streams from headwaters along the length of the stream continuum to larger streams and rivers. Females lay 75 to 100 eggs under large streambed substrates, primarily in riffles, and usually guard the nest until the eggs hatch. The aquatic larvae typically take two to three years to mature and metamorphose to a terrestrial form (Leonard et al. 1993). This salamander has evolved a dual life-history strategy, in which some individuals forego metamorphosis. These animals instead become sexually mature while retaining larval traits, including gills, which allow them to remain in the stream (becoming paedomorphs). They are voracious stream predators, feeding on anything they can swallow, including fish, invertebrates, and other giant salamanders (Parker 1994). The terrestrial phase of the giant salamander is also a voracious predator, taking small mammals and snakes as prey (Leonard et al. 1993). Welsh captured a large adult that regurgitated an adult chaparral mouse. Giant salamanders will also eat a wide range of invertebrate prey, including banana slugs.

The tailed frog breeds in headwaters to somewhat larger streams (e.g., Karraker and Beyersdorf 1997), occurring in highest abundances upstream of waters occupied by anadromous fishes. Larval tailed frogs require streams with cold, clear waters and clean, coarse substrates (Welsh 1993; Welsh and Ollivier 1998; Welsh and Lind in review.). Adults are found in the riparian zone near streams, where they forage at night along the stream banks. Adult tailed frogs are highly philopatric (Daugherty and Sheldon 1982a) and long-lived (Daugherty and Sheldon 1982b).

The foothill yellow-legged frog occurs farther down the stream continuum than the tailed frog or torrent salamander, preferring more open areas along larger streams and rivers where it often can be observed sunning on rocks and gravel bars. This species breeds in the slower-moving portions of these streams and rivers, attaching its eggs to rocky substrates in shallow, slow-flowing water near the stream margins (Zweifel 1955; Fuller and Lind 1992; Kupferberg 1996; Lind and Welsh, unpub. data).

The red-bellied newt is perhaps the closest to an endemic vertebrate in the redwood forest (see Twitty 1964 and chap. 5, this volume). Although it is biogeographically restricted to northwestern California within the range of coast redwood, it is not found exclusively in redwood forest habitats. It also occurs in mixed Douglas-fir–hardwood forests and adjacent meadowlands (Stebbins 1985; H. Welsh, personal observations). Red-bellied newts are active on the forest floor, mainly at night and only when litter moisture is high, throughout all but the summer months (H. Welsh, unpub. data). They migrate to breeding sites as early as the first week in February (Stebbins 1985; H. Welsh, unpub. data).

The western pond turtle is the only freshwater turtle in the redwood region. This turtle is found along large bodies of water, including lakes, ponds, and streams. It is spotty in occurrence, associated with areas having secure basking sites (e.g., logs and rocks surrounded by water) and undercut banks, where turtles can seek cover to avoid such predators as river otters, raccoons, and minks (Reese and Welsh 1998). It is a long-lived species and can reach fairly high densities where conditions are favorable. Pond turtles use upland forest habitats both for nesting and overwintering (Reese and Welsh 1997). On the whole, this species is uncommon in most areas of the redwoods, but populations do occur along parts of the major rivers of the northern California coast. The Mattole River has pond turtles in relatively low numbers.

The Oregon aquatic garter snake is a highly aquatic snake found along both small and large streams, ponds, lakes, and marshes, where individuals forage primarily on amphibians and fishes. At Hurdygurdy Creek in Del Norte County, juvenile snakes fed along stream margins on young-of-the-year steelhead and Chinook salmon or, when they become abundant in the summer, on tadpoles of the foothill yellow-legged frog (Lind and Welsh 1994). As adults these snakes shift prey preferences to larval and paedomorphic Pacific giant salamanders, which they pursue in the streambed substrates of even the fastest waters (Lind and Welsh 1990, 1994).

Birds

Bird species that depend on riparian/aquatic habitats in the redwood region are similar to those of riparian zones throughout northern California (e.g., osprey, bald eagle, belted kingfisher, American dipper, great blue heron, green heron, common merganser, killdeer, and spotted sandpiper). Most of these species are fish predators and therefore are positioned energetically at the top of the aquatic food web (e.g., fig. 6.5). Sustaining viable populations of these top predators

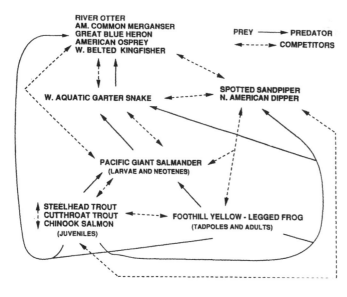

Figure 6.5. Vertebrate food web relationships at Hurdygurdy Creek, Del Norte County, California. (From Lind and Welsh 1990, 1994, unpub. data, and personal observations. Drawing by A. Lind.)

requires a healthy, productive riparian/aquatic ecosystem, a condition that is increasingly rare in the redwood region because of negative impacts from timber harvesting, water diversions, agricultural runoff, and introduced exotic predators. Many other birds species occur periodically or marginally in the riparian zone; these are addressed in chapter 5.

Mammals

Mammal species that utilize the riparian/aquatic habitats of the redwood region (e.g., river otter, mink, muskrat, raccoon, black bear) are similar to those found in riparian zones throughout northern California and southwestern Oregon. Most of these mammals eat fish or other aquatic vertebrates, as well as invertebrates, such as crayfish. Like the birds just mentioned, these species are positioned energetically high on the food chain (fig. 6.5); thus, sustaining their populations similarly requires productive, healthy aquatic ecosystems. Bats depend on open aquatic habitats for water (drinking on the wing) and aerial feeding (eating insects, many of which have aquatic larvae, and other flying arthropods). Proximity of aquatic and riparian areas to roost sites is essential for bats of the redwood region, species which include California myotis, little brown bat, long-legged bat, long-eared bat, Yuma myotis, hoary bat, and silver-haired bat (McKenzie et al. 1994).

Changes in Stream Ecosystem Processes Resulting from Timber Harvesting and Related Activities

The spatial and temporal scales of continuous timber harvest within the redwood region have led to widespread and persistent changes in stream conditions (table 6.4). The most pronounced change is the absence of, or reduction in, large trees within and adjacent to stream channels. Napolitano (1998) compared present-day log volumes in a historically (1864–1904) clear-cut and splash-dam logged, second-growth coastal redwood stream (north fork of Caspar Creek) (24 kg/m^2) with nine similar streams in old-growth redwood (49-268 kg m^2). (See Gates [1983] for a description of the highly destructive practice of splash-dam log transportation.) The pronounced difference in the abundance of large logs between Caspar Creek and the old-growth streams resulted in lasting channel changes, including channel incision, simplification of form, and reduction in sediment storage capability. Summarizing thirty years of research on the effects of timber harvesting on Caspar Creek, a 2,162-ha redwood-dominated watershed in the Jackson State Forest of Mendocino County, Ziemer (1998a) noted that logging had increased summer low flow (Keppeler 1998), subsurface and soil pipe flow (Keppeler and Brown 1998), and riparian tree mortality due to blow-down (Reid and Hilton 1998), and modified other riparian conditions.

For potential long-term effects on aquatic biota and their habitats, one of the

Table 6.4. Stream-Dwelling Fish Species in the Redwood Biome.

Anadromous	Euryhaline[a]	Freshwater
Pacific lamprey N (*Lampetra tridentata*)	prickly sculpin N (*Cottus asper*)	California roach N (*Hesperoleucus symmetricus*)
River lamprey N (*Lampetra ayresi*)	Coastrange sculpin N (*Cottus aleuticus*)	Speckled dace N (*Rhinichthys osculus*)
Pacific brook lamprey N (*Lampetra pacifica*)	Riffle sculpin N (*Cottus gulosus*)	Sacramento squawfish N (*Ptychocheilus grandis*)
Green sturgeon N (*Acipenser mediostris*)	Sharpnose sculpin N (*Clinocottus acuticeps*)	Hardhead N (*Mylopharodon conocephalus*)
White sturgeon N (*Acipenser transmontanus*)	Staghorn sculpin N (*Leptocottus armatus*)	Hitch N (*Lavinia exilicauda*)
Coho salmon N (*Oncorhynchus kisutch*)	Longfin smelt N, E (*Spirinchus thaleichthys*)	Tench I (*Tinca tinca*)
Chinook salmon N (*Oncorhynchus tshawytscha*)	Penpoint gunnel N (*Apodichthys flavidus*)	Golden shiner I (*Notemigonus crysoleucas*)
Pink salmon N, O (*Oncorhynchus gorbuscha*)	Saddleback gunnel N (*Pholis ornata*)	Carp I (*Cyrpinus carpio*)
Chum salmon N (*Oncorhynchus keta*)	Tidewater goby N, E (*Eucyclogobius newberryi*)	Klamath smallscale sucker N (*Catostomus rimiculus*)
Sockeye salmon N, O (*Oncorhynchus nerka*)	Starry flounder N (*Platichthys stellatus*)	Sacramento sucker N (*Catostomus occidentalis*)
Rainbow trout N (*Oncorhynchus mykiss*)	Shiner perch N, E (*Cymatogaster aggregata*)	Brown bullhead I (*Ictalurus nebulosus*)
Cutthroat trout N (*Oncorhynchus clarki*)	Bay pipefish N, E (*Syngnathus leptorhynchus*)	
Eulachon N (*Thaleichthys pacificus*)	Threespine stickleback[b] N2 (*Gasterosteus aculeatus*)	
striped bass I (*Morone saxatilis*)		
American shad I (*Alosa sapidissima*)		

[a]Lower estuarine zones may include several species of marine fish
[b]Sticklebacks can be resident or anadromous
N = native; I = introduced; E = esturarine; O = occasional.
Source: Moyle (1976); McGinnis (1984); Dill and Cordone (1997).

most significant modifications of riparian conditions from logging evidenced in the Caspar Creek watershed was the alteration of the large woody debris regime (Lisle and Napolitano 1998). The initial increase of large woody debris from blow-down in the riparian buffers increased habitat heterogeneity by storing sediments and forming pools (Nakamoto 1998). Nevertheless, the limited size spectrum and species composition of this large woody debris, as a consequence of logging in the nineteenth century (Napolitano 1998), suggested that the long-term result would be less favorable habitat conditions overall for aquatic species. Habitat suitability is expected to decline because the small, now hardwood-dominated woody debris decays quickly—much faster than that from conifers—and as input rates from depleted riparian sources in adjacent clearcuts and buffer zones decline (Lisle and Napolitano 1998; see also Reid and Hilton 1998).

Riparian forests on heavily logged landscapes are often dominated by deciduous trees (i.e., alders, big-leaf maple, and willows), with a lesser component of second- or third-growth conifers. Red alder often dominates the riparian zones of coastal streams in clear-cut watersheds (Dahm 1981). Lacking large and well-distributed, rot-resistant conifer wood, and with excess sediment filling the streambed, many streams in the redwood region have fewer and shallower pools and a less diverse array of physical conditions than reference streams in pristine settings.

The physical habitat conditions in streams constitute the ecological context for organisms that evolved there (Holt and Gaines 1992; Southwood 1977). Stream communities in logged versus unlogged streams are often different because the former have unsuitable habitat for many aquatic organisms. For example, many small, non–fish-bearing streams logged during the past 40 to 100 years, which are largely unprotected by current forest practice rules, are covered by 1–5 m of debris and sediments (Welsh, pers. obs.). These buried stream channels frequently flow subsurface for tens of meters, providing little or no habitat for many native aquatic organisms. Lewis (1998) found that although changes in forest practice rules between the 1970s and the 1990s reduced the amount of suspended sediment in Caspar Creek, logged areas still accounted for 89 percent more sediment than unlogged areas. Much of this sediment appeared to be mobilized from the unbuffered first-order watercourses (Lewis 1998).

Stream Classification and Its Problems

Stream ecosystems are supposed to be protected by regulations regarding timber harvest. The greatest scientific shortcoming in the current California Forest Practices Act (see chap. 8), from the perspective of riparian and aquatic resources, is its reliance on the conceptual approach to stream buffer protections used in the California Department of Forestry (CDF) stream classification system. CDF uses a system of stream categories in which class I represents fish-bearing streams, class II represents streams supporting aquatic life other than fish, and class III represents streams not supporting aquatic life. This system is in contrast to the geomorphological system based on watershed position (Strahler 1957), where first-order streams receive the highest classification—headwater tributaries, which combine to create second-order streams, which then combine to make third-order streams, and so on. The CDF system fails to recognize that a stream ecosystem and its vital processes are a functional continuum (fig. 6.2). The CDF system establishes differential protection measures along the stream continuum based on two factors, permanence of surface water and the presence of fish and other aquatic life. The presence of surface water has become the functional equivalent for the presence of aquatic life. Because headwater areas may have ephemeral surface flow (but perennial subsurface flow) and are not fish-bearing, they receive the least protection in terms of buffer zone width and retention of canopy cover and large woody debris. Perennial fish-bearing streams or rivers, under the CDF system, have the highest standards for these parameters.

Current buffer width designations have little basis in science. Rather, the different buffer widths for different stream classes reflect a bias in human valuation for game fishes over other riparian and aquatic biota, as well as ignorance of the stream continuum and the requirements of a healthy stream ecosystem upon which fish and other organisms depend. If relevant scientific data were considered, wider buffers would be provided on headwaters (CDF Class II and III or first- through third-order streams [Strahler 1957]) because they (1) tend to be transport reaches that provide crucial structural components such as LWD, (2) contribute a mixture of sorted coarse sediments of varying sizes downstream, and (3) are generally the source of the coldest waters. These headwater channels, if disturbed, are also potentially the greatest source of fine sediments, which can congest streambed interstices. Sediment-free interstices downstream are required for successful spawning by salmonid fishes; they also shelter the early life stages of stream macroinvertebrates and several species of stream amphibians (Welsh and Ollivier 1998). Consequently, headwater channels require significant riparian buffers to filter out fine-sediment runoff from the generally steeper terrain in which these channels are typically embedded.

The headwaters are where some of the stream biota thrive in the absence of fish predators. Some species appear to require areas with little or no predation in order to maintain viable populations on the landscape (e.g., see torrent salamander section below). Among such life-forms are several amphibians that require both aquatic and terrestrial environments in which to carry out their complex biphasic (aquatic and terrestrial) life histories. Among their terrestrial needs are cool, moist, stable microclimates in riparian forests alongside streams, where the adult life stages hide, forage, and seek mates. As currently constituted, even buffered areas along CDF Class I (fish-bearing) streams are primarily "edge" habitat (Laurance and Yensen 1990) and lack sufficient "interior core" areas where terrestrial microclimates are ameliorated and stabilized (Yahner 1988; Saunders et al. 1991; Brosofske et al. 1997). Recent research suggests that no-harvest buffers of 30–60 m are required to maintain suitable streamside and aquatic conditions for several cold-temperate adapted amphibian species (Brosofske et al. 1997; Ledwith 1996; Welsh and Hodgson, unpub. data).

The majority of large woody debris in a healthy stream enters the system in the headwaters and upper tributaries of the stream network, with less contributed along the larger, lower stream reaches (Maser et al. 1988). Adequate provisions for this ecosystem component are absent from the California Forest Practices Act for Class III, and insufficient for Class II streams. No leave trees (retained trees) are required or designated (dedicated) in the riparian management zone (RMZ) that might eventually contribute large woody debris greater than 61 cm to the Class II stream networks. Under current rules, the largest trees can (and probably will) be removed from the RMZ. Without provisions in the rules for some number of well-distributed, dedicated trees for recruitment of large woody debris, especially in size class 5 (>61 cm dbh; Mayer and Laudenslayer 1988), along the entire stream continuum, this critical structural element (table 6.1; Sedell et al. 1988) will continue to be lacking from the upper reaches of the stream networks of the redwood landscape.

By ignoring these considerations, the CDF system establishes and maintains a negative feedback system whereby downstream habitats can be progressively and continuously degraded because the headwaters are unprotected. The Mattole River Basin, and many other severely degraded watersheds on the North Coast, reflect this process of serial magnification of negative cumulative effects from poor timber-harvest practices (Mattole Sensitive Watershed Group 1996; MRC 1989). The result is a cascading disaster for aquatic and riparian resources where even portions of the stream that may initially support fish (CDF Class I or third-order and higher stream reaches) shrink and retreat with each harvest re-entry in a drainage basin. Tributaries are changed from Class I to II and II to III, as fewer and shorter portions of a stream system can support cold-water adapted fish or amphibians.

Effects of Forest Practices on Stream Biota

Most aquatic organisms have not been studied thoroughly enough to document impacts on their populations from human activities. For the several species discussed below, however, enough is known to make them potential indicators of the ecological condition of streams.

THE SOUTHERN TORRENT SALAMANDER. The southern torrent salamander may be more tolerant of stream canopy removal in the marine-influenced coastal redwood zone than elsewhere, based on its present distribution on altered landscapes (Diller and Wallace 1996). Nevertheless, declines on commercial timberlands may proceed undetected because the presence of individuals at some harvested sites is considered evidence of population persistence without any testing of this assumption (e.g., Brode 1995). Torrent salamander spatial distributions (metapopulation structure) and densities on managed redwood timberlands have never been compared with those of nearby pristine parklands. Data for this species from commercial timberlands need to be compared with data from reference sites to test the assumption that the torrent salamander is protected sufficiently by current forestry rules.

We found a mean density of 0.372 salamanders/m^2 (S.E. 0.0039) in five pristine streams in Prairie Creek Redwoods State Park in Humboldt County (Welsh and Ollivier 1998). Although we were not able to obtain comparable density data to compare with data from our reference sites, we did find relative abundance data (captures/unit of search effort) from Pacific Lumber Company lands in Humboldt County, California (Wroble and Waters 1989). Wroble and Waters (1989) reported an average of 0.052 southern torrent salamanders/hour of search time (S.D. 0.092) from seventeen streams on Pacific Lumber lands. For our comparison, we took a liberal approach (one that would tend to favor the timber company) by including data from all ten streams in our study (Welsh and Ollivier 1998), which included five affected streams (those with an infusion of fine sediments from a storm-triggered, road-building-related slope failure) and five pristine streams. We still found an average of 0.724 southern torrent salamanders/hour (S.D. 0.786) in the ten streams sampled at Prairie Creek—a highly significant difference (Mann-Whitney test; Z = 2.93, p = 0.003) (Welsh and Ollivier, unpub. data).

Diller and Wallace (1996) argued that torrent salamander distributions on commercial timberlands were determined by the gradient of the stream channel. Their argument was that stream channels with steeper gradients generate higher water velocities, which tend to flush fine sediments from the coarse substrates and thus create better microhabitat for this salamander. We agree with the general process model they describe (i.e., flushing of coarse substrates creates better microhabitat); however, we disagree that it follows that steeper channels are therefore required to support salamanders. Though this may be the case on commercial timberlands (Diller and Wallace 1996), interpretation of the data suggests that much of the suitable microhabitat along streams with gentle gradients has been compromised by excessive fine sediments in the streambed substrate on managed lands. To test this hypothesis, we examined our data from the ten reference streams in late-seral redwood forest at Prairie Creek Redwoods State Park (Welsh and Ollivier 1998). For those stream habitat units (e.g., pools, riffles, runs) with salamanders, we found a range of channel gradient from 2–14 percent (n = 26 stream habitat units; Welsh and Ollivier, unpub. data). We found no statistically significant correlation between channel gradient and salamanders (r = -0.213, p = 0.295).

Wroble and Waters (1989) proposed that low numbers of southern torrent salamanders on commercial timberlands are the result of a parent geology that is unconsolidated and naturally yields high levels of fine sediments, making streambed substrates unsuitable for torrent salamanders on these lands. This argument fails when considering that the streams with abundant populations of torrent salamanders in Prairie Creek Redwoods State Park also dissect an unconsolidated parent geology. To examine this question, we tested the hypothesis that parent geology influences the occurrence of populations of this salamander, using all available locality data from across the range in northwestern California (N = 83 sites; Redwood Sciences Laboratory, unpub. data). We classified all 83 localities as to parent geology, using U.S. Geological Survey maps for northern California. The presence of torrent salamanders was independent of parent geology type (consolidated vs. unconsolidated) (X^2 = 0.153, p = 0.696; Welsh and Ollivier, unpub. data).

Welsh and Lind (1996) argued that current forestry rules are inadequate to protect this salamander because they fail to protect headwaters areas. The continued isolation and fragmentation of headwater habitats, and the minimal protections afforded these sensitive habitats under current forestry rules, could have significant negative genetic consequences for populations on commercial forestlands (Welsh and Lind 1996). Torrent salamanders display some of the highest genetic diversity among sibling populations ever documented for a vertebrate species (Good and Wake 1992), suggesting that fragmentation and isolation have long been a part of their evolutionary history. Movement studies (Welsh and Lind 1992) suggest that torrent salamanders are highly sedentary, which could partly explain the high genetic diversity observed among populations. Nevertheless, it is highly unlikely that isolating processes occurred at the current rate or across so much of the range of this species in the past. Few data exist to determine how much fragmentation and isolation this species can tolerate before going extinct, but there is some circumstantial evidence.

Multiyear sampling of aquatic vertebrates from Caspar Creek in Mendocino County failed to detect torrent salamanders (Nakamoto 1998), although this drainage is well within the species' range and contains suitable habitat. Caspar Creek was splash-dam logged in the nineteenth century (Napolitano 1998); torrent salamander populations were likely extirpated by this activity and apparently have failed to recolonize in the intervening 100+ years. Evidence from the Mattole River suggests that anthropogenic disturbance has caused this species to decline across most of the watershed (Welsh et al., in prep.). Gene flow among the few remaining subpopulations is unlikely because of the limited amount of equable forest cover in the Mattole watershed. Further declines in late-seral habitats in the Mattole watershed will result in a high probability of the southern torrent salamander being extirpated from this watershed.

PACIFIC GIANT SALAMANDER. Although aquatic giant salamanders appear relatively impervious to some alterations of the near-stream environment (e.g., Murphy and Hall 1981; Murphy et al. 1981; Hawkins et al. 1983), their numbers can decline in response to habitat degradation resulting from sedimentation of the channel (see previous citations and Welsh and Ollivier 1998). They are sufficiently widespread and abundant in streams of the redwood region that their numbers might serve as an indicator of relative stream quality. Welsh et al. (1997) reported high numbers of Pacific giant salamanders along a 1-km length of Little Lost Man Creek in Redwood National Park, with an estimated mean density of approximately one salamander/m^2 ($\bar{x}$ = 0.956; S.E. 0.170). Welsh and Ollivier (1998) found a mean density of 1.33 salamanders/m^2 (S.E. 0.0069) for five pristine streams in old-growth redwood forest at Prairie Creek Redwoods, inclusive of Little Lost Man Creek. Nakamoto (1998) found similar densities of Pacific giant salamanders in the north fork of Caspar Creek and slightly fewer in the south fork, but found no differences before and after logging in either drainage. We are unaware of comparable data from commercial timberlands. More research on the response of the Pacific giant salamander to stream habitat modifications are needed, given the conflicting results obtained by Nakamoto (1998) and Welsh and Ollivier (1998).

THE TAILED FROG. The tailed frog is associated with late-seral forest conditions in northern California; the species appears to be negatively affected by timber harvesting there (Welsh 1990, 1993) and in Oregon (Corn and Bury 1989). Recent evidence indicates that the tailed frog has declined in the Mattole Basin as a result of anthropogenic changes to the landscape (Welsh et al., in prep.). Nakamoto (1998) reported tailed frog larvae in the north fork of Caspar Creek, but apparently too few were present for a comparative analysis. He indicated that tailed frog larvae were absent in the more heavily altered south fork, which was logged in the nineteenth century and then again in the 1960s. On the other hand, salmonid abundances in the south fork of Caspar Creek appeared to have returned to near prelogging levels (Burns 1972 cited in Nakamoto 1998).

We found mean densities of 0.929 tailed frog tadpoles/m^2 (S.E. 0.265) in one kilometer of Little Lost Man Creek in Redwood National Park (Welsh and Ollivier, unpubl. data) and 1.01 tailed frog tadpoles/m^2 (S.E. 0.0008) in five pristine streams in old-growth redwood forest inclusive of Little Lost Man Creek (Welsh and Ollivier 1998). Using the liberal analysis approach described above, we converted our data from Prairie Creek Redwoods (Welsh and Ollivier 1998; all ten streams) to captures/unit of search effort, then compared our data with those from seventeen streams on Pacific Lumber Company lands (Wroble and Waters 1989). Wroble and Waters (1989) reported 0.108 tailed frogs/hour (S.D. 0.097). Our surveys yielded 2.40 tailed frogs/hour (S.D. 1.58), a significantly greater abundance of tailed frog tadpoles than occurred on the Pacific Lumber landscape (Mann-Whitney test; Z = 4.30, p < 0.0001; Welsh and Ollivier, unpub. data). Given that Pacific Lumber Company lands lie well within the range of this species, and streams in redwood forest in this region typically support high numbers of this frog, our conclusion from this comparison is that the streams on Pacific Lumber Company lands are in exceptionally poor condition for tailed frogs as a result of timber harvesting.

THE FOOTHILL YELLOW-LEGGED FROG. The foothill yellow-legged frog is listed as a sensitive species by state and federal agencies based on dramatic declines of populations in California's Sierra Nevada foothills (see chap. 5, table 5.2). Populations of this frog appear to be stable in the redwood region, but they are highly susceptible to water diversions and channel alterations (e.g., gravel mining), which can interfere with their reproductive cycle, resulting in dramatic population crashes (Lind et al. 1996).

THE WESTERN POND TURTLE. The sedimentation of pools associated with timber harvesting, road building, and gravel mining probably have a negative effect on western pond turtle populations by reducing required habitats. Water diversions may also have negative influences on this turtle, particularly if they reduce the quality and amounts of shallow, warm-water microhabitats required as rearing areas by juvenile turtles (Reese and Welsh 1998).

Amphibians as Bioindicators

Their high relative abundances, sensitive physiology, and high site fidelity (relatively low mobility) make amphibians excellent candidates for bioindicators of ecosystem health. Welsh and Ollivier (1998) have tested this idea in streams of the redwood forest, where they documented significant adverse effects of increased fine sediments on three amphibian species (see fig. 6.6 a–c). Welsh et al. (in prep.) demonstrated a link between low stream temperatures, mature forest cover along the riparian zone, and the presence of cold-water adapted amphibians (i.e., the tailed frog and the southern torrent salamander) in the Mattole watershed of northern California.

Of the native stream amphibians of the redwood region, tailed frogs may be the most sensitive to anthropogenic perturbations. The larval stage, in particular, has great potential as an ecological indicator (Welsh and Ollivier 1998). Larval tailed frogs, along with the southern torrent and Pacific giant salamanders, are logical surrogates for gauging habitat conditions for anadromous fishes, such as coho and Chinook salmon, because they are long-lived, highly philopatric, and require very similar stream microhabitat conditions, such as clean gravels and cold water temperatures (Welsh and Ollivier 1998). Data on amphibian densities from reference stream habitats in pristine redwood forests have the potential to serve as a benchmark against which to compare densities in streams on managed lands. These data could also be used to generate indices of the relative condition ("health") of managed streams and to establish threshold targets for recovery efforts as proposed by Welsh and Ollivier (1998). Studies of stream amphibian densities are needed across all of the redwood parklands to establish the natural range of variability for these sentinel species.

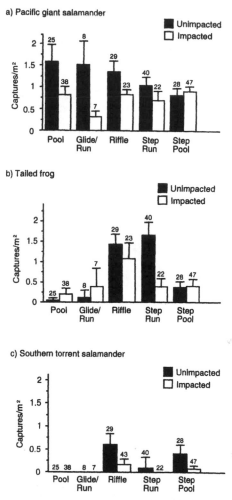

Figure 6.6. a–c. Densities of three species of amphibians with respect to stream habitat type and anthropogenic stress of fine sediment infusion (impacted). Data from five impacted and five unimpacted tributary streams in the Prairie Creek watershed, Prairie Creek State Redwoods. (From Welsh and Ollivier 1998.)

Research Needs

Several recent publications provide excellent reviews of current knowledge of stream ecology and terrestrial/aquatic linkages in the Pacific Northwest (Meehan 1991; FEMAT 1993; Spence et al. 1996; NRC 1996; Stouder et al. 1997; Naiman and Bilby 1998). Nevertheless, little integrated, holistic research on stream ecosystems has been carried out in the redwood region (but see Ziemer 1998b). This shortcoming can be explained by the enormous complexity of aquatic ecosystems, resistance within scientific disciplines to integrated interdisciplinary research (Regier 1978; Bella 1987), lack of funding for extensive research, and the private ownership of most of the region. Even the few pristine streams in public ownership (primarily state and national parks) are poorly studied. Although anadromous salmonids probably are the most thoroughly studied aquatic organisms in the redwood region, knowledge of their historic and current populations is limited. The observations of renowned Stanford University ichthyologist John Snyder (1925) still apply:

> It may not be out of place here to call attention to the well-known fact that stream fishing for trout, a major sport in California, is rapidly entering a critical stage. The extension of roads easily negotiated by the automobile, the building of high dams, the netting of steelheads in the rivers, water pollution, the use of water for irrigation, and many other things incident to a rapid growth in population are causing a marked and sudden depletion in the number of fish. It has been said that intelligent conservation must depend largely on our knowledge of the natural history of the species, and nowhere else is this more applicable. Very often our attempts at conservation serve among other things to bring to the surface our lack of definite knowledge of the habits and life history of the very fish that we are striving to protect. It is to be hoped that active support will be given to the Fish and Game Commission in every effort at careful investigation along this line.

We know little about the competitive and other interactions between salmonids and other aquatic organisms. Wilson and Halupka (1995) regarded anadromous fish as keystone species within vertebrate communities, yet nothing is known about the community effects of the drastic declines in salmonid populations throughout the redwood region. The loss of nutrients formerly brought in from the ocean by salmon (Bilby et al. 1996) undoubtedly has altered stream and riparian ecosystems, yet the magnitude or direction of change is unknown. Microbial dynamics in nutrient cycling and spiraling, although certainly a vital part of stream community structure and function, are virtually unstudied in the redwood region.

Research is urgently needed on exotic species introductions in the streams and rivers of the redwood region. Several predator species have been introduced into aquatic systems, and the potential influences on native species could be serious. For example, the predator Sacramento squawfish has been introduced into the Eel River system (Brown and Moyle 1997), as has the bullfrog (Kupferburg 1994), a voracious predator on other amphibians, reptiles, birds, and small mammals. The bullfrog has also recently been detected in the Mattole River (H. Welsh, pers. obs.).

Conclusions

In this chapter we have reviewed the available science on the processes and functions of riparian and aquatic ecosystems in the redwood region, along with the evidence of anthropogenic effects on these systems. Our impression is that, collectively, these studies support the notion that "the retention, restoration, and protection of aquatic and riparian processes and landforms that contribute habitat elements to streams and promote good habitat conditions for fish and other aquatic and riparian-dependent organisms" (Sedell et al. 1994) are essential to maintaining a functioning aquatic ecosystem and healthy populations of these resources across the landscape in perpetuity. Much work remains to be done to reverse the decline of the aquatic and riparian ecosystem in the redwood region. We are all stakeholders.

Appendix 6.1. Plant and Vertebrate Species Associated with Aquatic Ecosystems of the Redwood Region.

Common Name	Scientific Name	Common Name	Scientific Name
Alder	Alnus spp.	Long-eared bat	Myotis evotis
Red alder	Alnus rubra	Yuma myotis	Myotis yumanensis
Big-leaf maple	Acer macrophyllum	Hoary bat	Lasiurus cinereus
Willow	Salix spp.	Silver-haired bat	Lasionycteris noctivagans
Osprey	Pandion haliaetus	Pacific giant	Dicamptodon tenebrosus
Bald eagle	Haliaeetus leucocephalus	salamander	
Belted kingfisher	Ceryle alcyon	California giant	Dicamptodon ensatus
American dipper	Cinclus mexicanus	salamander	
Great blue heron	Ardea herodias	Southern torrent	Rhyacotriton variegatus
Green heron	Butorides virescens	salamander	
Common	Mergus merganser	Northwestern	Ambystoma gracile
merganser		salamander	
Killdeer	Charadrius vociferus	Rough-skinned	Taricha granulosa
Spotted	Actitis macularia	newt	
sandpiper		Red-bellied newt	Taricha rivularis
River otter	Lutra canadensis	Pacific tree frog	Hyla regilla
Mink	Mustela vison	Yellow-legged frog	Rana boylii
Muskrat	Ondatra zibethicus	Tailed frog	Ascaphus truei
Raccoon	Procyon lotor	Western toad	Bufo boreas
Black bear	Ursus americanus	Northwestern	Clemmys marmorata
California myotis	Myotis californicus	pond turtle	
Little brown bat	Myotis lucifugus	Oregon aquatic	Thamnophis atratus
Long-legged bat	Myotis volans	garter snake	

Chapter 7

Conservation Planning in the Redwoods Region

*Reed F. Noss, James R. Strittholt, Gerald E. Heilman Jr.,
Pamela A. Frost, and Mark Sorensen*

The reader is aware by now that saving the redwoods means much more than saving big trees. It means protecting the forest ecosystem in its natural condition, wherever such opportunities still exist. It means repairing damage that has been done by land uses that have disrupted natural processes and environments. And it means managing the overall landscape in a sustainable way to provide commodities, recreational opportunities, beauty, and other values for humans. Sustainability is interpreted here to include sustenance of the structure, composition, and function of the redwood ecosystem in perpetuity, not simply a sustained yield of wood or other products (Noss 1993). Because some species require refugia from human disturbances—and because the ultimate effects of our management practices are still unknown—sustainability is appropriately interpreted as a landscape or regional-scale property. It depends on protected areas (reserves), as well as areas where redwoods are harvested.

The first six chapters of this text review the history and basic biology and

Author contributions: Noss, lead author; Strittholt, codesigned and supervised focal area assessment; Heilman and Frost, focal area assessment; Sorensen, integration with planning.

ecology of the redwood forest. Chapter 8 reviews effects of management practices, especially silviculture, and describes some experiments in potentially sustainable forestry. In this chapter, we offer ideas and procedures that can be used to address such questions as where timber management, recreation, or other uses are appropriate and where strict protection of biodiversity should be emphasized. Given an assortment of possible landscapes in which to initiate conservation efforts, which ones warrant immediate investment? Answering these questions requires a "big picture" point of view, unconstrained by the relatively arbitrary legal, political, and regulatory boundaries that humans place on the landscape.

Deciding What to Protect

The need to evaluate sites (defined here as areas ranging from small parcels or single stands to large landscapes) in a systematic and rigorous way becomes increasingly obvious as we learn more about the distribution of biological elements and the consequences of past land-use decisions. The need becomes more urgent as resource extraction and development pressures accelerate and as funding for land conservation becomes more limited. Under such conditions, decisions about how to allocate scarce conservation dollars must be made using the best scientific information and analytic tools available. Many decisions will be irreversible.

One thing we have learned is that biodiversity is not distributed randomly or uniformly across the earth at any scale, but is concentrated in particular places. These places have been called "hot spots" (Myers 1988), and they can be recognized from a global scale (e.g., centers of endemism and species richness for various taxa) down to regional and ultimately within-landscape scales (e.g., serpentine outcrops with endemic plants, cold pool refugia in streams, and remnant old-growth stands in logged landscapes). Also, some parts of the landscape are of greater functional importance than others for maintaining biodiversity, even though they may have modest biodiversity themselves. For example, slopes prone to surface erosion and landslides are critical for controlling landscapewide erosion and sediment-delivery processes that determine the status of aquatic ecosystems and biota all the way to the sea (see chap. 6).

Past conservation decisions all too often have failed to result in a biologically meaningful system of reserves (Noss and Cooperrider 1994). Many protected areas have been established for their scenic or recreational values or because they were not useful economically. Other protected areas have been selected purely opportunistically—they were available at the right time for the right price. We do not fault those pioneering conservationists, who established the first redwood parks and other noteworthy reserves; their legacy speaks for itself (see chap. 2). Nevertheless, although early conservationists did the best they could, given the state of knowledge and information available at the time, many important areas went unprotected. The existing system of reserves (e.g., state and national parks) in the redwood region is arguably not extensive or connected enough to sustain the biota of the region, nor are regulations governing land uses on nonreserved lands adequate to prevent damage to aquatic ecosystems and other critical habitats (see chap. 6 and 8). Opportunistic, aesthetic, recreational, and other nonbiological approaches to reserve selection virtually guarantee that options to establish reserves will be exhausted before biological elements are adequately represented (Pressey et al. 1993; Noss and Cooperrider 1994).

The redwood region contains many potential sites for conservation, ranging from single stands of redwoods with extraordinary structural or compositional biodiversity to watersheds critical for controlling ecological processes such as mass erosion that determine aquatic ecosystem characteristics far from the site itself. New acquisitions of redwood sites should include lands of highest biological and ecological value—those areas that are irreplaceable (e.g., contain species or communities found nowhere else) or that otherwise contribute most meaningfully to conservation goals.

A broad temporal perspective should underlie all conservation decisions, so that sites can be evaluated not only in terms of their present biological value but also their potential value after restoration and their ability to remain viable (i.e., functioning as a natural ecosystem) over a long period of time. Large forests are usually better able to maintain their biodiversity over time than small pieces of forest (Diamond 1975; Harris 1984). Thus, protecting a large second-growth redwood forest might be preferable biologically to protecting a smaller parcel of old growth. Scenarios exist where a logged-over site in a strategic location may be of higher value for conservation than a pristine but isolated site, or where a collection of small, remnant forest patches contribute more to conservation than one or more larger patches. Exceptions occur to all general rules, so landscapes must be evaluated on the basis of multiple criteria. Despite these complications, a comprehensive, scientifically defensible, regionwide process of reserve selection and design is likely to result in a network of areas that captures more biodiversity and has a better chance of retaining its biological values over time, compared to a system resulting from site-by-site decisions.

At the heart of the conservation-evaluation process is a mechanism for viewing sites in a broader context and making fully informed comparisons among sites. Sites should be evaluated for their conservation potential at several scales. At the within-site scale, the focus is on the inherent qualities of sites—big trees, rare species, sensitive watersheds, and so on. At the among-site or landscape scale the focus is on the relationship of sites to surrounding landscape elements—as components of a larger constellation of sites in the landscape or as linkages between existing or potential reserves. Informing decisions at both these scales is the knowledge that the redwood forests are globally significant (see chap. 1).

Regional Conservation Goals and Guidelines

The following general goals of conservation planning (Noss 1992) will help guide conservationists as they evaluate options for conservation planning:

1. Represent in protected areas all kinds of ecosystems (natural communities) across their natural range of variation.
2. Maintain or restore viable populations of all native species in natural patterns of distribution and abundance.
3. Sustain or restore ecological (including physical) and evolutionary processes within a natural (historic) range of variability.
4. Create a network of conservation lands that is adaptable and resilient to a changing environment.
5. Encourage human uses of the landscape that are compatible with biological conservation goals, while discouraging those that are not.

These goals, derived from conservation biology, and the value judgment that biodiversity is worth protecting, are entirely consistent with the original "objects" of Save-the-Redwoods League, formulated decades earlier: rescuing from destruction "representative areas of our primeval forests," cooperating with state and federal agencies in establishing parks and reservations, purchasing redwood groves, fostering general understanding of the values of redwood and other forests, and supporting reforestation and conservation of forests (Save-the-Redwoods League, unpub.).

The goals of conservation planning suggest several, more specific guidelines for evaluating redwood sites for conservation. First, it is crucial that conservation efforts not be concentrated in a restricted portion of the redwoods region. Recognizing that species composition, species abundance patterns, forest physiognomy (architecture), genetics, ecological dynamics, and ecological context (i.e., surrounding vegetation) of redwood forests vary across the region in response to changes in climate, topography, soils, and other factors—including biogeographic history—conservationists should attempt to protect viable examples of redwood forests and other plant and freshwater communities across the northern, central, and southern sections (see chap. 3).

Second, the goal of maintaining viable populations of native species suggests that efforts be focused to some extent on those species of the redwood region whose viability is most at risk—the most imperiled plants listed in chapter 3, Appendices 3.4 and 3.5, and the most imperiled vertebrates and invertebrates listed in chapter 5, tables 5.2, 5.3, 5.4, and 5.5. Among the animals on these lists that inhabit redwood forests, several (e.g., northern spotted owl, Pacific fisher) have large home ranges or are otherwise area-sensitive; hence, large sites that can sustain large populations are preferable to smaller sites. Listed aquatic species in the region (e.g., coho salmon) also must be considered over large watersheds. Large reserves in key locations usually will provide more secure protection than such regulatory requirements as streamside buffers, which tend to be insufficient (see chap. 6). For most imperiled species in the region, however, maintaining the particular habitat conditions they require is the greatest need. Biologists and managers therefore must consider the often specialized life histories of these species.

Third, maintaining ecological and evolutionary processes generally requires protecting sites large enough to accommodate these processes. Some processes, such as floods, may result in changes on a large watershed scale. Fires sometimes cover very large areas, often crossing watershed boundaries. Evolution requires populations large enough to avoid rapid extinction (Frankel and Soulé 1981). Therefore, to maintain the full range of natural processes, the more intact (less fragmented), larger sites have higher conservation value than small, fragmented sites, all else being equal. Maintaining natural processes also requires sustaining ecologically functional (as opposed to minimally viable) populations of keystone species—species that play pivotal roles in the ecosystem, disproportionate to their abundance (Power et al. 1996). Salmon, especially those that spawn in the fall (Chinook and coho) during a time of protein scarcity for terrestrial predators and scavengers, may supply critical nutrients to the entire coastal ecosystem, terrestrial and aquatic. The ecological consequences of salmon decline in the region are unknown but potentially great. Other possible keystone species in the redwood region include the large carnivores (mountain lion, grizzly bear, wolf—the latter two species extirpated), beaver (which creates a diversity of habitats on the landscape), cavity-excavating birds, and undoubtedly others to be determined.

Fourth, the goal of creating a conservation network that is adaptable and resilient to environmental change, including global climate change, suggests that heterogeneous sites should be preferred over more homogeneous sites because sites with more microhabitat diversity allow organisms more opportunities to make small-scale shifts in distribution in response to environmental change. For instance, during a period of global warming, plants and small animals now inhabiting south-facing slopes may be able to shift distributions to north-facing slopes or to favorable microhabitats, such as shaded areas or seeps. Large areas generally can be expected to be more heterogeneous and contain more microhabitat refugia than small sites; however, a collection of small sites within dispersal distance of the species concerned may serve this purpose just as well. The need for a reserve network to accommodate environmental change argues for connectivity (e.g., landscape linkages) among reserves, so that individuals can move in response to long- or short-term habitat change. Wide-ranging species that consist of several partially linked subpopulations (i.e., metapopulations; see discussion of forest carnivores in chapter 5), also depend on regional-scale connectivity. For aquatic- and riparian-dependent species, connectivity is determined by drainage networks; a single large watershed is inherently more connected and often more biologically diverse than multiple small watersheds of the same aggregate area (C. Frissell, pers. comm.).

Finally, the goal of encouraging uses compatible with biological conservation and discouraging other uses is more helpful as a criterion for management than for site selection. Nevertheless, sites or landscapes with lower road density generally are preferred for conservation because they offer better control of human access.

Principles of Conservation Biology

The guidelines offered in the previous section are consistent with the emerging empirical generalizations ("principles") of conservation biology, which include the following. Some of the evidence supporting these principles was provided by Noss et al. (1997).

1. Species well distributed across their native range are less susceptible to extinction than species confined to small portions of their range.
2. Large blocks of habitat, containing large populations, are better than small blocks with small populations.
3. Blocks of habitat close together are better than blocks far apart.
4. Habitat in contiguous blocks is better than fragmented habitat.
5. Interconnected blocks of habitat are better than isolated blocks.
6. Blocks of habitat that are roadless or otherwise inaccessible to humans are better than roaded and accessible blocks.
7. The fewer data or more uncertainty, the more conservative (i.e., causing less reduction or disruption of natural habitats) a conservation or development plan should be.
8. Maintaining viable (i.e., undegraded, fully functioning) ecosystems is usually more efficient, economical, and effective than a species-by-species approach.

Although considerable empirical evidence and practical experience supports these generalizations, each has exceptions. To be useful for conservation planning, such principles must be interpreted and applied to particular cases cautiously and only by competent individuals familiar with the region and taxa involved. Uncritical interpretation of general principles can lead to flawed decisions. Each principle is based on the assumption that all else is equal; for example, in stating that protecting a large site is preferred to protecting a small site, we assume that habitat quality in the two sites is equal. Because all else is usually not equal, general principles must be interpreted in light of landscape-specific, site-specific, and often species-specific information. For instance, small remnants of old-growth redwoods in altered landscapes are important refugia for many species, such as many invertebrates and bats (Zielinski and Gellman 1999; see chap. 5). We should not rule out protection of these remnants by insisting that protected sites always be large.

Developing a Conservation Plan

Conservation planning is a process of evaluating the capability or suitability of different areas in a landscape for different land uses, then allocating each area to the best possible category of use. "Best" is interpreted from the standpoint of explicit conservation goals, such as those stated earlier. People with divergent goals likely will have different opinions about what constitutes best land use; ideally, such differences can be worked out through the political process or, more directly, through land acquisition.

Reserve Network Design

In its simplest and traditional form, conservation planning consists of selecting and designing nature reserves (protected areas). It becomes more complex when networks of reserves are proposed or when the design includes a variety of management zones, each accommodating a different range of human uses. A UNESCO biosphere reserve, for example, includes core areas designated for strict protection surrounded by buffer zones and transition zones (also called "zones of cooperation") that allow more intensive human uses, and sometimes other zones, such as areas for manipulative research or habitat rehabilitation (UNESCO 1974; Hough 1988; Noss and Cooperrider 1994). Noss (1992) revised this basic design to include habitat corridors (connectivity zones or landscape linkages) connecting reserves both within and among landscapes. These concepts have been applied by scientists and conservation activists to many regions (e.g., the Klamath-Siskiyou ecoregion, which lies adjacent to the redwood region; fig. 7.1).

Conservation planning relies on a site assessment methodology to help guide decisions about land purchases or designations. Species richness, endemism, rarity, habitat quality for species of interest, seral stage, sensitivity of habitats to human disturbance, and level of representation of habitat types in reserves are among the biological criteria used to select sites for conservation (Noss and Cooperrider 1994; Noss et al. 1999). These criteria complement nonbiological criteria, such as scenic value, recreational potential, and level of conflict with other land uses, which until now have been the dominant criteria in decisions regarding land protection. Sites that score high in an evaluation algorithm would be assigned to a zone, such as a core area, where perpetuation of biological values is the primary management goal (Noss et al. 1999). If the current protection status of a highly scored area is low and threats to its integrity are

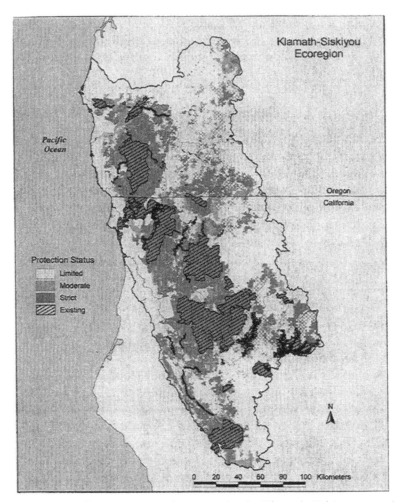

Figure 7.1. Draft conservation alternative for the Klamath-Siskiyou ecoregion (see Noss et al. 1999).

imminent, then that site is legitimately of high priority for acquisition or designation as a core area.

A comprehensive conservation plan takes into consideration more than core areas centered on sites of highest biological value. It attempts to address all sites in a landscape, envisioning how they might be knit together into a functional, ecological network that sustains natural processes and biota while permitting a range of human uses. The result of such a planning process might be conceptualized as a series of gradients, with protection increasing with proximity to core areas, and intensity of human use increasing outward from cores (Harris 1984; Noss and Harris 1986). For practical purposes, both for map-making and for deciding what happens where on the ground, lines around core areas need to be rigorously defined spatially. Otherwise, incompatible uses may encroach. These lines can be flexible over time, however, with reserve additions, deletions, and land trades altering the boundaries of core areas, but always with conservation objectives firmly in mind. Lines around other zones (e.g., buffers) can be fuzzier. Compatible and incompatible uses should be identified, but the list can be flexible and revised as more is learned about the effects of various activities (i.e., adaptive management; Holling 1978; Walters 1986).

Reserve network design, then, starts with core areas of recognized high biological value or sensitivity, then builds outward to encompass other lands necessary to maintain ecological integrity at a broader spatial scale. One key principle is to buffer core areas and other sensitive features spatially from incompatible uses on adjacent lands. In many instances, buffer zones need not be formally designated as such. Sometimes it might be perfectly adequate to use positive incentives to encourage low-intensity uses (e.g., recreation, uneven-age forest management) near core areas. In other instances, easements on lands intended as buffer zones might be acquired by an agency to fulfill conservation objectives.

Flexibility in meeting conservation goals is also required for corridors. The issue is not whether an identifiable corridor exists in the landscape, but whether the landscape provides functional connectivity for animals (and sometimes plants) sensitive to habitat fragmentation (Noss and Cooperrider 1994; Schumaker 1996; Beier and Noss 1998). Functional connectivity is usually a species-specific property measured by successful movement, population interchange, or gene flow. For many species, the best way to achieve functional connectivity is to minimize movement barriers and mortality risks in the portions of the landscape lying between patches of suitable breeding habitat; this may or may not result in a discrete swath of habitat that we recognize as a corridor. The important question is whether the species of concern will recognize the swath as a corridor.

Whether we call habitat that provides connectivity a "corridor" is largely a question of scale. Spotted owls, for example, are not known to follow physical corridors of forest cover, but dispersing individuals are vulnerable to predation by great horned owls, which thrive in fragmented landscapes with abundant edge habitat (Thomas et al. 1990). Therefore, a relatively intact landscape matrix (i.e., managed forest in which patches of old growth remain) may, at a regional scale, provide an avenue for dispersal and thus function as a corridor—especially in contrast to intensively logged landscapes that make up the rest of the matrix. For species sensitive to human persecution or mortality associated with vehicles, low road density might be the best measure of functional connectivity (Noss and Cooperrider 1994). Dispersal of most aquatic animals is necessarily confined to linear stream networks; thus, by definition they are dependent on aquatic and riparian corridors. Riparian corridors are also beneficial in that they mediate the flow of water, sediments, and nutrients from terrestrial to aquatic ecosystems.

Among the general suggestions for maintaining functional connectivity in a landscape are these: (1) all else being equal, wide swaths of suitable habitat (e.g., old forest) are better than narrow corridors, which may suffer from edge effects; (2) landscape linkages longer than normal dispersal distances for a target species should be sufficiently wide to provide for resident individual home ranges or have "stepping-stone" habitat patches, among which animals can move; (3) animals, besides seeking appropriate forage and cover, usually follow a path of least resistance when moving through a landscape; thus, ridgelines, adjacent parallel slopes, and riparian networks are natural movement corridors for many large, flightless animals; (4) planners should base connectivity designs on the needs of species most sensitive to fragmentation (Noss and Cooperrider 1994; Beier and Noss 1998).

Connectivity therefore must generally be addressed species by species, but focusing on those species most likely to require safe movement routes at broad spatial scales. In the redwoods region, what little is known about the dispersal and other movement needs of sensitive terrestrial fauna, such as forest carnivores (see chap. 5), suggests that high levels of overhead cover provide increased protection from predation and reduce the energetic costs of traveling between foraging sites, as well as providing favorable microclimatic conditions (e.g., Powell and Zielinski 1994). The current condition of the redwoods region is more fragmented than it was before human settlement and can be expected to have reduced functional connectivity for species associated with older forests, as has been shown in other study areas for American marten (Hargis and Bissonette 1997). Therefore, maintaining or restoring wide corridors with high levels of canopy closure within and among redwood protected areas is a sound conservation strategy. We recommend, however, that particular redwood forests and other communities that are naturally isolated generally should not be connected artificially.

Conservation Planning for the Redwoods

To put available financial resources to best use, Save-the-Redwoods League and other conservation organizations must be able to locate sites, from small forest patches to landscapes, throughout the range of redwood that offer the best opportunities for long-term maintenance of the redwood ecosystem with all its components. This is a large task. The League has long recognized the need for forging alliances and partnerships to create a unity of action among diverse interests. By working with government agencies, landowners, private companies, conservation interests, and others, the League hopes to create a synergy and long-term effectiveness in the redwood conservation movement that no single organization could accomplish alone. Part of what the League can offer to other organizations—including government agencies—is scientific and tactical advice on conservation planning.

A consensus has emerged within the conservation community regarding the usefulness of working at broad geographic scales to conserve biodiversity (Ricketts et al. 1999). The ecoregional approach to conservation planning has been accepted by many organizations the world over. The issues related to the conservation of redwood forests are complex and differ significantly across the redwood region. Socioeconomic situations within the northern, central, and southern sections of the range are likewise different and are reflected in local attitudes concerning the redwoods and environmental conservation in general.

The current conservation network in the redwoods region does not capture the full range of biodiversity present in redwood forests. We made a first attempt at addressing this issue by overlaying maps of protection status on the current distribution of redwood forests (table 7.1). Four categories of management or protection status are recognized by the national Gap Analysis Program (GAP) of the U.S. Geological Survey Biological Resources Division (Crist et al. 1998):

Status 1. An area having permanent protection from conversion of natural land cover and a mandated management plan in operation to maintain a natural state within which disturbance events (of natural type, frequency, intensity, and legacy) are allowed to proceed without interference or are mimicked through management.

Status 2. An area having permanent protection from conversion of natural land cover and a mandated management plan in operation to maintain a primarily natural state, but which may receive uses or management practices that

Table 7.1. Representation Summaries for the Redwood Region by Section Compared to the Klamath-Siskiyou Ecoregion.

	Private/Tribal	Public	Protected GAP 1	Protected GAP 1+2
Redwood, entire region	82.54	17.46	4.82 (19.75)	11.68 (56.71)
Redwood, northern section	81.34	18.66	5.76 (22.16)	12.43 (66.98)
Redwood, central section	88.31	11.69	1.36 (7.04)	6.72 (37.64)
Redwood, southern section	72.02	27.98	10.75 (22.99)	21.27 (43.96)
Klamath-Siskiyou, ecoregion	36.04	63.96	12.83 (16.57)	16.68 (20.69)

Note: First two columns provide general ownership percentages. For columns three and four, top figure indicates percentage of region or subregion protected, lower figure (in parentheses) indicates percentage of old growth protected. See text for explanation of GAP 1 and 2 categories.

degrade the quality of existing natural communities, including suppression of natural disturbance.

Status 3. An area having permanent protection from conversion of natural land cover for the majority of the area, but subject to extractive uses of either a broad, low-intensity type (e.g., logging) or localized intense type (e.g., mining). It also confers protection to federally listed endangered and threatened species throughout the area.

Status 4. There are no known public or private institutional mandates or legally recognized easements or deed restrictions held by the managing entity to prevent conversion of natural habitat types to anthropogenic habitat types. The area generally allows conversion to unnatural land cover throughout.

Only areas in GAP status 1 or 2 are considered protected areas. Redwood forests are best represented in protected areas in the southern section (although old-growth redwoods are most thoroughly protected in the northern section), lowest in the central section, and probably nowhere sufficient to represent the full range of physical habitats and plant communities (table 7.1). In this way, the redwood region is like virtually every region in the world (Noss and Cooperrider 1994). Furthermore, the current reserve network for redwood forests is not large and connected enough to guarantee long-term viability for many species. The conservation planning approach offered here makes a start at solving these problems by organizing the best available information on biological and ecological values within a broad regional context. The process involves a series of successively more focused investigations, used to establish regional priority areas (i.e., focal areas—large landscapes, each of which contains a collection of redwood sites) where more detailed assessment and site-level conservation strategies can be applied. The general sequence of planning is as follows:

- Delineate the redwood regional area of interest, including subregions (sections; see chap. 3). This step provides the boundaries for subsequent investigations.
- Identify the location, extent, and general character of remaining redwood forests.
- Identify landscapes of potentially high biological significance and value for redwoods conservation at the subregional (section) scale. These are potential focal areas.
- Apply a watershed overlay to identify least-disturbed catchments as additional focal areas for terrestrial and aquatic biodiversity.
- Identify existing and potential threats to and conservation opportunities for focal areas.
- Prioritize focal areas for site-level planning.

- Develop conservation plans for focal areas based on site-level assessments, identification of key ecological and local community issues, threats to biodiversity, opportunities for local partnership building, and identification of conservation, monitoring, and adaptive management actions. Plans should include core areas, corridors (where appropriate), and other zones.

We expect this process to result, fairly quickly, in specific plans of action for high-priority focal areas. Several of the tasks identified in this process (i.e., watershed assessments, partnership-building) are outside the scope of this chapter and book, but they are no less critical. The League and its partners can use the experiences gained in implementing the early focal area plans to refine the process, and can then develop plans for additional areas until—ideally—the entire range of the redwoods is covered. Focal area boundaries could be adopted by the League and its partners as conservation management units in which to manage, monitor, and evaluate the effectiveness of conservation efforts over time.

This focal area identification and assessment methodology, while essentially objective and mechanistic, requires the participation of numerous individuals with knowledge and experience in redwoods ecology and conservation. There is considerable room—indeed, a requirement—for flexibility and creativity in applying this approach.

Focal Area Identification and Assessment Model

The focal area identification and assessment model is intended to implement the initial steps of the conservation planning approach just described. (See Strittholt et al. 1999 for details of the model and its application across the redwood region.) The model is essentially a formula for assigning conservation scores to potential focal areas. Conservation scores are a measure of potential value for conservation of native biodiversity. Focal areas with moderate to high scores in the initial evaluation can then be subjected to more intensive assessments, including field surveys and species-specific analyses, as time and funding allow. Four basic features distinguish this approach:

1. It is based on accepted empirical generalizations and principles of conservation biology (e.g., Moyle and Sato 1991; Wilcove and Murphy 1991; Noss and Cooperrider 1994; Meffe and Carroll 1997; Noss et al. 1997), such as the need for large, multiple, well-distributed, well-connected, and unfragmented reserves to maintain biodiversity.
2. It is hierarchical. The analysis begins by dividing the redwood region into three sections (northern, central, southern) as defined in chapter 3. The ecological mosaic (e.g., redwoods, hardwoods, grassland, chaparral, coastal

scrub) differs among sections, as well as among potential redwood focal areas in each section. The analysis then proceeds from section to focal areas to individual sites (i.e., watersheds, parcels, or other relatively discrete units) within focal areas, with the level of spatial resolution and biological detail increasing at each step.

3. It assesses focal areas at three spatial scales: (a) patches—discrete, mappable vegetation units of a defined composition and structure, such as redwood forest stands spatially distinct from other such stands, or more mechanistically defined units of analysis, such as a regular array of grid cells; (b) neighborhood—the immediate surroundings of a patch; and (c) watershed—the hydrologic unit (e.g., sixth-level watershed) in which a group of patches exists. Use of a computer-based geographic information system (GIS) makes multiscale analysis feasible.

4. This conservation planning exercise is focused largely on the biology and ecology of the redwood forest ecosystem. It emphasizes conservation of the redwood seral stages, especially old growth, that are most depleted in the region and the species associated with redwood forest that are most imperiled. Nevertheless, opportunities to protect other species, plant communities, and freshwater habitats associated with redwood forests in the landscape mosaic should be exploited whenever possible. As documented in chapters 3 and 5, most of the rare species of the redwood region are associated with habitat types other than redwood forest; and as explained in chapter 6, the condition of the terrestrial landscape largely determines the health of associated aquatic ecosystems.

Assessment Criteria

The focal area assessment applies GIS modeling to evaluate the biological conservation potential of large landscapes within each of the three redwood sections defined on the basis of vegetation characteristics (see chap. 3). Sections are delineated to the closest watershed subbasin for mapping purposes (fig. 7.2). Treating the three sections individually has practical benefits, but an overarching perspective that spans sections is also important. For example, the ranges of some species of concern, both aquatic and terrestrial, do not conform to sections determined by the distribution and composition of redwood forests.

The method applied here does not say anything specific about the conservation and management options for an identified focal area or the sites within it. That task remains for a much more detailed assessment at the within-focal-area and individual-site levels. Similarly, management recommendations for matrix lands outside sites targeted for reserves within as well as outside focal areas are beyond the scope of this assessment, but should be developed after further study.

The first step in a focal areas assessment is identifying potential focal areas across the region of interest. Our method was inclusive of all redwood forests

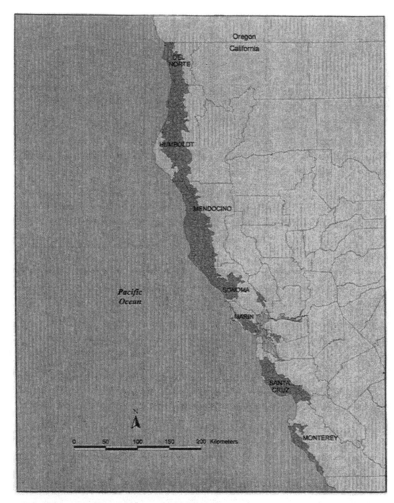

Figure 7.2. Watershed subbasins (6th level) based on 1:24,000 mapping for the redwoods study region (data obtained from California Fish and Game).

that have been mapped in the most recent inventories. Potential focal areas are delineated by GIS modeling of the mapped forest patches. The potential focal areas are defined as areas that score high after all evaluation criteria are analyzed and considered. These focal areas have no predetermined size. Rather, their sizes and shapes are based on the results of the model and the level of generalization most comfortable to the conservation practitioner. If the subbasin (sixth-level watershed) is the unit of organization, focal areas could consist of single sub-basins or, more commonly, clusters (e.g., 6–7 units) of subbasins.

Our method applies available, spatially explicit data at an intermediate map scale (1:100,000). Larger scaled data (e.g., 1:24,000 roads and hydrography) are incorporated into some model criteria. The modeling approach uses ARC/INFO, ERDAS, and FRAGSTATS GIS software and assesses present as well as potential value of focal areas by considering existing attributes (e.g., forest age and species composition) along with such long-term considerations as position in the watershed and restoration potential.

The technique applies ten criteria to rank potential focal areas in each section in terms of conservation value, that is, the contribution the focal area could make to the conservation goals stated earlier. We first analyzed each criterion separately, then integrated them into an overall ordinal index (the overall index is subdividable, however, and any criterion can be applied by itself as a stand-alone analysis). Ordinal scores for each criterion are expressed as 1–5 (further discrimination could be employed, as desired). We designed the model to weight old-growth redwood forest higher than other criteria. Weightings are largely subjective but have the benefit of applying professional judgment. We express all results (identification and ranking of potential focal areas) in an ordinal fashion.

Nine of the ten criteria can be categorized as a patch function, neighborhood function, or watershed function (see earlier description), depending on the scale they address, although some criteria address two or more scales. (A tenth criterion, restoration and management potential, is based on expert opinion.) For example, road density can be calculated within a patch of forest using a regular grid cell array (e.g., 1 km × 1 km, an arbitrary but convenient scale for landscape-level assessments), generalized using a GIS moving window filter over a larger neighborhood, or calculated according to watersheds at several scales. The following is a discussion of each criterion, its rationale for inclusion in the model, and a brief technical explanation. We depict the basic model as a flow chart (fig. 7.3). After the nine quantitative criteria are examined and combined to delineate areas of high conservation value (high-ranked focal areas), the spatially explicit model results are reviewed by individuals who are knowledgeable about socioeconomic and other practical realities of each area. Focal areas are then prioritized further on these grounds.

Criterion 1, Location of Largest Late-Successional Patches (Patch Function)

This criterion is based on the conservation biology principle that, all else being equal, large patches are better than small patches (see earlier discussion and chap. 5). "Large" and "small" are context specific; in some landscapes, the largest remaining patches of natural redwood forest may be only a few hectares in size, but they nevertheless are valuable (see Zielinski and Gellman 1999). We are most interested in late-seral (especially old-growth) stands of redwoods because such stands and the species dependent on them have been reduced far more than early-seral stands since European settlement (see chap. 2 and 5); these communities generally are more imperiled. We used the most current and accurate map of late-seral vegetation data for the region, provided by Peter Morrison of the

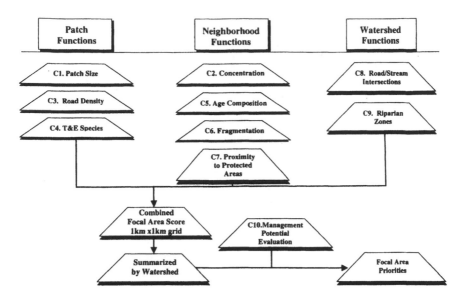

Figure 7.3. Flow chart of GIS-based focal areas model. (C1–C10 correspond to Criteria 1–Criteria 10 as described in the text.)

Pacific Biodiversity Institute. In this layer, late-seral forest patches were represented as two categories: old-growth redwood and residual redwood. Other iterations of the model included old-growth and residual Douglas-fir as well.

Because required patch sizes cannot be specified in the abstract, but depend on landscape context, we assigned ordinal scores to potential focal areas based on the range of patch sizes (i.e., frequency distribution) encountered in each subregion (1 = smallest patches, 5 = largest patches).

Criterion 2, Concentration of Late-Successional Patches (Neighborhood Function)

The second criterion is based on the principle that patches close together are better than patches far apart (i.e., they combine to form a larger functional patch from the standpoint of organisms and ecological processes that move freely among patches). As with Criterion 1, the desirable patch density could not be specified deductively. First, we determined neighborhoods by buffering all existing old-growth patches as defined in Criterion 1 and then excluded areas greater than 500 meters. The results produced discrete neighborhoods. The old-growth patch data used in Criterion 1 were organized and analyzed according to these neighborhoods. FRAGSTATS (a fragmentation analysis software) was run on the late-seral patches found within each neighborhood. We assigned ordinal scores relative to the range (frequency distribution) of patch mean nearest neighbor (1 = lowest mean nearest neighbor, 5 = highest mean nearest neighbor).

Criterion 3, Road Density (Neighborhood Function)

All else being equal, the lower the road density, the more desirable an area for conservation, in terms of protecting aquatic ecosystem integrity, reducing invasion
potential of exotics, offering security to species sensitive to human disturbance, and
other reasons (see Noss and Cooperrider 1994:54–57; Trombulak and Frissell
2000). To calculate road density, we created a 1 km × 1 km grid cell array across
the entire region using ARC/INFO and summarized road density using 1:24,000 scale
data. The result was a map layer showing road density for each grid cell. An ordinal score was assigned to each cell, based on the range of existing densities (1 = highest road density, 5 = lowest road density). Note that application of this criterion
depends on an up-to-date, comprehensive roads database—including unpaved logging roads. The roads data were based on USGS 7.5-minute quadrangles that show
many, but not all, unimproved (unpaved) roads. The largest problem with these
data sets is the wide spread of completion dates; some quadrangles, for example,
were dated in the 1960s and others in the 1990s. Nonetheless, a 1:24,000 analysis
is far superior to coarser-scaled roads analyses (e.g., 1:100,000).

Criterion 4, Location of Imperiled Species (Patch and Neighborhood Function)

Known locations of imperiled species—especially places where many of these
species are concentrated—are well accepted as areas for conservation emphasis
(Noss and Cooperrider 1994). For this application, we included all imperiled
species from the California Natural Diversity Data Base (CNDDB) found within
the redwood regional boundary. After establishing a 1 km × 1 km grid over each
section, we scored each cell by the number of element occurrences (rare species
locations) encountered, combined with a weighting based on level of rarity and
threat (from natural heritage program criteria of The Nature Conservancy). For
example, a grid cell might have two occurrences of rare species, one critically
imperiled globally and one more common globally but imperiled in California.
We considered the two together to determine the value for that grid cell, applying
a score of 50 for the globally imperiled species and a score of 10 for the species
imperiled in California but more common elsewhere. The result was a map layer
showing grid cells containing the most known imperiled elements. We then
assigned ordinal scores to grid cells according to their position in the range of values (1 = lowest imperiled element occurrences, 5 = highest occurrences).

An important caveat for interpretation of this criterion is that the only reliable database on imperiled species in the study region—the CNDDB—is, like
most such databases, incomplete. Furthermore, it is usually impossible to separate areas that truly lack populations of imperiled species from areas where no
one has conducted the necessary surveys. The precise locations of imperiled
species populations within grid cells become important in the more detailed
analyses within focal areas and sites, which should be conducted subsequent to
the analyses described here.

Criterion 5, Forest Neighborhood Age (Neighborhood Function)

An ideal, long-term conservation strategy for redwoods would maintain an ecological mosaic with all seral stages of redwood and associated habitat types included in natural patterns of distribution and juxtaposition. Such a mosaic can be achieved only at a landscape or regional scale. At the patch or stand level, and considering the historic trend of loss of old-growth redwoods, late-seral redwood stands have greater conservation value today than younger forests (again, all else being equal). Furthermore, the age of surrounding forest patches is likely to influence the population viability of late-seral species within a given forest patch. Using the best vegetation data available, we described forest-age composition in each potential focal area. We mapped average forest age using a moving window of 1 km × 1 km. Areas of relatively older forests were preferred to younger areas. We then scored the results based on relative age (1 = youngest forest neighborhood, 5 = oldest forest neighborhood).

Criterion 6, Forest Fragmentation (Neighborhood Function)

Studies in many regions have implicated anthropogenic habitat fragmentation as one of the greatest threats to biodiversity (Noss and Csuti 1997). After making an intermediate map layer showing approximate tree sizes in forest patches (4 size classes: 1–11" dbh, 11–24" dbh, 24–36" dbh, > 36" dbh) and nonforest, we applied FRAGSTATS to these patches for "landscape units" that were defined by 1:24,000 roads data. We assigned a total fragmentation score to each landscape unit by combining ordinal scores for each of four fragmentation metrics (mean core area per patch, mean nearest neighbor, area weighted mean shape index, and interspersion and juxtaposition). We tried to consider only anthropogenic fragmentation, not mosaics created by site factors (e.g., differences in soils or slope) or natural disturbances. We applied composite ordinal scores to each landscape unit (1 = highest fragmentation, 5 = lowest fragmentation). Interpretation of maps to provide scores did not take into consideration the probability that some species occurring in late-seral redwoods are dependent on other seral stages or vegetation types seasonally to meet critical life-history needs. For example, winter wrens, varied thrushes, and other birds in northern California are known to shift habitat associations seasonally (B. Marcot, USDA Forest Service, pers. comm.). This issue should and can be addressed in further, site-level examination of prioritized focal areas.

Criterion 7, Potential Connectivity to Existing Protected Areas (Neighborhood Function)

Redwood patches or collections of patches that are adjacent or close to existing protected areas have greater potential for enhancing the ecological integrity of both the existing reserves and the added sites. The same can be said, in some cases, for patches that have high potential to be physically linked to redwood parks, or other secure redwood forests, by strips of habitat through which red-wood-associated species will move. This criterion also can be used to evaluate the extent to which a focal area lies within a natural connection between pro-tected areas or other secure habitat. Riparian networks and ridge systems (if not compromised too severely by roads) are among the natural connections in land-scapes, as they offer a path of least resistance to many dispersing animals (Noss and Cooperrider 1994). We conducted this analysis by determining the prox-imity of every position in the study area to the existing protected areas. The range of scores was expressed ordinally (1 = lowest potential connectivity to pro-tected areas, 5 = highest potential connectivity to protected areas).

Criterion 8, Road/Stream Intersections (Watershed Function)

Places where roads cross streams are common sources of sediment. Using the CALWATER watersheds for California, we calculated the number of road/stream intersections based on 1:24,000 USGS data, both for roads and streams. We summarized the results by watershed subbasin and then assigned ordinal scores (1 = highest number of road/stream intersections, 5 = lowest number of road/stream intersections). Follow-up studies of potential focal areas should map logging roads in greater detail. We were surprised to find this crite-rion nonredundant with Criterion 3 (Road Density) (see Strittholt et al. 1999).

Criterion 9, Forested Riparian Zones (Watershed Function)

In many landscapes, forested riparian zones help protect the ecological integrity of streams; this is strongly the case for the redwood region (see chap. 6 and 8). To conduct this analysis, we buffered the 1:24,000 streams layer by 100 m and compared it to the vegetation data layer. The goal here was to quantify the amount of older forests along streams summarized on a watershed by watershed basis. Ordinal scores were assigned to each watershed subbasin (1 = low levels of older forest along streams, 5 = high levels of older forest along streams).

Aquatic Biodiversity Value

One of the criteria we hoped to use was an aquatic biodiversity value to comple-ment the largely terrestrial-focused analysis of CNDDB data and other criteria in our assessment model. After careful consideration of available aquatic species data, we opted to exclude this criterion from the model because the data lack equal coverage and discrimination across the redwood region. This criterion should be pursued further for future applications or refinements of the model.

A separate, follow-up study to identify aquatic diversity areas across the red-wood region is urgently needed to complement the approach taken here. Effective aquatic conservation will require a combination of (1) specially pro-tected watersheds, (2) protection of intact floodplains and other critical river segments, and (3) a set of blanket, landscapewide conservation practices that minimize disturbance of riparian zones, wetlands, and erosion-prone hillslopes (see Moyle and Sato 1991; Doppelt et al. 1993; Moyle and Yoshiyama 1994; Frissell and Bayles 1996).

Criterion 10, Management and Restoration Potential

A site cannot simply be acquired and expected to take care of itself. Some level of temporary or permanent management by humans is almost always necessary to maintain or restore ecological integrity. Restoration potential refers to how easily, and in what time span, a site might be restored to a semblance of naturalness. A logged-over site on a steep slope that has lost much of its topsoil and is filled with exotic plants would generally have a low restoration potential. This criterion also considers management intensity, the level of human effort required to maintain the site in the desired condition in perpetuity. All else being equal, small sites require more management expenditure per unit area than large sites, especially because they are more affected by external influences (Noss and Cooperrider 1994). Many aspects of restoration and management potential, however, are site— or landscape—specific. Each focal area ranked high under the first nine criteria could be reviewed, using expert opinion, for its relative restorability and manageability. This criterion could have a major influence on priority-setting for focal areas. It could be addressed either outside the mechanistic model or included in the model and weighted appropriately. Without weighting, the range would be 1 = very difficult to restore or manage, 5 = relatively easy to restore or manage.

The data requirements for conducting focal area assessments using the first nine criteria reviewed above are listed in table 7.2. The resolution and specificity of data required to make a useful assessment vary from case to case, depending on the interests of those conducting the analysis. Use of this assessment model provides a convenient way for data gaps and data-gathering priorities to be identified.

Table 7.2. Data Layers Used in the GIS-based Focal Areas Modeling.

Data Layer	Scale/Resolution	Source
Historic extent of coastal redwoods	1:130,000	Pacific Biodiversity Institute
Current extent of redwoods	1:130,000	CA Dept. of Forestry and Fire Protection
Current vegetation, 1994	30m × 30m	LEGACY
Current old-growth redwoods	30m × 30m	Pacific Biodiversity Institute
Element occurrences	1:24,000	CA Natural Diversity Data Base
Watershed subbasins	1:24,000	CA Fish and Game
Roads	1:24,000	USGS 7.5 minute quadrangles
Hydrography	1:24,000	USGS 7.5 minute quadrangles
Ownership	1:100,000	Teale Data Center
Protected areas	1:100,000	CA GAP
Administrative boundaries	1:100,000	CA GAP

Focal Area Selection and Monitoring

Results from each of the nine to ten analyses of conservation criteria just described can be integrated by summing raw or weighted criterion scores into an overall index. This process results in a ranking of focal areas by biological conservation priority, given the explicit emphasis on redwood-associated, old forest–dependent, area-dependent, and fragmentation-sensitive species. The focal area ranking can be used as one basis for determining where the League and other conservation organizations should take specific actions to preserve or restore redwood forests. Alternative rankings that reflect different objectives or emphases can be easily achieved by applying different criterion weightings.

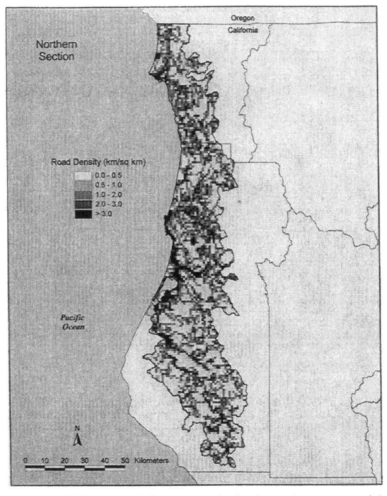

Figure 7.4. Road density patch-function results for the northern section of the redwoods region calculated for a 1 km × 1 km grid cell array.

Alternative weightings could be used, in part, to select for species that use different kinds of habitats or habitat configurations (B. Marcot, USDA Forest Service, pers. comm.). Geographic variation in site characteristics among the northern, central, and southern redwood forests may also suggest the need for different weightings of formula components in each section.

The complete focal area assessment for the redwood region is described in Strittholt et al. (1999). As a sample of our findings, we illustrate the results from the northern section for one patch function (road density; fig. 7.4), one neighborhood function (forest age composition; fig. 7.5), and one watershed function

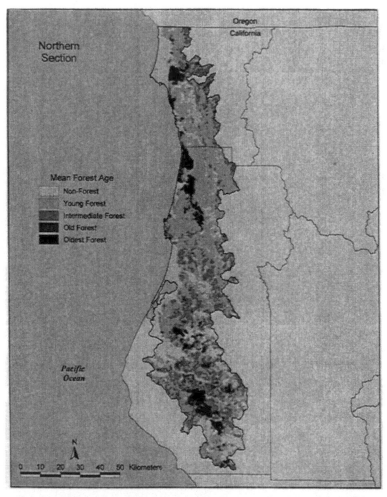

Figure 7.5. Forest age composition neighborhood-function results for the northern section of the redwoods region, calculated using a 1 km × 1 km rectangular averaging moving window.

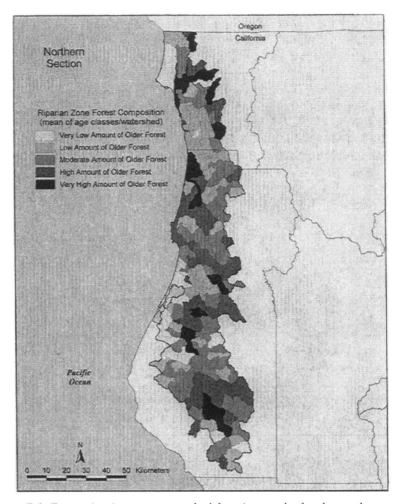

Figure 7.6. Forest riparian zone watershed function results for the northern section of the redwoods region.

(forested riparian zones; fig. 7.6). In addition, we present results for the composite rankings based on the first nine criteria organized both by polygon (fig. 7.7) and by watershed subbasin (fig. 7.8).

Biological and ecological values are not the only criteria by which focal areas should be evaluated. The League and its partners can apply additional criteria for identifying areas of high conservation or management priority. Different partners may value redwoods mainly in terms of public access for active recreation, scenic values, historical and archeological values, passive recreation values, educational or scientific values, or others. Ranking criteria can be developed to reflect these values.

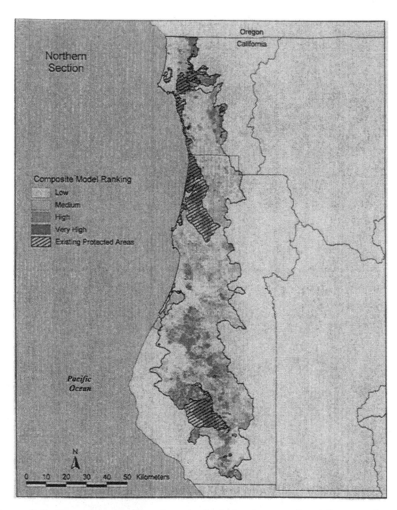

Figure 7.7. Composite ranking for criteria 1–9 using equal weighting and organized by 1 km × 1 km grid cell.

Focal area boundaries could eventually be established for the entire range of the redwoods. These ecologically defined boundaries can serve both as a logical units for localized planning and as administrative units for managing and tracking the League's master plan activities. As such, focal areas can be used as a reference for the following:

- Monitoring, recording, and reporting redwood conservation efforts over time;
- Development of partnerships with local groups;
- A geographic reference framework for tracking stakeholders and stakeholder programs;

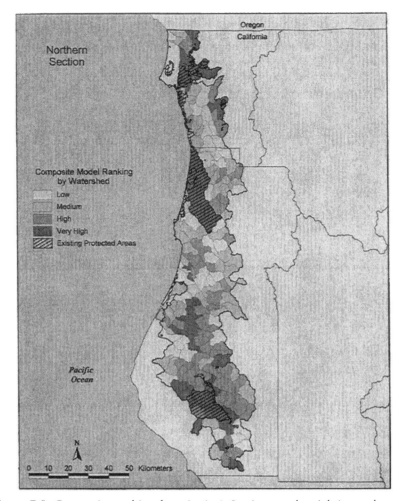

Figure 7.8. Composite ranking for criteria 1–9 using equal weighting and organized by watershed subbasin.

- Monitoring of land use and land cover change over time;
- Monitoring of urbanization and demographic trends over time.

Site-Specific Assessment

The focal area identification and assessment model can be used at the site level for general reference and to provide a broad context for local decision-making. Nevertheless, conservation design and management planning for particular sites require a more detailed evaluation. This assessment must include intensive field surveys by competent biologists. Additional considerations at this later stage of conservation planning should include:

- The biological distinctiveness and condition (e.g., naturalness, habitat quality, integrity) of individual sites within the focal area and the focal area as a whole;
- The precise locations, size, and potential viability of resident populations of imperiled species and other taxa of interest;
- More information on the restorability of the area and the effort (cost and time) required for restoration to natural condition;
- The specific management practices and uses appropriate for the area;
- The current management and development status of sites within the focal area; and
- Threats to particular parcels from logging or development.

Conclusions

The conservation planning principles and procedures offered in this chapter are meant to provide for a future landscape where the viability of the redwood ecosystem with all its attendant species and processes is assured. The landscape that ultimately arises from this process will not be an ideal landscape from a scientific or conservation standpoint. Data gaps, mapping errors, errors in assessing site values, ongoing logging and road building, human population and development pressures, unavailability of certain sites for conservation purchase, and—perhaps foremost—funding limitations will result in a compromised landscape. We are confident, however, that a science-based landscape design can be far superior to a landscape achieved by default, or one where all conservation decisions are made separately and opportunistically, without awareness of the big picture.

The process proposed here parallels in many respects the planning process being conducted by scientists and conservationists in the adjacent Klamath-Siskiyou ecoregion, another globally significant ecoregion (Ricketts et al. 1999; Noss et al. 1999; see fig. 7.1). In that project, conservation priorities were determined through an intensive research effort involving three tracks: (1) special elements, such as rare species hot spots, critical watersheds, and old-growth stands; (2) representation; and (3) focal species (e.g., fisher). Maps depicting locations of special elements, physical habitats, and vegetation types currently receiving little protection in reserves, and high-quality habitat for focal species, were overlaid in GIS analysis to develop a conservation plan that can be implemented in several steps. Conservation groups involved in the effort, which include World Wildlife Fund and several regional, grassroots groups, hope that the existence of this alternative plan, and the rigor of the underlying analyses, will help persuade the public agencies (chiefly, the U.S. Forest Service) that manage most of the region to revise their management plans to provide better protection to natural values. Alternately, the plan may be used in proposals for new national parks and other protected areas.

The Klamath-Siskiyou ecoregion is mainly (64 percent) public land. In contrast, the redwoods region is only 17 percent public land (varying by section; see table 7.1), and the vast majority of remaining conservation opportunities involve private lands. Therefore, the focal area–driven process described in this chapter is arguably better suited to the region than the more top-down, regional planning process followed for the Klamath-Siskiyou. Nevertheless, because considerations of landscape context, connectivity, and other broad values are part of the evaluation procedure offered here, we suspect that the overall design will converge on the design options developed for the Klamath-Siskiyou. That is, the (near-optimal) redwoods design will protect locations of rare species, old growth, and other special elements; it will result in most redwood site types and plant communities being protected; and it will provide a high probability for particular area-sensitive focal species to persist. Although the landscape that emerges from conservation activities over the next few decades will fall short of the ideal, it will likely be superior to what would have emerged without such guidance.

The contiguity of the Klamath-Siskiyou and redwood regions raises another important issue: inter-regional conservation opportunities. The North American conservation assessment conducted by World Wildlife Fund (Ricketts et al. 1999) highlighted a string of interconnected, globally significant ecoregions bordering the Pacific Ocean from Alaska to Mexico. These ecoregions—which collectively form the American Pacific Rim—were recognized as being in the highest class of conservation priority in this analysis. They include (from north to south): Northern Pacific Coastal Forests, Northern Cordillera Forests, Queen Charlotte Islands, British Columbia Mainland Coastal Forests, Central Pacific Coastal Forests, Klamath-Siskiyou Forests, Northern California Coastal Forests, Sierra Nevada Forests, California Interior Chaparral and Woodlands, California Montane Chaparral and Woodlands, and California Coastal Sage and Chaparral. This amazing string of ecoregions raises the conservation significance of the redwoods another notch. It also suggests priorities for future conservation planning—identifying and protecting connections among these ecoregions so that cross-regional processes, such as movements of wide-ranging animals and adjustments of communities to climate change, can be maintained.

The conservation planning strategy for the redwoods offered here is intended to advance redwood conservation through a rational and strategic process that makes use of many kinds of information. The strategy relies heavily on the value of redwood forests as reservoirs of regional biodiversity, besides such other values as scenic and recreational potential. The strategy provides a basis for identifying priority areas for conservation of redwood forests, not only to protect exemplary groves, but to sustain or restore whole ecosystems. Ultimately, only within intact ecosystems will the redwoods endure.

Chapter 8

MANAGING REDWOODS

Dale A. Thornburgh, Reed F. Noss, Dean P. Angelides,
Craig M. Olson, Fred Euphrat, and Hartwell H. Welsh Jr.

Previous chapters have made a variety of points relevant to the management of redwood forests. Four general lessons stand out: First, primary old-growth redwood forests are immensely valuable for biological, aesthetic, and other reasons and have become rare in the region after a century and a half of logging. As these forests have declined by more than 95 percent since European settlement, so have many of the species associated with them. Some of the structures, ecological processes, and biological assemblages of old-growth forests, such as those of the forest canopy (see chap. 3 and 4), are dimly understood and difficult to replicate in younger forests. Therefore, protecting as much as possible of the remaining old growth is a sensible policy.

Second, as will be elaborated further in this chapter, stands of second-growth redwoods (and, in some cases, third growth) that retain structural legacies from former old-growth stands, or that have regained some of these structural quali-

---

Author contributions: Thornburgh, lead author and organizer; Noss, writing and editing throughout; Angelides and Olson, management practices on commercial forests; Euphrat, box on Big Creek; Welsh, appendix on forest practices and riparian/aquatic ecosystems.

ties over time or through management, are also valuable; many "old-growth" species seem able to persist in these stands. Second- and third-growth redwood forests currently dominate the redwood region. These stands will become more valuable as they age and as more of the region is converted to nonforestland uses.

Third, the remaining redwoods on private lands, a small proportion (estimated as < 1 percent) of which are old growth, cannot all be placed in protected areas. Funds for land acquisition by Save-the-Redwoods League and other conservation interests are limited and must be devoted to those stands and landscapes of highest conservation value (see chap. 7). Difficult decisions must be made. Moreover, redwood produces a beautiful wood that is legitimately used by people for many purposes. Sustainable production of this commodity is desirable.

Finally, however, most previous logging of redwood forests has not been sustainable and has destroyed many biological and other values. New silvicultural systems, based on better understanding of redwood forest ecology would go a long way toward avoiding degradation of forests and encouraging sustainability. Some of these systems can be used to hasten the development of old-growth structural conditions and associated species in second- and third-growth forests and plantations. These points are the premises of this chapter.

This chapter reviews management practices in redwood forests, from prescribed burning, used to restore and maintain natural qualities in old-growth stands in parks, to new silvicultural techniques aimed at sustainable forestry on private lands. The main emphasis here is on silviculture, ranging from traditional even-aged and selection management to new, alternative approaches. New approaches strive to mimic natural disturbance regimes in order to maintain the total biodiversity of the ecosystem, produce clean water and air, store carbon, and provide products useful to society.

Management of Redwood Parks

Many people assume that management of parks and other nature reserves is an oxymoron, believing these areas are best left alone to take care of themselves. Paradoxically, however, in landscapes where natural processes have been disrupted, some kind of management is usually necessary to maintain the natural qualities of the ecosystem. This is true for the redwood parks, where, among other problems, the natural disturbance regime—especially fire—no longer operates as it once did.

Management of Old-Growth Reserves

As explained in chapter 4, recurring fire is a natural feature of redwood forests. The relationship between redwood forests and fire is complex, however. Depending on site conditions, either too much or too little fire (in intensity or frequency) can place redwood trees at a disadvantage relative to competing tree and shrub species (see box 4.3). In the old-growth redwood forests of Redwood National Park, and in the various redwood state parks, managers are attempting to reintroduce the natural fire cycle through controlled burns. The intent is to restore and perpetuate the natural structure and composition of the ecosystem.

Studies indicate that the pre-European, mean fire-return interval in some southern redwood forests was eight years (Brean and Svensgaard-Brean 1998), although it varied among sites in the region. Greenlee and Langenheim (1990) found intervals ranging from 17 to 175 years in southern redwood forests (see chap. 4). Since 1978, state park personnel have burned 100–140 ha of southern redwood forests annually, in a shifting pattern so that every site is burned approximately every eight years, in an attempt to return these forests to a more natural condition. After a century of fire suppression, however, large amounts of duff and litter have accumulated around the bases of trees. Combustion of this litter could potentially kill even large trees. To prevent mortality to the large trees, this layer is removed before a prescribed burn, accompanied by thinning of the understory. Although promising, the ecological consequences of these treatments in Big Basin Redwood State Park and elsewhere have not been evaluated (W. J. Berry, Cal. Dept. Parks and Rec., pers. comm.).

In the central redwood forest, at Humboldt Redwoods State Park, a fire-history study indicated a fire rotation of 26.2 years before European settlement (Stuart 1987). Again, however, considerable variation existed (see chap. 4). Many of these fires were probably set by the Sinkyone Indians who inhabited the region. No major fires have occurred since 1940, resulting in the accumulation of greater-than-normal levels of large and live fuels and increasing the probability of a large, intense fire. The California State Parks agency seeks to burn 2,000–2,800 ha of the central redwood forests annually (W. J. Berry, pers. comm.). In Humboldt Redwoods State Park, the objective is to prescribe-burn 300 ha of old growth and a smaller area of second-growth forest each year. This reintroduction of fire, meant to simulate the natural fire cycle, is viewed as essential to reduce fuels, expose the understory, prepare seedbeds, release seeds, and control nonnative plants (Brean and Svensgaard-Brean 1998). Prescribed burns have resulted in lower levels of litter fuels and live sclerophyllous vegetation, with no serious charring of old-growth redwood or Douglas-fir (Stuart 1985). Some of the older second growth has been underburned to accelerate the development of old-growth structural conditions.

Fire is thought to have a moderate ecological role in the northern redwood forest (Olson et al. 1990). Nevertheless, a study in Redwood National Park adjoining Prairie Creek State Park suggested a mean fire return interval of eight years before European settlement (S. Underwood, Redwood National Park, pers. comm.). Other studies have documented a great range of intervals, depending on distance from the coast and other factors (see chap. 4). Light ground fires that do not open the canopy favor western hemlock regeneration but usually eliminate older hemlocks from the stand. Redwood, grand fir, and tanoak appear to maintain their status with or without fire (S. D. Veirs, pers. comm.); in the absence of fire, redwood can regenerate on downed logs or mineral soil. At Redwood National Park, the goal is to reintroduce the natural role of fire while limiting the risk to resources on adjacent lands. The area of old-growth forest prescribed-burned in the park has generally increased each year. In 1997 and 1998, park personnel burned 35 ha. Plots have been established to monitor the effects of the burn, which include the amount of burnt bark and cambium on large and small trees and the effects of fire on understory tree seedlings and herbaceous vegetation (S. Underwood, pers. comm.). Preliminary observations suggest that old-growth stands managed with frequent, low-intensity fires show reduced biodiversity, including understory flora, canopy trees, and evidence of animals (D. Thornburgh, unpub. data), though fires of higher intensity produce a more diverse canopy structure and higher biodiversity. No burning has occurred in the northern Redwood State Parks: Prairie Creek, Del Norte, and Jedediah Smith.

Another management issue in the northern redwood region, as elsewhere, is exotic plants (see chap. 3, box 3.2). Numerous exotic plant species, including English and Cape ivy and Spanish heath, have become established in the old-growth stands of Redwood National Park, threatening to upset the biological community. The park's exotic plant management plan proposes eradication by hand and biocontrol (Fritzke and Moore 1998). The relationship between exotic plant invasion and fire in redwood forests has not been well documented.

Management of Second-Growth Reserves

About 18,800 ha in Redwood National Park, or 63 percent of its total area, is second-growth redwood–Douglas-fir forest. Most of these stands regenerated from old-growth forests that were harvested in the 1950s and 1960s using the seed-tree method for redwood regeneration and aerial seeding for regeneration of Douglas-fir. Some cutting also took place in the 1970s. In 1987, these stands had a conifer density of 3,303–4,525 stems/ha, with 3,158 hardwood stems/ha (Mastrogiuseppe and Lindquist 1996).

A study was initiated in the park in 1968 to determine the efficacy of thinning second growth for the restoration of old-growth characteristics. The treatments thinned selected stands to densities in the range of 250–2,500 stems/ha (Harrington 1983). After thirty years, the understory shrubs huckleberry, Oregon grape, salal, and rhododendron were at the same density as in the old-growth stands. Understory birds also had equivalent abundance in thinned plots and old-growth forests (Menges 1994). Growth of trees in thinned plots was four times the growth in unthinned plots. Redwood National Park initiated further thinnings in 1978, which have accelerated the growth rates of trees and increased understory shrub biomass (S. Veirs, pers. comm., 1998). Thinned plots are developing some of the characteristics of old growth more rapidly than an untreated second-growth stand. The results of these studies suggest that, at least for the characteristics measured, thinning is a useful tool in the restoration of old-growth forests.

Old second-growth forests also occur in some of the state parks. Most of these stands are unmanaged. As noted earlier, however, portions of Humboldt Redwoods State Park have been underburned to hasten the development of old-growth structure. Also, in state parks in the southern redwoods section, second-growth stands are thinned to reduce the tree density from 2,000–3,000 stems/ha to 150–200 stems/ha before the understory can be burned. The structural characteristics of old-growth forests appear to be developing more quickly when stands are thinned.

A portion of the Arcata Community Forest is dedicated as a nature reserve. The area that is now in the reserve was logged in 1890 by handsaws, log-bark peeling, and slash burning, yarded by oxen, and left to regenerate. The reserve today is a dense forest of a few residual, old redwood trees and a mixture of 108-year-old redwood, Douglas-fir, western hemlock, grand fir, and Sitka spruce. The stand structure is being altered by a shifting pattern of relatively small patchy disturbances caused by infrequent windstorms. Storms with winds severe enough to topple large trees occur almost every year, with a high of thirteen storms in 1981. Blowdowns have created numerous canopy gaps, which provide resources for establishment of grand fir, western hemlock, and redwood in the understory and increased height growth of individual trees in lower and mid-canopy positions. These wind disturbances are slowly converting this stand into a typical northern redwood old-growth forest, with abundant coarse woody debris and an upper canopy dominated by redwoods.

The primary anthropogenic impact to the Arcata Community Forest reserve today is human recreational use, which is heavy, accompanied by a slow invasion of such nonnative plants as English ivy and holly (though this invasion has been slowed by volunteers hand-pulling the exotic plants). Most recreational use is confined to hiking trails; however, trampling impacts are noticeable in the forest understory, particularly in the riparian zones along small streams. Frisbee golf playing, illegal camping, and off-trail hiking are the major sources of impact. Some of the affected areas have been planted with sword fern and oxalis and covered with slash to discourage off-trail use. Other second-growth nature reserves experience so much human impact that the owners have given up trying to keep them natural; an example is the redwood forest near the Eureka Zoo.

The small size of these nature reserves in the context of an increasingly urbanized landscape creates uncertainty with respect to their ability to maintain a natural state over the long term. The heavy recreational use, noise, exotic plants, feral cats, and other influences stress habitat conditions and probably populations of some species within the forests. The surrounding land uses, such as urban/suburban development and short-rotation, even-aged forestry, are slowly turning these reserves into isolated islands of old trees with heavily altered understories. Nearby residential development already has limited the ability of reserve managers to employ prescribed burning as a management tool. It is too early to forecast long-term effects of these various landscape changes on the redwood forests, but they are unlikely to be positive.

In large second-growth reserves, such as some of the Redwood National Park units and smaller nature reserves, the forests are increasingly unlikely to develop natural cohorts of multistoried conifers, considered to be typical of old growth (see chap. 4). In most cases, stand development in today's patches of variable-density seedling, shrub-dominated, young forests will not lead to the same kinds of stands and habitats as those produced by natural succession (Tappeiner et al. 1997). Reasons for altered stand development include the following, in various combinations: (1) introduction of nonlocal genotypes; (2) lack of fire, resulting in dominance by different species; (3) establishment of exotic species, both flora and fauna; (4) timber harvest, which has changed species composition and seed supply, favoring the development of shrubs and hardwoods capable of vigorous sprouting; (5) establishment of stands after logging that are often more heavily and uniformly stocked with western hemlock and Douglas-fir than were the original stands; (6) change in climate or weather patterns since the time when the present old-growth forests developed; and (7) suburbanization of scattered blocks of second-growth, with accompanying lawns, roads, and domestic animals.

Silviculture in Redwoods: Incentives and Disincentives

If redwood forest is to remain a primary land cover in the redwood region, silviculture must be both ecologically responsible and economically viable. Approaches to timber management in redwood forests are dictated by current technologies, economic factors, forest practice rules, social concerns, and the condition of the forests themselves. The cost of managing private lands for harvestable redwoods has been increased by environmental and timber harvest regulations, including the Endangered Species Act and various state laws. It is often argued that increasing regulation has driven small landowners to convert their land from forest to residential development. Although this is true in some circumstances, the major factor in most conversion probably has been increasing land values in the vicinity of urban areas. Outside of urban areas and major highways, little conversion has occurred. Nevertheless, conversion of redwood forests to subdivisions is on the increase and undoubtedly is more destructive environmentally than properly conducted timber harvest; furthermore, the loss is essentially permanent (Anderson 1998).

Historically, silviculture has focused on site-specific or stand-specific prescriptions. It has become increasingly clear, however, that to meet conditions of ecological sustainability, managers must look beyond the individual stand and develop prescriptions that incorporate a mix of stand-specific treatments that will ultimately achieve desired future conditions on the landscape scale. The "big picture" view so necessary for conservation of the biological values of redwood forests (see chap. 7) therefore is also necessary for improved management of redwoods for timber. A big-picture approach requires the flexibility to apply a broad range of stand-specific management treatments and a decision-analysis process adequate to determine the appropriate mix of management activities across the landscape and through time. The financial viability of private forest ownership depends on the health of the region's timber economy, a rational regulatory structure, and most important in the long term, a healthy forest ecosystem.

Because most of the remaining redwood forests are on private lands—managed both by large corporations and small nonindustrial landowners—these owners must have some certainty that their long-term investments in forests are secure; otherwise, they might invest in opportunities such as residential development (where feasible). For a small owner, the cost of complying with regulation can be 25–50 percent of revenues. Blencowe (1998) suggests that streamlining the state regulation process, through "fast-track" Timber Harvest Plans for forest owners who are certified as practicing sustainable forestry, would help to preserve forestland in the redwood region. Standards for certification of forests in the Pacific states are not fully developed and remain controversial, however, especially for harvest of residual old-growth trees and issues concerning landscape ecology.

The California Forest Practices Act of 1973 mandated that nonfederal landowners and managers practice forestry in ways that will protect land productivity and public resources. The Timberland Productivity Act of 1982 added provisions for sustainable harvest practices. The California Environmental Quality Act of 1970, the Porter Cologne Water Quality Act of 1969, and the California Endangered Species Act of 1984 all aim at protecting public resources and are implemented through state regulation, particularly the California Forest Practice Rules. Together, these laws and rules are supposed to govern silvicultural practices by requiring that forest practices protect soil productivity and water quality, that practices be sustainable, and that specified stocking levels are achieved and maintained.

Despite the originally good intentions of the laws and rules governing forest practices, many observers believe that the rules are too weak—especially for the protection of riparian/aquatic ecosystems—and that enforcement is often lacking in the redwood region (see appendix 8.1). Evidence for the cynical view is abundant: most of the streams in the redwood region have been declared "impaired" by sediment, listings of species under endangered species laws are on the increase, and public conflicts over forest management issues are escalating. The incentives and disincentives that influence forest practices and forest management decisions in the region apparently need further development to achieve true sustainability of forest ecosystems.

Traditional Silvicultural Systems

Traditional silvicultural systems currently being applied in the redwood region can be lumped into two general categories: even-age (e.g., clear-cutting) and uneven-age (e.g., selection) systems. The current California State Forest Practice Rules make this distinction; however, both systems can result in highly variable landscape patterns and stand structure depending on how they are implemented. No definitive research is available to compare timber yields or wildlife habitats provided by even-age versus uneven-age systems in redwood forests. Nevertheless, the two systems normally result in distinctly different patterns and structures; thus, they provide habitats for different assemblages of species through time.

Even-Age Systems

Clear-cutting is the most widely used system of silviculture in the redwood region. Removal of all trees except the very smallest during harvest permits easier site preparation, slash disposal, and control of species composition and stocking (Helms 1995). Clear-cutting also provides an open, sunny environment for the new stand, which promotes rapid growth of redwoods. Even after stands achieve crown closure, redwood trees continue to grow and develop rapidly, making continued management investment to enhance their growth attractive. Typically, second-growth redwood stands are harvested at sixty to eighty years; however, the culmination of mean annual increment (the point of maximum average productivity over time) does not occur until the stands are more than one hundred years old.

Because of abundant sprouting redwoods, hardwood stumps, logging slash, and residual vegetation on highly variable terrain and site conditions, hand planting of seedlings in clear-cuts to create "row plantation monocultures" is usually impractical. Grasses, forbs, shrubs, and hardwood trees invade most clear-cuts soon after logging. The rapid development of vegetation provides food and cover for many animal species adapted to early seral vegetation.

The shelterwood and seed-tree systems of harvesting and regenerating redwood stands is similar to the clear-cut method in that most trees except the very smallest are initially harvested, but in this case an overstory of seed trees is left to provide seed and shelter for the next stand of trees. California Forest Practice Rules specify the number and spacing of seed and shelterwood trees to be retained in this system. Often five to ten years before harvest, some trees are removed around the intended seed trees to help them become more wind-firm. Once a new stand is established under the seed trees, the seed trees are removed, usually within five to ten years following the harvest of the over-story.

Seed tree and shelterwood systems can be used to facilitate regeneration of species other than redwood. Indeed, the seed tree system usually has been more successful for regeneration of Douglas-fir than of redwood. Redwoods have unpredictable seed crops, seed viability is low, and shelter is not needed to enhance sprout development (Helms 1995). Furthermore, redwood seedlings are extremely vulnerable to infection by damping-off and *Botrytis* fungi during their first year (Hepting 1971).

The removal of seed trees after the new stand is established usually causes some damage to the young trees; often 10–25 percent of the remaining trees are damaged. Retention of the overstory throughout the next rotation, either for seed tree or shelterwood systems, will provide habitat elements (i.e., large trees and subsequent coarse woody debris) that might otherwise be missing from a managed landscape. Such an approach was recommended (Oliver et al. 1996) and is currently being implemented on the Jackson Demonstration State Forest.

Bruce (1923) developed "normal" yield tables for fully stocked redwood stands. Such yield tables indicate how even-age redwoods might grow under intense management, but up to the present time few stands have been managed in this manner. Lindquist and Palley (1963) developed empirical yield tables (growth projections) for well-stocked (though not necessarily fully stocked) young-growth redwood stands over the range of site-quality conditions found in the redwood region. They projected that redwoods grown 100 years yielded more than 3,500 board-feet per acre (35 m^3/ha) per year on high-productivity lands, but barely more than 500 board-feet (5 m^3/ha) per year on low-productivity lands.

As noted earlier in the review of park management, thinning of second-growth stands can accelerate development of old-growth characteristics. Thinning also has silvicultural value. Typically, even-age systems require intermediate treatment to ensure high productivity over time. Precommercial thinning, combined with control of invading hardwoods and shrubs ("brush") is often needed during early stand development to reduce competition between noncommercial species and the young stand of redwoods. Following clear-cutting or a seed-tree cut, red alder from seed and Pacific madrone and tanoak from sprouts are the hardwoods that compete most commonly with redwood in early stand development.

Redwood responds well to commercial thinning (Carr 1958; Oliver et al. 1996). One or two commercial thins are often applied at some point after the stand is thirty years old to remove some of the growing stock so the remaining trees have more room to grow. The overall yield over the length of the rotation is increased by this practice, and the average tree diameter of the stand cut at the end of the rotation is greater. Nevertheless, the ultimate effects of commercial thinning on even-aged redwood stands are not well known. With a light thinning, sprouting is low and less vigorous and generally will not develop as a significant understory. On the other hand, a thinning that greatly opens the canopy will generally result in vigorous sprouting, which should then develop as a younger age class. Successive thinning of the larger trees in a stand ("thinning from above") would then likely result in a change to an uneven-age structure (Helms 1995).

Uneven-Age Systems: Single-Tree Selection

Uneven-age systems can involve removal of individual trees or groups of trees. The uneven-age single-tree selection system seeks to maintain all age classes and a multilayered canopy, along with a constant supply of wood from year to year over a smaller area than possible with even-age management. At each entry, trees are removed within each diameter class to retain a target distribution. Harvest of older trees is compensated for by growth of young trees, enhanced by opening the canopy. Often all trees above some diameter, perhaps 1 m dbh, are removed. The single-tree selection system is based on the concept of self-thinning; that is, a certain number of small trees die because of shading and competition and a certain amount of regeneration occurs under the closed canopy. Because it is moderately shade tolerant, however, redwood does not self-thin like many other tree species. Small redwoods can persist in low-light situations, but the growth of individuals is less vigorous than under more open-canopy conditions.

In single-tree selection, stands are entered and trees are removed periodically, generally every five to twenty years. Factors such as economics, residual basal area, and growth rate need to be considered when determining the appropriate return interval. Frequent entries will result in greater soil compaction and residual stand damage, whereas return intervals that are too long are likely to result in holding costs of the residual growing stock (i.e., forgone revenue). A minimum volume to remove to offset logging and roading costs was estimated as 5,000 board-feet per acre (50 m³/ha) in two studies in the redwood region (Adams 1980; Kennedy 1983).

Because small redwood trees survive and persist within closed stands, at each entry within a single-tree selection system numerous small trees need to be cut and removed to maintain the target diameter distribution. This practice adds considerable expense to the logging process compared to group selection or even-age systems (Kennedy 1983). In addition, Jacobs (1987) found that because redwood is unable to reproduce by seed under dense forest canopies because of low light and increased probabilities of infection by damping-off fungi, a light selection cut will favor the regeneration of grand fir and western hemlock over redwood.

At Jackson Demonstration State Forest, clusters of trees are removed as an approach to selection management. This technique is intermediate between single-tree selection and group selection systems. The groups of trees removed are usually clusters of sprouts growing around stumps. This practice has the benefit of minimizing the problems of felling and yarding trees; in addition, it opens the stand more, aiding the establishment of young redwoods. Such an approach eliminates some of the problems associated with classic single-tree selection management in redwood forests.

Uneven-Age Systems: Group Selection

The group selection system is similar to even-age silviculture, except that the clear-cut areas are small, usually 0.04–1.2 ha. At each entry, several small groups of trees are removed and regenerated, whereas the remainder of the stand might receive a commercial thin. The area to be included in the group cuts at each entry is computed by multiplying the total stand area by the intended return interval and dividing by the intended rotation length. Biological and economic considerations influence the area selected. No definitive research has been conducted in the redwood region to establish optimal group sizes for this method.

Several recent studies have examined the growth and yield of second-growth stands under uneven-age silviculture (Adams et al. 1996; Helms and Hipkin 1996; Piirto et al. 1996). In a preliminary analysis, Helms and Hipkin (1996) found no marked effects on tree growth from widely different selection systems. Adams et al. (1996) concluded that uniform single-tree selection lacked the ability to assure the rapid regeneration growth of redwood afforded by the group selection system.

Current Silviculture Practices of Small-Forest Owners

Most landowners with small-forest holdings in California are interested in maintaining aesthetics and wildlife habitat. Consequently, they use some type of "green-tree retention" cuts (i.e., heavy seed tree or shelterwood cuts) or some other selection system—uniform, group, or irregular. A common approach is to enter a 40–60-year-old forest that has developed after the original logging of the old growth and remove 40–65 percent of the volume. The stand is then planted with a mixture of tree species. The next entry, which involves cutting part of the overstory and the younger, second canopy layer, may occur in twenty to thirty years. Trees are again planted. This practice creates a three-storied stand that produces revenue and is aesthetically pleasing. Nevertheless, unless managed specifically for biological values, these stands can have low populations of some species, such as spotted owls, that are found in a natural, late-seral redwood forest (L. Diller, Simpson Timber Company, pers. comm.).

Some small-forestland owners apply small to medium-size clear-cuts or seed-tree cuts. The resulting stands usually grow into the low-biodiversity stem exclusion successional stage and are then sequentially recut before biodiversity increases through natural stand structural changes (L. Diller, pers. comm.).

The Current and Future Landscape

The traditional silvicultural systems reviewed above, along with the diverse land-ownership pattern and the variable responses of landowners to regulations, have shaped the landscape of the redwood region. Currently, some 93–95 percent of the redwood forest is second and third growth (L. Fox pers. comm.). These young to middle-aged forests are a mosaic of dense, stem-exclusion stands dominated by sprouting redwoods and some partial selection-cut forests and recently clear-cut forests in the early shrub-seedling stage. Redwood trees grow well in this environment because of their sprouting ability and rapid regeneration from seed in disturbed forests in contrast to their slow growth under the old-growth canopy. A small proportion of these young forests is dominated by early successional tree species, such as red alder. Most of central Humboldt County between the only two old-growth reserves, the Headwaters Forest and Redwood National Park (a distance of approximately 35 miles), consists of this pattern.

The present pattern of young forests in irregular patches generally favors animal species that require early successional habitats to meet some or all of their life-history needs. These species include black-tailed deer, elk, black bear (in part), foxes, mountain lions (responding to deer), bobcats, various rodents, and raptors. The present mosaic is less suitable for species of plants and animals associated with late-seral forest (Diaz and Bell 1997). Nevertheless, some of the older second-growth forest patches support populations of red tree vole, flying squirrel, and fisher (see chap. 5).

Even the northern spotted owl, which is generally dependent on old-growth forests throughout its range, appears to be persisting in the landscape mosaic of reforested clear-cuts and mature second-growth forests in the redwood region. This mosaic provides a mixture of prey reservoirs, forage areas, and roosting and nesting structures for the owl (O'Dell 1996). In 1993, the California Department of Fish and Game estimated a population of 4,450–8,500 northern spotted owls in California, with many of the owls found in managed second-growth redwood forests (Lucas 1998). The rapid growth rate of redwood trees, combined with the persistence of structural legacies from former old-growth stands in many second-growth forests, explains in large part the survival of spotted owls and some other old-growth associated animals in these forests (Noon and Murphy 1997). Another factor benefiting the owl is that one of its major prey species, the dusky-footed woodrat, thrives in the landscape mosaic created by recent forestry activities (Sakai and Noon 1993; Giusti 1999).

The suitability of the present landscape for many other species associated with old-growth redwood forests is unknown. It appears, however, that many canopy lichens, mosses, liverworts, fungi, terrestrial mollusks, insects, and other taxa characteristic of the old-growth redwood forest are not found, at least as viable populations, in the modified landscape. Species that rely on snags and downed coarse woody debris, such as many salamanders, are reduced in these forests (see chap. 5). The long-term effects on biodiversity of eliminating 95 percent of the old-growth redwood forests remain to be seen.

The landscape pattern of the redwood region is beginning to change again. In some areas near cities and along the public highways that transect the region, the human population is growing rapidly, principally through people moving into new subdivisions. This trend is expressed by an increase in buildings, roads, lawns, pastures, fences, domestic animals, and other artifacts of civilization. The average size of individual parcels of private forestland is declining in response to high inheritance and estate taxes, among other influences. Some of the large timber companies are selling parcels of their large forest ownerships adjacent to newly suburbanized areas because of conflicts with the public over silvicultural practices. A major increase in the human population in the redwood region is expected within the next twenty years, which inevitably will lead to more land in rural residential uses. These trends will probably increase fragmentation of the forest, with corresponding effects on fragmentation-sensitive species.

Most of the new forest owners say their priorities are aesthetics, wildlife, investment values, and the like; fewer forest owners have timber production as their primary goal. Almost half, however, think they will harvest some timber from their land in the next decade. These owners generally do not belong to forestry organizations or read forestry magazines. Hence, they seldom are knowledgeable about forestry or know where to look for advice on forest management. The result is often lost opportunities for the landowner and declines in the health and sustainability of the redwood forest (Sampson 1998).

Future Silvicultural Management of Private Forestlands

In the late 1970s, forestry practices began to change dramatically in the redwood region. Increasingly, the public regarded clear-cuts as ugly and environmentally destructive. A developing environmental consciousness among a large and politically effective segment of the population led to pressure on forest managers to be more "ecological." For example, in 1979 the citizens of the city of Arcata passed an initiative to manage their forests for utilization of the resources in accordance with the principles of ecological forestry and perpetual sustained yield for both consumptive and nonconsumptive uses.

Soon, new organizations were founded to promote sustainable forestry in northwest California. The Institute for Sustainable Forestry, a community-based organization, was founded in 1991 to promote stewardship forestry through education and demonstration, and has developed a program of certification and labeling of sustainably harvested forest products.

The Pacific Forest Trust was founded in 1993 to promote stewardship forestry on private forestlands. Their main objectives are to restore, enhance, and protect private, productive forestlands. The Trust serves as the coordinating body in the Pacific states for developing forest certification standards consistent with international Forest Stewardship Council standards and criteria. Working with several small-forest owners in the redwood region, the trust has the first paid-for, forest carbon-storage program in the United States. The management plan allows cutting of annual growth for economic benefit, while promoting structural and tree species diversity, overall biodiversity, clean air and water, recreation, aesthetics, and carbon storage. The trust also has developed conservation easements for numerous small-forest ownerships.

With the listing, under the federal Endangered Species Act, of the northern spotted owl in 1990, the marbled murrelet in 1992, and coho salmon of the central California Coast Evolutionarily Significant Unit (ESU) in 1996 as threatened species, the management of private, industrial forestlands has had to include provisions for their habitat requirements. Commercial harvesting practices intended to accelerate creation of late-seral habitat components have been developed by state and private land managers. Practices include retention of large trees and multistoried canopy structure, and creation and maintenance of streamside ecosystem structure and function. The following are some of the major new silvicultural approaches being applied in the redwood region.

Shifting Mosaic of Variable-Size Patch Cuts with Variable Thinnings

Second-growth redwood–mixed conifer–hardwood forests in some landscapes are being managed by creating a shifting mosaic of variable patch sizes on a 100–200-year rotation. This type of silviculture attempts to mimic the natural stand dynamics that occurred in the primary forests of this region, which were characterized by a shifting pattern of relatively small patchy disturbances, such as the death of individual canopy trees or groups of trees, forming gaps of various sizes and shapes. These gaps provide resources for establishment of young trees in the understory and increased height growth of individuals in lower and mid-canopy positions (Spies 1997).

This management regime attempts to maintain a high photosynthetic surface to take full advantage of available light, water, and nutrients. This is done by maintaining a deep and irregular canopy composed of all the tree species normally found on the site. This practice seeks to maximize the amount of fixed carbon converted to stored carbohydrates (wood). The optimal age and size structure of the canopy trees for storage of carbon is determined by the point where the amount of CO_2 fixed and stored is equal to the amount of CO_2 given off in respiration.

When stands reach this optimum age, size structure, and species mixture, the stands are thinned by removing the trees with small crowns and cutting some dominant canopy trees to create a mosaic of variable-sized, open patches. Never is more than 90 percent of annual growth cut in a ten-year period. A natural mixture of tree species is planted, using local seed sources, or naturally regenerated in these patches. Some of the patches must have sufficient sunlight to allow growth of shade-intolerant species, such as sugar pine and Douglas-fir, as well as moderately shade-tolerant species, such as redwood.

As the trees in regeneration patches grow and their crowns fill out, thinnings are applied to forestall early canopy closure and the development of the stem-exclusion successional stage, which is low in biodiversity. These thinnings favor a diversity of overstory and understory plant species and development of snags and coarse wood debris. They also optimize the growth of some trees, maintain the dense, variable vertical canopy, and provide a sustained flow of wood products and revenue (Carey and Curtis 1996). Usually, because of the uncertainty of this silvicultural system, 20–40 percent of the landscape is retained in uncut reserves, including riparian zones, special wildlife habitats, and inoperable areas. An attempt is made to reduce road densities by recontouring entry roads and planting trees and other vegetation.

More than a thousand hectares of forest in the redwood region are being managed using this silvicultural system; some have been for at least twenty years. Several have been certified as being managed sustainably. Monitoring of operations to assure that they meet sustainability criteria has been limited, however. Populations of spotted owls, wood rats, red tree voles, flying squirrels, winter wrens, nesting great blue herons, and other species of interest have been monitored in some of these forests. The willow flycatcher has been found breeding in young stands planted with Douglas-fir (Anderson 1998). Fishers have been reported to use older second-growth forests, where small patch cuts and group selection cuts duplicate the natural windthrow and fire regimes of older forests (L. Diller, pers. comm.), which are characterized by a diversity of tree sizes and shapes, light gaps and associated understory vegetation, snags, fallen trees and limbs, and limbs close to the ground (Ruggiero et al. 1994).

Mendocino Redwood Company, a new company that recently purchased 93,000 ha of Louisiana Pacific's redwood lands in Mendocino and Sonoma Counties, is in the process of developing a forest management plan based on principles of sustainable forestry. They will not cut old-growth trees, apply herbicides, or create clear-cuts. They intend to maintain the structural attributes of natural ecosystems on their managed stands, including a mixture of dominant, intermediate, and small trees. Large snags and downed logs will be maintained in each stand. The stands will be managed to create horizontal habitat heterogeneity, with canopy gaps, dense clumps of trees, and a variable understory. When stands are harvested for timber, varying quantities of biological legacies will be left to provide quick successional recovery of the ecosystem, including soil organisms, tree symbionts, and decomposers (Franklin 1995). Although this type of management for a complete redwood ecosystem currently has limited application and monitoring in the redwood forest, it is consistent with the paradigm for forest management suggested by many conservation biologists (Noss and Cooperrider 1994; Carey and Curtis 1996; Meffe and Carroll 1997).

Single-Tree Selection with Late-Seral Habitat Components

The Pacific Lumber Company, which developed a Habitat Conservation Plan (HCP) and Sustained Yield Plan (SYP) pursuant to Section 10(a) of the U.S. Endangered Species Act, has proposed two late-seral selection regimes intended to provide late-seral habitat components and stream shading and buffering. The first regime targets a J-shaped distribution of tree diameters, maintaining trees up to 40" (1 m) dbh with a postharvest basal area of 240 sq ft (55 m²). Stands would be entered no more often than every twenty years. The second late-seral regime is similar to the first, but seeks to maintain larger trees, up to 48" (1.2 m) dbh, and higher residual basal area, 300 sq ft (69 m²). Stands must have at least 276 and 345 sq ft (63 and 79 m²) of basal area, respectively, for the two regimes before entry and retain at least 20,000 board-feet per acre (200 m³/ha) in volume. Pacific Lumber has proposed 30-foot (10 m) no-cut buffers adjacent to streams.

Approved by the U.S. Department of Interior just before this book went to press, Pacific Lumber's HCP/SYP was hotly contested. The streamside buffers are considered inadequate by many aquatic biologists who have reviewed the plan (see appendix 8.1 and chap. 6). Furthermore, Pacific Lumber defines as "late seral" and "old growth" redwood forests that most forest ecologists would call "young." The HCP/SYP defines late-seral forests as "made up of stands with overstory trees that on average are larger than generally 24" [0.6 m] and may have developed a multistoried structure. It occurs in stands as young as 40 years but more typically in stands about 50 to 60 years old and older" (Pacific Lumber Co., unpub.). By well-accepted criteria and definitions, however, redwood and Douglas-fir forests less than 100 years in age would be considered young forests, not late-seral (see chap. 4, box 4.1). Pacific Lumber's late-seral and old-growth management strategy will fail to maintain true late-seral and old-growth forest in the planning area (Noss, unpub. review of Pacific Lumber HCP/SYP).

Managers for the Jackson Demonstration State Forest (JDSF) have proposed several silvicultural prescriptions intended to maintain uneven-age stands with large trees. The goal of the regimes is to create multistoried stands dominated by large trees as soon as possible. For their upland all-aged, large-tree emphasis regime, the targeted residual stand condition is 23–46 m² of basal area in trees greater than 24" (0.6 m) DBH and 35–46 m² of basal area in trees less than 24" DBH, depending on site quality. The prescription also calls for retention of all snags and downed logs, depending on safety considerations.

In addition, JDSF managers have proposed two approaches to maintain late-seral components within water and lake protection zones (WLPZs). The first approach is similar to the upland all-aged, large-tree emphasis regime, but stipulates that at least 75 percent canopy closure will be maintained within 25 feet (8 m) of the stream, and at least 50 percent canopy closure beyond 25 feet. The second targets a J-shaped diameter distribution for trees less than 24" (0.6 m) DBH, while maintaining a uniform distribution and number of trees greater than 24" DBH. The managers also aim to maintain 75 percent canopy closure within 8 m of streams and 50 percent beyond.

Another example of single-tree selection, but with group selection applied in some areas, is provided by Big Creek Lumber (box 8.1). This company has

Box. 8.1. Big Creek Lumber

When it comes to logging redwood in California, there are few examples that people point to as exemplary, both economically and environmentally. But Big Creek Lumber, a family-owned timber grower, miller, and retailer, has won the admiration of a broad coalition. The company has become a leading example of local control that allows timber and residential landowners to coexist amicably.

Begun in the 1940s by the McCreary brothers, Big Creek now owns and operates 10,000 acres of the Santa Cruz Mountains, the Coast Range south of San Francisco. Their mill in Davenport, on the Pacific Coast north of Santa Cruz, uses logs from these lands as well as logs from another 50,000 acres, what forester Mike Jani calls the Big Creek "client base." The entire timber supply is from private lands, and the mill saws up to 15 million board-feet (150,000 m³) of timber per year.

Big Creek's "vertical integration" is one key to its success. Big Creek manages the forest from the soil to the market. Working with owners of parcels as small as one hundred acres, Big Creek's foresters prepare management plans for sustainable harvests well into the future. This cuts down on future paperwork, assures the landowners that their land is well cared for, and creates a log base for the mill. Big Creek can count on this supply, so they do not need to approach land with a liquidation philosophy; they know that their timber supply is as predictable as the growth of trees. Their retail outlet sells half of their total production and specializes in environmentally certified redwood lumber. The lands and foresters of Big Creek have been certified by Scientific Certification Systems (SCS) under Forest Stewardship Council standards and guidelines to husband the lands owned by the company and also those of the client base.

(continues)

It is interesting that Big Creek views their major block of production, those of the unconsolidated landowners, as "clients." In many ways, these lands are the vendors, those who keep the supply flowing for the mill. The landowners, however, look to Big Creek for environmental guidance, forest management, and regulatory compliance as well, so the lands are both suppliers and clients. Jani says the process works because their client base has a like-minded philosophy about such things as stream protection, growing stands for old-growth characteristics, and a 100-year rotation.

Big Creek has chosen not to expand beyond its production capacity. Its expansion into new markets has allowed the business to grow and mature. increasing the company's profitability without increasing its production or its client base's cut. The selection management required in Santa Cruz County has increased overall volumes in the area, and logs that Big Creek cannot use travel hours and hundreds of miles to more hungry mills.

Big Creek's management is most often single-tree selection. As required in Santa Cruz County, no clearing in a dominantly redwood stand can exceed one-half acre. Cut trees must have leave trees within 75 ft (25 m), and at least half of the trees in the stand greater than 12″ (0.3 m) dbh must be retained. There are a few exceptions to this approach. In a few Douglas-fir stands, Big Creek conducts group selection, with clearings up to 1 ha; in unentered old growth, they cut lightly or not at all (remnant old-growth trees, however, may be cut, which has engendered some controversy).

Big Creek is proud of their habitat work. Near streams, they maintain greater than 75 percent canopy retention, and areas with marbled murrelet habitat are not entered during the breeding season; critical habitat trees are retained. The company conducts a comprehensive murrelet survey. Using the results of this survey, they will attempt to selectively harvest some unoccupied stands with old-growth characteristics. Big Creek is also interested in marketing its environmental resources, through easements and sale of fee lands, as lands that can be removed from their timber base.

Big Creek recently placed 146 ha into a permanent conservation easement. The property contains significant stands of rare old-growth and mature second-growth redwoods. Situated near Butano State Park, Año Nuevo State Reserve, Big Basin Redwoods State Park, and the West Waddell Creek State Wilderness Area, this forest is an integral part of a network of protected land in San Mateo County. Butano Creek, home to steelhead trout, is protected through restrictions on harvest to reduce sedimentation. Forest management will be focused on maintaining and enhancing old-growth habitat required by threatened and endangered species, such as the marbled murrelet.

Invasive exotic plants are a constant concern in Big Creek's management program. Pampas grass and French broom are controlled with an annual mowing program, without herbicides. Jani says the key to control of these plants is keeping the canopy intact. and their cutting methods are designed to limit the expansion of these aggressive, nonnative plants.

Resistance from environmentalists is a common deterrent to logging in the Santa Cruz Mountains, but generally good relations with the community, based on a consistent track record, has kept Big Creek in the woods rather than the courtroom. They regard their certification by SCS as a demonstration to the community of their good intentions and environmental stewardship.

Big Creek views itself as an important player in the future of Santa Cruz County and a model for the continuous production of redwoods in an increasingly urban and restrictive environment. Pressure to build more private residences in the woods, to keep timber operations further segregated from residences, and to reduce timber management options are unrelenting in the area, the forest just over the hill from Silicon Valley. Big Creek works with the county Board of Supervisors to reduce the effects of urbanization and to make logging rules more accommodating to the people of the community.

achieved the advantage of having its operations certified as environmentally responsible.

Short-Rotation Plantations with Variable Green-Tree Retention

The Simpson Timber Company manages even-aged plantations on short rotations. In an effort to protect and enhance spotted owl habitat as part of their HCP, managers are applying low to medium levels of green-tree retention. This system includes maintaining intact patches of trees greater than 0.2 ha in clear-cuts, enhanced WLPZ protection, and scattered groups of 10 to 12 trees and snags depending on site-specific conditions (Hibbard 1996).

The variable green-tree retention system has been applied to second-growth stands that developed after the cutting of the primary forests. These stands, which contain remnant snags, cull trees, and downed coarse woody debris, are clear-cut at about fifty years of age, with most of the merchantable residual logs and snags removed. Some snags and downed logs are left for wildlife, as well as a variable number of second-growth trees of all species left for wildlife and streamside protection. Today, this retention varies widely from one clump of five trees per 2 ha to 25 percent of the volume left as streamside cover and irregular clumps of trees in each clear-cut. Clumps of trees to be retained are selected by a wildlife biologist. Slash and second-growth understory vegetation are not burned. Two-year-old bareroot seedlings of redwood and Douglas-fir are planted in the untreated slash and ground vegetation. The seedlings are protected from deer browsing by vexar tubes. If early successional trees and shrubs, such as red alder, tanoak, blue blossom, and manzanita, begin to outcompete the planted conifers, the competing vegetation is treated with a basally sprayed herbicide. Some of these stands may later be precommercially and commercially thinned to optimize stand volume and individual tree growth.

Structural Retention Cuts

Structural retention silvicultural systems are being used by several consulting foresters and one large timber company in redwood forests. In 30–100-year-old, second-growth redwood–mixed conifer–hardwood forests, the stands are entered once every ten years to remove 30–40 percent of the volume, leaving the best-formed trees regardless of spacing (J. Able, forestry consultant, pers. comm.). It is assumed (not proven) that the volume removed will be replaced on the remaining leave trees within a ten-year period. The stands are always kept in a mixture of tree species.

This retention system is a "thinning from below," or natural thinning method, which works well for redwood and associated tree species. Trees with less than 30 percent live crown are thinned out, leaving redwoods with greater than 30 percent live crown more light and growing space. Redwood trees respond by filling out their crowns with greater growth of branches and more needles on each branch; this creates a much greater photosynthetic surface with less respiration cost, thus increasing growth. Furthermore, most of the growth is in larger trees, which increases their economic value. This type of partial cut eliminates heavy blowdown, excessive damage from bears and other animals, and thinning "shock" to the leave trees.

Growth in stands harvested by partial-retention cuts shows no signs of slowing (D. Thornburgh, pers. obs.). Eventually, these stands must be regenerated by a heavier cut or other disturbance, but can probably produce wood profitably and continue growing for another 100 to 200 years. These stands, which have a dense upper canopy but very little understory of shrubs or small trees, appear very similar to a natural, self-thinned, upper-slope old-growth forest with frequent understory burns. A disadvantage is that this type of silviculture requires frequent entries with logging equipment (e.g., every ten years) and a high road density of 10–12 miles/square mile (6.25–7.42 km/km^2). Many of these stands seem to have low species diversity and lack spotted owls and understory mammals (D. Thornburgh, pers. obs.). Where managers retain old-growth "heritage" trees, as well as downed wood, snags, and other structures, biodiversity is higher (F. Euphrat, pers. obs.).

Other Approaches

Among the measures to enhance wildlife habitat in redwood forests is noncommercial stand manipulation to create wildlife habitat components. In many traditional silvicultural systems, as well as some alternative or "ecoforestry" systems, a major objective is to capture tree mortality—that is, to cut trees that would soon die naturally. This practice results in fewer snags and downed logs—structures that provide nesting and foraging habitat for numerous species of animals (see chap. 5)—within the managed forest. Large, dead wood also plays an essential role in stream ecosystems (see chap. 6)—for example, by helping to create pools and thereby improving habitat for salmon.

Because trees are not left to die in traditional managed forests, other strategies must be employed to provide crucial habitat components. Girdling of trees to create snags provides habitat for cavity-excavating birds; these cavities, in turn, are used by many other birds, mammals, and invertebrates. Snags ultimately fall, providing downed logs. Because redwood is resistant to fungi and other decomposers, girdling of associated species, such as Douglas-fir and grand fir, will produce snags usable by additional species. Little is known of the suitable density and size distribution of snags to leave for wildlife in the redwood region. Bingham and Sawyer (1988) found 957 m^3/ha or 200 t/ha of downed logs within an 80-ha area of upland old-growth redwood forest. Whether this volume is representative of all old-growth redwood stands is unknown, but it is comparable to that found in studies in old-growth Douglas-fir and western hemlock forests in coastal Oregon and Washington (Harmon et al. 1986; Graham and Cromack 1982).

Summary of New Silvicultural Practices

All of the approaches reviewed above seek to leave more large trees and corresponding structures—live and dead—on the landscape than in traditional even-age silviculture. The precise amount and juxtaposition of leave areas (retained patches), leave (uncut) trees, and structures needed to sustain a healthy redwood forest are unknown. It will take time to understand fully the ultimate ecological effects of new silvicultural approaches, which underscores the need to treat these approaches as experimental (i.e., as adaptive management; see following section).

Throughout the redwood region, many forest owners and foresters are moving in the direction of sustainable forestry as outlined in the Montreal Process and the Santiago Agreement in 1995, which were signed by the U.S. government (see *Journal of Forestry* 93[4]:18–21). The Santiago Agreement recognized that "forests are essential to the long-term well-being of local populations, national economies, and the earth's biosphere as a whole" and endorsed six criteria of sustainable forestry: (1) conservation of biological diversity; (2) maintenance of productive capacity of forest ecosystems; (3) maintenance of forest ecosystem health and vitality; (4) conservation and maintenance of soil and water resources; (5) maintenance of forest contribution to global carbon cycles; and (6) maintenance and enhancement of long-term multiple socioeconomic benefits to meet the needs of societies.

Large and small timber companies alike are taking criteria such as these into consideration in their operations. They are attempting to manage their lands not only for timber but for biodiversity, clean water and air, carbon storage, alternative forest products, recreation, and aesthetics. Large timber companies in the region have expanded their wildlife and fisheries staff and are preparing HCPs and other broad-spectrum plans. These plans remain controversial and may ultimately fail to meet the criteria for ecological sustainability, where the natural structure, function, and composition of the forest is sustained in perpetuity (Noss 1990, 1998; Christensen et al. 1996). Nevertheless, if they can be strengthened and made adaptable to changing conditions (i.e., they are not constrained by biologically unrealistic "no surprises" clauses; Noss et al. 1997) they will have a far better chance of meeting biological objectives than traditional silviculture.

Adaptive Management and Monitoring

To move silviculture toward ecological sustainability requires additional and continued changes in current practices. Among the changes that forest ecologists agree are necessary, long rotations, structural retention, and structural restoration stand out as particularly well supported by current theory and empirical data (Kohm and Franklin 1997). Nevertheless, many questions remain unanswered: How long should rotations be? Precisely what kind of structure should be retained—for example, what size and decay classes of logs and in what proportions? How should leave trees be spaced—separately, in clumps, or in a mixture of patterns? How many are needed? For how long or over how many rotations should leave trees be retained? What specific silvicultural practices will best restore second-growth forests or plantations to natural structure? How will populations of sensitive species respond to the lag time between the initiation of restoration and achievement of the desired conditions?

These questions have no general answers. For the redwood forests, different plant associations, site conditions, stand histories, landscape contexts, and other factors will call for different practices. Faced with the inevitable uncertainties about the effects of alternative silvicultural regimes, the best that forest managers can do is to use their ecological judgment as a starting point and then apply adaptive management—learn by doing—in an intelligent, reasonably controlled, responsive way (Holling 1978; Walters 1986; Noss and Cooperrider 1994).

Adaptive management requires, first, humility—the admission by forest managers that they do not have all the answers and that everything they do is an experiment with an uncertain outcome. In practice, "management" and "humility" have been countervailing concepts, and this needs to change. Beyond humility, adaptive management requires a commitment of scientific oversight to forest management in perpetuity. Information coming in from on-site monitoring (e.g., on population trends of target species, responses of a stand to harvests and to management treatments such as prescribed burning) is combined with remote sensing information showing the condition of the broader landscape, along with data from relevant research projects, to inform and revise site-specific and regional forest management plans.

Adaptive management relies on measurable indicators that correspond to the elements of forest biodiversity, health, and sustainability that forest managers—and society generally—find valuable. Such broad values cannot be measured directly. Only by measuring indicators can managers gauge the effects of their management treatments. Although few indicators have been adequately tested or validated, many reasonable ones have been suggested. A commonsense approach is to develop indicators that correspond to trends of interest in a particular forest landscape (table 8.1). These indicators can be measured to track

Table 8.1. Indicators That Might Be Used to Monitor Recovery of Redwood Forests.

Desired Trend	Indicators	Scale and Type of Measurement
Shift from younger to older age classes of trees; recovery of old-growth stands and old individual trees	Rotation period of stand-replacing disturbances (natural and human-caused); diameter and age class distributions of surviving trees in stand and trees removed from stand; mean and range of tree ages within defined seral stages across landscape; diversity of tree ages or diameters in stand; area of landscape occupied by old growth and other seral stages; amount of late-successional forest habitat in all patches and per patch	Landscape (remote sensing) and stand (direct measurements)
Shift from simplified secondary forests and plantations to structurally complex, all-aged natural forests	Abundance and density of key structural features (e.g., snags and downed logs in various size and decay classes); spatial dispersion of structural elements within stand; physiognomy, including foliage density and layering (profiles), canopy openness, and horizontal patchiness of profile types; percentage of stand in gaps of various sizes and ages since formation; diameter and age class distributions of surviving trees in stand and trees removed from stand; diversity of tree ages or diameters in stand; abundance of species dependent on particular structural features	Direct stand-level measurements for most indicators; remote sensing for some (e.g., gaps)

(continues

Table 8.1. *(continued)*

Desired Trend	Indicators	Scale and Type of Measurement
Shift from small, isolated patches of forest to large blocks of continuous forest	Forest patch size frequency distribution for each seral stage and community type and across all stages and types; size frequency distribution of late-successional forest interior patches (minus defined edge zone, e.g., 100-200 m); fractal dimension (a measure of boundary length and complexity); patch shape indices (e.g., deviation from roundness); patch density; fragmentation indices (e.g., from FRAGSTATS software); relative abundance and demographic characteristics of species requiring large patches of forest	Landscape-scale measurements using remote sensing (with ground-truthing); surveys of area-dependent species
Separate, isolated patches of forest replaced by continuous or connected forest	Patch density; fragmentation and connectivity indices; inter-patch distance (mean, median, range) for various patch types; juxtaposition measures (percentage of area within a defined distance from patch occupied by different habitat types, length of patch border adjacent to different habitat types); structural contrast (magnitude of difference between adjacent habitats, measured for various structural attributes); presence of habitat corridors or other movement routes for fragmentation-sensitive species; relative abundance and demographic characteristics of species with poor dispersal abilities or otherwise isolation-sensitive	Landscape-scale measurements using remote sensing (with ground-truthing); surveys of isolation-sensitive species
Recovery of natural fire cycles and other aspects of the natural disturbance regime	Frequency, return interval, intensity, timing (seasonality or periodicity), patch size (areal extent), predictability, variability, and other characteristics of fires and disturbances; patch size frequency distribution for each seral stage and community type; abundance and density of	Landscape (remote sensing) and stand-level measurements; surveys of disturbance-sensitive species

Desired Trend	Indicators	Scale and Type of Measurement
	key structural features (e.g., snags and down logs in various size and decay classes); physiognomy, including foliage density and layering (profiles), canopy openness, and horizontal patchiness of profile types; percentage of stand in gaps of various sizes and ages since formation; relative abundance and demographic characteristics of species sensitive (either positively or negatively) to fire and other kinds of disturbance	
Reduction of road networks and associated impacts	Road density (mi/ mi^2 or km/km^2) for different classes of road and all road classes combined; percentage of landscape in roadless area (for different size thresholds, e.g., 1,000 ha and above, 5,000 ha and above); miles or kilometers of roads constructed, reconstructed, and closed (seasonally and permanently) each decade; amount of roadless area restored through permanent road closures and revegetation each decade	Landscape-scale measurements using remote sensing (with ground-truthing); engineering data
Eradication or effective control of exotic species that invaded following road construction, site disturbance, and dispersal by vehicles, other equipment, and humans	Ratio of exotic species to native species in community (species richness, cover, and biomass); invasion rates and pattern of spread of exotic species; demographic characteristics of particular exotic species and native species sensitive to predation or competition from exotics	Stand-level measurements; landscape-level measurements for exotic species tha can be sensed remotely
Decreased air pollution, including low-level ozone, acid fog, acid precipitation, and particulates	Direct measures of air and precipitation contents; biomass increment and other measures of tree productivity; input/output budgets of ions (as indicators of change in soil pH and nutrient status and of tree nutrition); level of direct damage to leaves and other tissues; status of pollution-sensitive and pollution-tolerant species	Stand-level measurements; remote sensing of patterns of mortality and morbidity

(continues

Table 8.1. (*continued*)

Desired Trend	Indicators	Scale and Type of Measurement
Reduced negative impacts of recreational use of forests (hiking, hunting, fishing, camping, off-road vehicle use, etc.)	Access indicators (see roads indicators above; also density of airstrips, boat landings, other access points); size and proportion of area closed to human use; measures of erosion, ground-level vegetation density and condition; measures of exotic species invasion (see above); visitor-days for various types of recreation; abundance and demographic characteristics of species sensitive to human harassment or simply human presence; visitor attitudes	Stand and landscape measurements; surveys of sensitive species and visitor attitudes

Note: These are only examples of many potential indicators for monitoring and assessing the biodiversity and ecological integrity of forests.
Source: Adapted from Noss (1999).

the movement of particular stands and their components in the desired direction of ecological recovery. Different treatments can be compared with each other according to the response of indicators within each treatment.

The indicators to be measured in a particular forest must be narrowed considerably from the expansive list in table 8.1. Criteria for selecting indicators appropriate in a given case include (1) a validated relationship of the indicator to the phenomenon of interest, (2) convenience and cost-effectiveness of the indicator for repeated measurement, (3) ability of the indicator to provide an early warning of change or trouble ahead, and (4) ability of an indicator to distinguish changes caused by human activity from "natural" changes (Noss 1990). Although funding and staffing limitations will restrict the number of indicators that can be measured, relying on one or a very few indicators is precarious. Forest managers should try to generate reasonable hypotheses about the controlling factors that maintain the communities and species of concern, based on available empirical data and theory, and select indicators with verified or highly probable relationships to those factors. Indicators should be validated periodically through focused research that quantifies and verifies their relationships to ecological factors of interest.

Forest management and restoration should not be overly prescriptive, in the sense of aiming for a well-defined, desired future condition. Rather, adaptive management progresses by measuring the responses of indicators carefully, relating those responses to the particular management practices that produced them, and continually guiding forests in the desired direction through adjustment of management. "Knowing when we get there" is elusive because managers will never understand exactly what determines forest health and integrity nor will they be able to separate definitively human impacts on forests from the vagaries of nature (Noss 1999). The more carefully managers track the responses of forest ecosystems to alternative management practices, however, the more will be learned from these experiments. If this knowledge is combined with a true concern for the redwood forest and its many inhabitants, the forest ecosystem will be conserved.

Appendix 8.1. California Forest Management and Aquatic/Riparian Ecosystems in the Redwoods

The California Forest Practices Act governs the most significant resource management activities on redwood landscapes in California: it determines what and where various forestry methods are permitted. These practices can have profound effects on the aquatic and riparian ecosystems of the redwood region (fig. 8.1; see chap. 6). Touted by some as the most stringent forestry rules anywhere, the regulations now appear inadequate to protect and maintain the natural structure and function of aquatic/riparian ecosystems. In 1997, for example, the California Department of Forestry (CDF) declared five redwood region watersheds "cumulatively impacted."

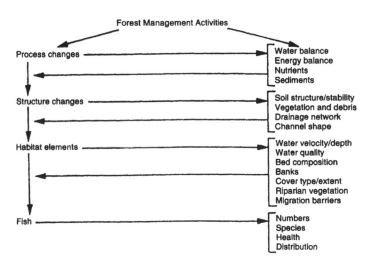

Figure 8.1. Linkages between timber management and biotic productivity. From Chamberlin et al. (1991).

The California forest practice rules were designed for the purpose of maximizing timber harvests while fulfilling minimal requirements for "consideration to" the public-trust resources of fish, wildlife, and water quality. One serious problem is the system of stream classifications used to determine allowable forestry practices adjacent to stream channels (class I = fish-bearing, class II = supports aquatic life, class III = does not support aquatic life). Basing management prescriptions on where fish occur reflects a bias for game fish over other aquatic life-forms and the integrity of the ecosystem as a whole (see chap. 6). Streams are a continuum (Vannote et al. 1980); what happens at the top of a watershed or catchment basin flows down through the system, influencing ecological processes and biotic interactions from the headwaters to the river mouth. Conditions in upstream channels determine the conditions in downstream channels. Ironically, the focus on where fish dwell ignores much of the ecosystem that supports and sustains them; consequently, not even fish are well protected by these regulations.

The timber industry, the California Board of Forestry (which oversees the timber industry), and the CDF have been slow to acknowledge the adverse effects of timber harvesting under current forestry rules (cf. Bella 1997). The nine members of the Board of Forestry are appointed by the governor, but must include four representatives of the timber industry. The denial of adverse influence of forest practices on riparian/aquatic ecosystems is sustained by the political leverage of the industry and its influence on local and state governments, despite scientific studies demonstrating negative effects of large clear-cuts and site conversions, poorly designed roads, harvesting on overly steep slopes, poorly designed stream crossings, and inadequate riparian protections (Meehan 1991; NRC 1996; Spence et al 1996; Stouder et al. 1997).

The following examples illustrate the inadequacy of current forestry rules for protecting riparian/aquatic systems. Under current rules, timber operators are responsible for maintaining roads for only three years postharvest and are required to erosion-proof only new or reconstructed roads. Yet the erosion of old logging roads has been shown by numerous studies to be a major preventable cause of siltation in many watercourses (Furniss et al. 1991; Harr and Nichols 1993; McGurk and Fong 1995). Under current rules, harvesting trees from the edges of the inner gorge of class II and III streams (those not supporting fish) is allowed. Nevertheless, these are the trees that contribute most to bank stability along headwater watercourses. This practice is analogous to pulling out the cornerstone of a building, and then when an earthquake occurs and the building collapses, citing the earthquake as the cause of the problem.

The current cultural/political climate surrounding forestry practices is exemplified by a recent situation in the Mattole watershed, an area with a substantial redwood component in its headwaters. A large portion of the Mattole watershed is listed as a Tier One Key Watershed under the President's Plan for the Forests of the Pacific Northwest (FEMAT 1993). This designation means the Mattole still retains populations of declining native species, such as coho and Chinook salmon, and therefore is a potential refugium for these threatened stocks. In the 1970s, the Board of Forestry was petitioned by citizens of the Mattole, who were concerned over cumulative adverse effects from timber harvesting (see MRC 1989). In response, the Board first proposed to set special timber-harvesting rules for the Mattole. Under threat of lawsuits from the timber industry, however, the Board decided instead to make a "sensitive watershed rule" whereby citizens could petition for special rules in a given watershed to address situations like that of the Mattole by presenting evidence of effects from past harvesting. Sensitive watersheds then must have some resource at risk that would not be protected under regular forest practice rules. The sensitive watershed rule went into effect in 1994.

In 1996, a petition was filed by local citizens nominating the Mattole under the new rule. The nomination was accompanied by copious evidence documenting increased sedimentation to tributary streams, lethal water temperatures, declining fisheries (coho and Chinook), declining amphibians, decreasing and ineffective amounts of late-seral forest (<8 percent of 1943 levels), erosion of old logging roads, widespread landsliding associated with past harvest activities, and other problems (Mattole Sensitive Watershed Group 1996). A panel to review this evidence was selected by the staff of the Board of Forestry in consultation with Board members. This panel was charged with evaluating the petition, assessing new evidence for and against it, and making a recommendation to the Board on the petition's merits. The sensitive status of the Mattole was acknowledged by five out of six state resources agencies that testified at public hearings. Only CDF denied the sensitive status of the watershed, despite an earlier letter from their own director identifying the Mattole as one of the most impaired watersheds in California. Three scientists on the review panel recommended acceptance of the nomination. Not surprising, however, given the balance of power that had been arranged by the Board, the vote was seven to six against the petition. With scientists kept in the minority, the Board of Forestry rejected the petition for sensitive watershed status.

To preserve and enhance the aquatic/riparian systems of the redwood region, more ecologically sensitive forest practice rules and an informed body to implement them are needed. A first step would be the appointment of a Board of Forestry representing the wide range of interests that depend on healthy and productive forest ecosystems. The Board also needs to develop a process to incorporate scientific evidence into its decision-making process (cf. Bella 1987). Such changes are necessary if California is to reverse the severe downward trends in fish and wildlife populations and water quality on commercial redwood timberlands and other heavily managed landscapes (Meehan 1991).

To sustain and promote riparian/aquatic ecosystem structures and functions

A. PHYSICAL FACTORS

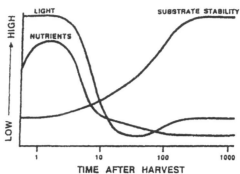

B. RIPARIAN VEGETATION

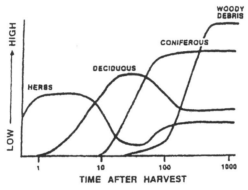

Figure 8.2. Recovery trajectories after timber harvesting of riparian structure in mature forest: *(A)* Physical factors; *(B)* Riparian vegetation. Time is expressed as years on a logarithmic scale. From Spence et al. (1996), after Gregory et al. (1987.)

before, during, and after timber harvest (fig. 8.2), revisions to the Forest Practice rules should (1) reduce excessive, human-accelerated watershed erosion and the amount of fine sediments entering stream systems by providing larger buffers on class II streams and buffers on class III streams where they are not currently required; (2) require larger buffers on all stream courses to moderate stream temperatures so streams can support native, cold-water–adapted fauna (fig. 8.3a) and to assure a future legacy of large trees that will eventually fall into stream channels, trapping and sorting streambed sediments and providing habitat diversity for stream fauna (fig. 8.3b); and (3) provide a constant source of nutrients from streamside forests from the headwaters downstream by buffering all classes of streams (fig. 8.3b).

Perhaps even more crucial than the protection of existing aquatic ecosystems

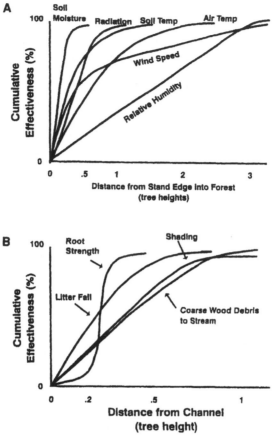

Figure 8.3. *(A)* The effects of riparian buffer width (distance expressed in site-potential tree heights) on microclimate. *(B)* The effects of riparian buffer width on four key ecological processes. From FEMAT (1993).

in the redwood region is the reestablishment of late-seral riparian communities. In streams where this is still possible (i.e., industrial forest lands), this condition will require new restrictions on timber harvest in the riparian zone to reestablish natural succession in these sensitive areas (fig. 8.2). As succession proceeds through the decades and centuries, the physical and chemical processes of streams will respond to changes in riparian form and function, and the aquatic community will be shaped by these evolving conditions.

Only a long-term solution, one that addresses both sustainability of ecosystems and human needs (table 8.2), can halt the ongoing decline in habitat quality and quantity in the riparian/aquatic component of the redwood ecosystem. Science does hold useful answers (e.g., Sedell et al. 1994, 1997; Reeves et al.

Table 8.2. Essential Components of Ecosystem Management.

Attribute	Description
Sustainability	Ecosystem management entails managing in such a way as to ensure that opportunities and resources for future generations are not diminished. Sustainability should not be evaluated based on the delivery of specific goods and services, but rather on the maintenance of the ecosystem structures and processes necessary to provide those goods and services.
Goals	Ecosystem management requires clearly defined goals. These goals should not focus exclusively on individual commodities (e.g., board feet of timber, catch of fish, visitor days). They should be explicit in terms of desired future trajectories or behaviors for components and processes necessary for sustainability.
Sound ecological models and understanding	Ecosystem management is founded on sound ecological principles, emphasizing the role of ecosystem structures and processes. It must be based on the best science and models currently available.
Complexity and connectedness	Ecosystem management recognizes that ecological processes are connectedness complex and interwoven and that this complexity and connectedness may confer particular properties (e.g., stability, resistance, resilience) to ecosystems.
Recognition of dynamic nature of ecosystems	Ecosystem management recognizes that environmental change and biological evolution are inherent properties of ecosystems and that attempts to maintain particular ecosystem "states," rather than ecological capacities, are futile over the long term in a changing environment.
Context and scale	Ecosystem management acknowledges that ecosystem processes operate over a wide range of spatial and temporal scales and that their behavior (including their response to human perturbations) at a given location is strongly influenced by the surrounding landscape or system and by the legacy of past events.
Humans as ecosystem components	Ecosystem management acknowledges that humans are components of ecosystems, as well as the source of most significant challenges to sustainability. Humans who are a part of ecosystems will, of necessity, define the future of those ecosystems. Thus, ecosystem management applied alone, without consideration of social and economic systems (and their sustainability), is insufficient to ensure resource sustainability.
Adaptability and accountability	Ecosystem management recognizes that current models and paradigms of ecosystem structure and function are provisional and subject to change. Acknowledging limits to scientific understanding and adapting to new information as it becomes available are central to successful ecosystem management.

Note: Based on recommendations of Ecological Society of America. Christensen et al. (1996).

1995; NRC 1996; Spence et al. 1996; Stouder et al. 1997; Swanson et al. 1997; Naiman and Bilby 1998), but with current political and cultural climates, a major shift in how the public relates to ecosystems and their long-term sustainability must take place before broad support will exist to take advantage of this knowledge (Frissell et al. 1997). The challenge is to apply this knowledge now, in the face of political and cultural adversity, and to educate and persuade the doubters.

Chapter 9

LESSONS FROM THE REDWOODS

Reed F. Noss

The content of this volume demonstrates that the redwood forest can be examined from a variety of perspectives and scales—from the millions of years of history recorded in the fossil record to the natural processes and actions that affect the redwood forest today; from the broad regional scale of redwood's Pacific coastal distribution to the canopy architecture of a single redwood tree. People whose work and thoughts center on one spatial or temporal scale commonly have difficulty communicating with people thinking and working on other scales. We hope this book stimulates a more integrative kind of thinking about redwood forests.

The public often is concerned with single stands of redwoods—those they know and love. Many ecologists, on the other hand, think predominantly of communities, landscapes, and regions. This does not mean ecologists do not care about individual, beautiful redwood groves, but their professional concerns are broader. Within science, there are also many different perspectives. A biologist who studies beetles or fungi operates on a scale quite different from that of the student of salmon, elk, or mountain lions. Scale aside, perspectives differ among people who view redwood forests as a source of timber, those who are most interested in the forest as habitat for rare animals, and others—who go to the forest to seek spiritual fulfillment.

In this book, thirty-two authors and I have tried to bring together these various scales and perspectives. The redwood forest can be a source of commodities and a place to find oneness with nature, though perhaps not always in the same place at the same time. Primitive peoples seemed to have little trouble uniting the spiritual and the practical into one view of the world, but today, people find it difficult to reconcile contemplation of nature with the whine of chainsaws and the incessant beeping of bulldozers. Yet this book contains glimpses of many experiences, from intricate accounts of the life histories of tiny creatures to prescriptions for sustainable forestry—even using chainsaws. We have something to learn from all these perspectives and scales.

Consider what we know about redwoods from a broad, historical perspective. Here, in the coast redwood, *Sequoia sempervirens,* we have a species that some call a "living fossil," a "relic." We know that the lineage of redwoods is ancient, with the earliest ancestors growing from the soil more than 250 million years ago. All the modern genera in the redwood family, including *Sequoia,* were present during the Cretaceous period (ca. 144 to 66 million years ago), when *Tyrannosaurus rex* and its enormous prey walked the earth. The world is a very different place today from when the Taxodiaceae were one of the most broadly distributed groups of plants on earth. *Tyrannosaurus* is gone, but the redwoods remain. Barely.

One might surmise from the ancient lineage of redwood and its shrunken range today that the species is somehow senile, doddering, unfit for the modern world of six billion humans and their minivans, airplanes, spacecraft, computers, and other wonders of industry. Redwoods are no match for modern machinery—they are easily felled by modern logging equipment. Nonetheless, the species is hardly feeble. Consider the size of these trees. Rivaled only by its closest relative, the giant sequoia, the coast redwood is the tallest and among the most massive organisms on earth today. Bigger and heavier than the blue whale and among the fastest growing and most vigorous of trees, the redwood defies its status as a living fossil.

Yet, paradoxically, the redwood forest is vulnerable. Although extinction of the coast redwood species in the near future is doubtful, the sensitivity of the redwood ecosystem is undeniable. From lichens to salamanders, beetles to salmon, the old-growth redwood forest will not be able to perpetuate itself under current conditions of escalating logging, road building, subdivision, and other threats. The outstanding collection of redwood parks, combined with more ecologically sensible management of redwood stands on many private lands, has helped forestall collapse. Nevertheless, aquatic communities, old-growth associated species, and many other components of the redwood ecosystem are still in decline. The achievements of redwood conservationists so far is a commendable start, but the mission is far from accomplished.

This is a book primarily of redwood science. Botany, zoology, ecology, paleontology, forestry, conservation biology, and other natural sciences form the core of its content. Can science save the redwoods? No, but it can help. Consider these findings about redwood forests reported in this book and what they suggest for conservation and management of the ecosystem:

- The waxing and waning of redwood's distribution over millions of years has been dramatic, but has taken place extremely slowly in response to geologic activity and associated changes in climate. Then, in less than two centuries after the arrival of Europeans to the region, more than 95 percent of the ancient redwoods disappeared. There is a mismatch between natural rates of change and changes related to the activities of modern humans.

- The redwood forest is, in fact, an assortment of many different plant associations responding to a plethora of site conditions. Several tree species, such as Douglas-fir and tanoak, often share dominance with redwood. These different associations have different species composition and ecological relationships. Protection of redwoods in parks and other reserves has not sampled the various associations equally. Some types of redwood forest are unrepresented. For example, 10.75 percent of the redwood forests in the southern section is in the highest category of protected areas, compared to 5.76 percent in the northern section and only 1.36 percent in the central section.

- Terpene evidence suggests that the southern redwood forests were founded by a colonizing population different from those that founded the redwood forest north of San Francisco. The central and northern redwood forests were founded by two other, incompletely mixed populations. Mixing genes from different populations in breeding programs can result in reduced adaptation to local conditions. Conversely, inbreeding depression has been documented in redwoods, with inbred individuals showing reduced growth under stressed conditions, compared to noninbred individuals. Breeding and restoration programs must be pursued carefully in light of these considerations.

- Holocene sediments containing redwood pollen also frequently contain charcoal, an indication of fire, and human artifacts. This evidence, together with the oral history of the Yuroks and other native peoples of the redwood region, suggests that these people burned many of the redwood forests fairly regularly. How strong was the influence of Indian burning compared to lightning-set fires? What does this influence suggest for management of redwood parks today? We still lack definitive answers to such questions.

- Natural fire frequencies vary substantially among redwood forests in different areas, ranging from 6 to 600 years. Generally, fire-return intervals are longest near the coast and shortest inland. Fire frequency also varies naturally within a landscape, depending on exposure and moisture conditions. Fire therefore has a variable influence on redwood stands, ranging from very strong to virtually nonexistent.

- The sprouting ability of redwoods is legendary and highly unusual for a conifer. Sprouts arise from a large burl, or lignotuber, at the base of the tree, and also laterally from stems. Sprouting serves the species well in environments prone to windstorms and fire, and also allows trees to grow back rapidly after cutting.

- Old-growth redwood canopies possess a surprisingly high diversity of bryophytes, fungi, lichens, invertebrates, and even terrestrial vertebrates (e.g., the clouded salamander). Researchers have only begun to study these communities. What are we losing with forestry plans that do not allow redwood trees to reach even 100 years of age?

- The fauna of late-seral redwood forests represents, in large part, the more vulnerable end of a continuum of biological resilience in coastal temperate rain forests, with many species showing narrow adaptations and tolerances.

- Although no species or subspecies of vertebrate is restricted entirely to redwood forests, some taxa are restricted to the redwood region and many more are found mainly there. The survival of some species is probably tied to the persistence of these forests.

- Several species commonly associated with old-growth forests in the Pacific Northwest, including the northern spotted owl, can be found in second-growth redwoods. Apparent explanations for this phenomenon include the rapidity with which young redwoods can reach a reasonably large size, and the presence in second-growth stands of biological legacies (snags and downed logs) from the former old-growth forest. As redwood forests are managed on shorter rotations, and as the biological legacies decay over time, habitat suitability for old-growth species can be expected to decline.

- Recent molecular genetic research on species in the redwood region, especially amphibians, has documented "cryptic species"—species that look alike (to us) but differ genetically and are probably genetically isolated from one another. Considerable research remains to be done on many taxa. If this phenomenon is widespread, the species richness and endemism of the redwood region may be considerably higher than generally believed.

- The invertebrates of the redwood forest and region are very poorly known. Many species appear to have extremely narrow distributions. Some species have been collected just once and never been seen again. They may be extinct.

- The ecosystems in the most degraded condition in the redwood region are likely riparian/aquatic ecosystems. Among other problems, they collect the sediments from logged, roaded watersheds and deteriorate when large trees are no longer available in the riparian zone to create shade, fall into streams and create pools, and stabilize streambanks. Forest practices rules in California are grossly inadequate to protect riparian and stream ecosystems. In particular, the vitally important headwaters areas (first- through third-order streams) receive the least protection in terms of buffer zone width and retention of canopy cover and large woody debris. Wider buffer zones and increased protective measures across watersheds are needed.

- Funds for land acquisition and other conservation actions in the redwood region are limited. Therefore, new acquisitions should be based on scientifically defensible reserve selection methods. Needed are protection of redwood forest types that are not currently represented well in reserves, protection of landscapes with rare species and communities (including those outside redwood forest), better connectivity among late-seral redwood forests, and increased protection of critical watersheds, floodplains, and river segments.

- Most or all traditional silviculture systems are unsustainable ecologically: they fail to sustain the natural composition, structure, and function of the redwood ecosystem. To complement reserve-based conservation, forestry must be improved to pass rigorous tests of ecological sustainability.

- A variety of incentives are available to encourage sustainable forestry, but more are needed. Several forestry experiments are promising, but it is too early to say that they are ecologically sustainable. Adaptive management is required.

This list is but a sampling of the research findings, questions, and recommendations that emerge from this book. We have learned much about the redwood forest and region, but much more remains unknown. Nature is like that.

The recommendations that emerge from this book are summarized in the introductory chapter, and I repeat them here. For an ecosystem as globally significant as the redwoods, are the following too much to ask?

1. Protection of biologically significant stands of redwoods, both old growth and second growth, representing the natural range of variation of redwood forest types and within a configuration of reserves adequate to maintain ecological integrity over time. Arguably all remaining old growth should be protected.
2. Restoration of many areas of degraded redwood forest to something resembling natural conditions. Grow more old growth!
3. Truly sustainable management of appropriate redwood stands for timber and other values.

This book is essentially a case study of an endangered ecosystem (see Noss et al. 1995; Noss and Peters 1995). Many more case studies of this sort are needed. Within the redwood region alone are a number of globally imperiled natural communities, some on the verge of total loss (table 9.1). A comprehensive conservation strategy for the redwood region must consider the special needs of all these communities and the species that compose them. The same is needed for all other regions of the earth; for them, we hope this book encourages similar work.

Table 9.1. Globally Imperiled Natural Communities of the Redwood Region.

Community	Status
central dune scrub	G2
central maritime chaparral	G2
coastal brackish marsh	G2
coastal terrace prairie	G2
fen	G2
grand fir forest	G1
maritime coast range ponderosa pine forest	G1
Mendocino pygmy cypress forest	G2
Monterey cypress forest	G1
Monterey pine forest	G1
Monterey pygmy cypress forest	G1
northern coastal bluff scrub	G2
northern foredune grassland	G1
northern interior cypress forest	G2
northern maritime chaparral	G1
serpentine bunchgrass	G2
valley needlegrass grassland	G1

G1= critically imperiled, with five or fewer occurrences or < 800 ha extant.
G2 = imperiled, with fewer than 20 occurrences or < 4,000 ha extant.
Source: Information from California Natural Diversity Data Base (electronic).

Glossary of Technical Terms

***Ad valorem* tax system:** Taxed in proportion to the value of the property.

Additive genetic variation: The proportion of genetic variation that responds to natural selection, mass selection, or pick-the-winner selection. This is the basis of a parent's breeding value, or general-combining ability.

Aerobes: Living things that require oxygen for metabolism.

Allelopathic: Negative, chemical effects of certain plants on nearby, potentially competing plants.

Alluvial: Deposited by flowing water, as in alluvial sediments (alluvium).

Angiosperms: Flowering plants, which have seeds enclosed in an ovary.

Anthropogenic: Caused or created by humans.

Axil: The upper angle between the axis and branch of an appendage; for example, between stem and leaf.

Axillary or axillary bud: Within an axil.

Basal sprouts: Sprouts emerging from the base of a plant.

Biosynthetic metabolism: Chemical reactions (metabolism) within a cell or organism that result in the formation of organic compounds from elements or simple compounds.

Bole: Trunk of a tree.

Bud collars: Buds from which sprouts emerge, as with redwood (i.e., axillary meristem gives rise to a central bud and two or more accessory buds).

Butt swell: Enlargement of the base of the trunk of a tree. Or, what happens when you spend too much time sitting to write a book.

Chronostratigraphic: Pertaining to a time sequence inferred from layers (strata) of sediments.

Circumboreal: Distributed throughout the northern regions of the Northern Hemisphere.

Collateral accessory bud: Smaller buds associated with a central bud in the growing tissues (meristem) of a seedling.

Cotyledon: Seed leaf; a modified leaf present in the seed, often functioning for food storage.

Culls: Trees that foresters reject as unmerchantable.

Detritivores: Organisms that feed on dead organic matter, including scavenging animals, litter feeders (e.g., earthworms), and decomposers (fungi and bacteria).

Ecotone: The boundary or transition zone between two biotic communities or habitats.

Endomycorrhizae: Fungal symbionts living inside the fine roots of a plant.

Endophytic: Living inside a plant.

Epiphyte: A plant, lichen, or fungus growing on a tree or other plant.

Epizootic: Growing on an animal.

Exudate: An exuded substance.

Fairy ring: Usually a ring of fungi; for example, a circle of mushrooms corresponding to the outward, belowground growth of the fungal body (mycelium). Also, redwood sprouts growing outward from a stump.

Floristics: The study of the numerical distribution of plants.

Foliose: Leafy.

Frass: Insect feces.

Heliotherm: An ectothermic animal that gains warmth from sunlight, such as many reptiles.

Hexaploid: Having six sets of chromosomes (i.e., three times the normal complement of chromosomes in the adult form in diploid species).

Holarctic: Of or designating the biogeographic region that includes the northern areas of the earth and is divided into Nearctic and Palearctic.

Horizon: The uppermost zone of mineral soil, below the layer of organic debris; the zone of greatest biological activity.

Hyphae: The filaments of a fungus that make up the mycelium (body).

Hypsithermal: The period of warmer temperatures in North America during the middle Holocene, generally dated as 9,000-4,000 years ago, corresponding to a wedge of warm, dry, Pacific air.

Isomesic: Equally moist.

Isotope: Atom of an element that differs from other atoms of the same element in the number of neutrons in the atomic nucleus. Some isotopes are unstable and emit radiation.

Lignotuber: The burl located at the base of a tree, as with redwood, from which sprouts arise.

Macrofossil: Fossils that can be seen and often identified with the naked eye (as opposed to fossil pollen and other microfossils).

Madrean: South-temperate; for example, originating in southern California, the American Southwest, or Mexico.

Meristem: The undifferentiated plant tissue, including a mass of rapidly dividing cells, from which new tissues arise.

Midden: A refuse heap from a primitive human habitation.

Monoecious: Having male and female sexual structures on the same individual plant.

Monoterpenoids: A class of volatile essential oils found in some plants, including redwood.

Mycologist: One who studies fungi.

Mycorrhizae: The symbiotic association of the mycelium of a fungus with the roots of certain plants.

Neoendemic: An endemic taxon (i.e., a species or subspecies restricted to a particular area) that evolved in recent history.

Nonadditive genetic variation: The proportion of genetic variation that does not respond to simple mass selection, or to natural selection, and that causes the average performance of offspring of specific pairwise crosses to depart from the performance values predicted by the breeding values of the parents.

Nurse tree: A tree that provides support, shade, or other benefits to another plant.

Ontogeny: The course of development of an individual organism.

Paleoendemic: An endemic taxon (i.e., a species or subspecies restricted to a particular area) that evolved in the distant past.

Phylogenetic: Pertaining to the evolutionary development of a species or lineage.

Physiognomy: The structure and shape of a plant community.

Radiocarbon ages: Ages determined by radiocarbon dating.

Reiterated trunk: New trunks arising in the canopy of a tree after breakage.

Rhizines: Threadlike or rootlike structures arising from the lower surface of a lichen and serving for attachment.

Rhizome: An underground stem from which upright stems or fronds arise.

Rhizosphere: The soil zone of increased microbial growth and activity that surrounds the roots of a plant.

Sag ponds: Ponds forming in surface depressions.

Saprobe: A plant or fungus that lives on decaying organic matter.

Sequiterpenoids: A class of volatile essential oils found in some plants, including redwood.

Sierra Madre Occidental: The major mountain range in northwestern Mexico.

Sporocarp: A structure enclosing spores, found in fungi, bryophytes, and ferns.

Squamulose: Having minute scales.

Symbiont: One of the organisms in a symbiotic relationship (living together).

TCT pollen: Pollen of the closely related families Taxodiaceae, Cupressaceae, and Taxaceae.

Tectonic: Pertaining to or resulting from structural deformities and movements of the earth's crustal plates.

Terpenes: Essential oils.

Thermoperiod: The rhythmic fluctuation of temperature, as in a day-night cycle.

Udic: Soils that are dry for fewer than thirty days in most years.

Ultramafic parent materials: Rocks, such as serpentine, having high concentrations of certain minerals (e.g., nickel, magnesium, iron) but deficient in important nutrients, such as calcium. These substrates often support endemic plants.

UNESCO: United Nations Educational, Scientific, and Cultural Organization.

Xeric: Of, characterized by, or adapted to an extremely dry habitat.

Species List

This section provides common and scientific names for species mentioned by common name in the text. Not all species of extinct plants mentioned in chapter 2, or various taxa listed by both common and scientific name in chapter tables and appendices, are repeated here.

Animals

Allen's hummingbird *(Selasphorus sasin)*

American marten *(Martes americana)*

anthaxia *(Anthaxia aeneogaster)*

aphid *(Amphorophora morrisoni)*

aquatic garter snake *(Thamnophis attratus* ssp.)

badger *(Taxidea taxus)*

bald eagle *(Haliaeetus leucocephalus)*

banana slug *(Ariolimax columbianus)*

bark beetle *(Taenioglyptes pubescens)*

bark moth *(Cydia cupressana)*

beaver *(Castor canadensis)*

bee moth *(Galleria mellonella)*

belted kingfisher *(Ceryle alcyon)*

black araucaria *(Lindingaspis rossi)*

black bear *(Ursus americanus)*

black-horned borer *(Callidium pallidium, C. sempervirens)*

black-tailed deer *(Odocoileus hemionus columbianus)*

blue grouse *(Dendragapus obscurus)*

bobcat *(Felis rufus)*

Botta's pocket gopher *(Thomomys bottae)*

brown creeper *(Certhia americana)*

bush rabbit *(Sylvilagus bachmanii)*

caddis fly (order Trichoptera)

California giant salamander *(Dicamptodon ensatus)*

California (mountain) kingsnake *(Lampropeltis zonata)*

California prionus *(Prionus californicus)*

California red tree vole (also California tree vole) *(Arborimus pomo)*

California red-backed vole
(Clethrionomys californicus)

California red-legged frog (Rana
aurora draytonii)

California sea lion (Zalophus
californianus)

California sea otter (Enhydra lutris)

California slender salamander
(Batrachoseps attenuatus)

carpenter ants (Camponotus sp.)

chestnut-backed chickadee (Poecile
rufescens)

cliff swallow (Petrochelidon pyrronhota)

clouded salamander (Aneides ferreus,
A. vagrans)

collembolids (order Collembola)

common garter snake (Thamnophis
sirtalis)

common kingsnake (Lampropeltis
getula)

common raven (Corvus corax)

cone moth (Commophila
fuscodorsana)

coyote (Canis latrans)

cutthroat trout (Oncorhynchus clarki)

cypress puto (Puto cupressi)

cypress twig borer (Phloeosinus
cristatus)

dampwood termite (Zootermopsis
angusticollis)

Del Norte salamander (Plethodon
elongatus)

Dunn's salamander (Plethodon dunni)

dusky-footed woodrat (Neotoma
fuscipes)

earthworm (Lumbricus terrestris)

eulachon (Thaleichthys pacificus)

fisher (Martes pennanti)

(northern) flying squirrel (Glaucomys
sabrinus)

foothill yellow-legged frog (Rana
boylii)

fringed myotis (Myotis thysanodes)

giant garter snake (Thamnophis gigas)

gopher snake (Pituophis melanoleucus)

gray fox (Urocyon cinereoargenteus)

great blue heron (Ardea herodias)

great horned owl (Bubo virginianus)

green chafer (Dichelonyx validum)

green sturgeon (Acipenser
transmontanus)

grizzly bear (Ursus arctos horribilis)

ground beetle (order Coleoptera,
family Carabidae)

hairy woodpecker (Picoides villosus)

harbor seal (Phoca vitulina)

harvestman (order Opiliones, family
Phalangiidae)

hermit thrush (Catharus guttatus)

hermit warbler (Dendroica occidentalis)

isopod (order Isopoda)

large cedarborer (Semanotus ligneus)

longfin smelt (Spirinchus thaleichthys)

long-tailed weasel (Mustela frenata)

lungfish (order Dipnoi)

marbled murrelet (Brachyramphus
marmoratus)

marten (Martes americana)

millipede (class Diplopoda, order
Julida)

mink (Mustela vison)

mountain beaver (Aplodontia rufa)

mountain lion (Puma concolor)

mountain quail (Oreortyx pictus)

nematode (phylum Nematoda)

northern alligator lizard (Elgaria
coerulea)

northern flicker (Colaptes auratus)

northern red-legged frog (Rana
aurora aurora)

northern spotted owl (Strix
occidentalis caurina)

northwestern garter snake
(Thamnophis ordinoides)

northwestern salamander (Ambystoma
gracile)

oribatid mite (order Acarina, suborder Sarcoptiformes)

(river) otter *(Lutra canadensis)*

pacific jumping mouse *(Zapus trinotatus)*

Pacific lamprey *(Lampetra tridenta)*

Pacific slender salamander *(Batrachoseps pacificus)*

Pacific-slope flycatcher *(Empidonax difficilis)*

Pacific tree frog *(Pseudacris regilla)*

piñon mouse *(Peromyscus truei)*

powderpost beetle *(Eiemicoelus gibbicollis)*

pygmy nuthatch *(Sitta pygmaea)*

raccoon *(Procyon lotor)*

red crossbill *(Loxia curvirostra)*

red tree vole *(Phenacomys longicaudus)*

red-bellied newt *(Taricha rivularis)*

red-legged frog *(Rana aurora)*

red-shouldered hawk *(Buteo lineatus)*

redwood bark beetle *(Phloeosinus sequoiae)*

redwood mealybug *(Spilococcus sequoiae)*

redwood scale *(Aonidia shastae)*

ringneck snake *(Diadophis punctatus)*

ringtail *(Bassariscus astutus)*

Roosevelt elk *(Cervus elaphus roosevelti)*

rough-skinned newt *(Taricha granulosa)*

roundheaded borer *(Anoplodera mathewsii, Dicentrus bluethneri, Leptura obliterate, Phymatodes nitidus)*

rubber boa *(Charina bottae)*

ruffed grouse *(Bonasa umbellus)*

rufous hummingbird *(Selasphorus rufus)*

sagebrush lizard *(Sceloporus graciosus)*

San Francisco garter snake *(Thamnophis sirtalis tetrataenia)*

Santa Cruz long-toed salamander *(Ambystoma macrodactylum croceum)*

sapwood borer *(Serropalpus substriatus)*

sharp-tailed (sharptail) snake *(Contia tenuis)*

short-tailed weasel (ermine) *(Mustela erminea)*

small cedar borer *(Atimia confuse)*

southern alligator lizard *(Elgaria multicarinata)*

southern torrent salamander *(Rhyacotriton variegatus)*

spider mite *(Oligonychus ununguis)*

spined woodborer *(Ergates spiculatus)*

(western) spotted skunk *(Spilogale gracilis)*

spotted towhee *(Pipilo maculatus)*

springtail (order Collembola, family Sminthuridae)

steelhead (rainbow trout) *(Oncorhynchus mykiss)*

Steller's jay *(Cyanocitta stelleri)*

stonefly (order Plecoptera)

striped skunk *(Mephitis mephitis)*

sturgeon *(Acipenser* sp.)

tailed frog *(Ascaphus truei)*

termite (order Isoptera)

tidewater goby *(Eucyclogobius newberryi)*

tip moth *(Argyresthia cupressella)*

Townsend's warbler *(Dendroica townsendi)*

Trowbridge's shrew *(Sorex trow bridgii)*

varied thrush *(Ixoreus naevius)*

Vaux's swift *(Chaetura vauxi)*

violet-green swallow *(Tachycineta thalassina)*

western fence lizard *(Sceloporus occidentalis)*

western horntail *(Sirex areolatus)*

western pond turtle *(Clemmys marmorata)*

(western) racer *(Coluber constrictor)*

western rattlesnake *(Crotalus viridis)*

western skink *(Eumeces skiltonianus)*

western tanager *(Piranga ludoviciana)*

western terrestrial garter snake *(Thamnophis elegans)*

western toad *(Bufo boreas)*

wild pig (boar) *(Sus scrofa)*

willow flycatcher *(Empidonax traillii)*

Wilson's warbler *(Wilsonia pusilla)*

winter wren *(Troglodytes troglodytes)*

wolverine *(Gulo gulo)*

wood warbler (family Parulidae)

wrentit *(Chamaea fasciata)*

Yuma myotis *(Myotis yumanensis)*

Fungi and Lichens

alectorioid lichens (*Alectoria* sp.)

angel's wings *(Pleurocybella porrigens)*

bark canker *(Pezicula livida)*

branchlet blight *(Chloroscypha chloromela)*

coral fungi *(Clavaria, Clavulinopsis,* and *Ramariopsis)*

cyanolichens (lichens with Cyanobacteria [Cyanophyta] partners)

damping-off *(Pythium)*

leaf blight *(Cercospora sequoiae, Cytispora pinastri, Mycosphaerella sequoiae, Pestalotiopsis funerea, Phyllosticta)*

mushrooms *Alboleptonia, Hygrocybe, Hygrophorus, Leptonia, Mycena,* and *Nolanea)*

root rot *(Armillaria ostoyae, Heterobasidion annosum, Phymatotrichopsis omnivore)*

rots of living heartwood *(Ceriporiopsis rivulosa, Fomitopsis rivulosa, Meruliporia incrassata, Oliogporus sequoiae)*

stem canker *(Botryosphaeria dothidea, Seiridium)*

twig dieback *(Leptostroma sequoiae, Macrophoma, Physalospora)*

Plants

alder (*Alnus* sp.)

alum root *(Heuchera pilosissima)*

Australian fireweed *(Erechtites arguta, E. minima)*

avocado (*Persea* spp.)

baldcypress *(Taxodium distichum)*

beach (shore) pine *(Pinus contorta contorta)*

bear-grass *(Xerophyllum tenax)*

bigleaf maple *(Acer macrophyllum)*

bishop pine *(Pinus muricata)*

bishop's cups *(Mitella ovalis)*

black cottonwood *(Populus balsamifera* ssp. *trichocarpa)*

black huckleberry *(Vaccinium ovatum)*

blue blossom *(Ceanothus thrysiflorus)*

blue gum *(Eucalyptus globulus)*

bluebell *(Campanula prenanthoides)*

boykinia (*Boykinia* sp.)

bracken fern *(Pteridium aquilinum* var. *pubescens)*

bull thistle *(Cirsium vulgare)*

California bay *(Umbellularia californica)*

California black oak *(Quercus kelloggii)*

California polypody *(Polypodium californicum)*

Canada thistle *(Cirsium arvense)*

canyon live oak *(Quercus chrysolepis)*

Cape ivy (German ivy) *Senecio milkanioides)*

cascara *(Rhamnus purshiana)*
chain fern *(Woodwardia fimbriata)*
clintonia *(Clintonia andrewsiana)*
coast lily *(Lilium maritimum)*
coast liveoak *(Quercus agrifolia)*
cotoneaster *(Cotoneaster lacteus,*
  C. pannosus)
Coulter pine *(Pinus coulteri)*
dawn-redwood *(Metasequoia*
  glyptostroboides)
deer fern *(Blechnum spicant)*
Douglas iris *(Iris douglasiana)*
Douglas-fir *(Pseudotsuga menziesii)*
elm *(Ulmus* spp.)
English ivy *(Hedera helix)*
eucalyptus *(Eucalyptus* sp.)
fairy bells *(Disporum hookeri, D.*
  smithii)
false hellebore *(Veratrum*
  fimbriatum)
false Solomon's seal *(Smilacina*
  racemosa)
festoon *(Selaginella oregana)*
fir *(Abies* spp.)
forget-me-not *(Myosotis latifolia)*
Formosan cypress *(Chamaecyparis*
  formosensis)
French broom *(Genista*
  monspessulanus)
fringecups *(Tellima grandiflora)*
giant sequoia *(Sequoiadendron*
  giganteum)
ginkgo *(Ginkgo biloba)*
golden chinquapin *(Chrysolepis*
  chrysophylla)
gorse *(Ulex europaeus)*
grand fir *(Abies grandis)*
hairy (hairyleaf) manzanita
  (Arctostaphylos sp.)
Harding grass *(Phalaris aquatica)*
hazel *(Corylus cornuta)*
Himalayan blackberry *(Rubus discolor)*

holly *(Ilex* spp.)
huckleberry *(Vaccinium* sp.)
Japanese cedar *(Cryptomeria japonica)*
Labrador tea *(Ledum glandulosum)*
lady fern *(Athyrium alpestre, A.*
  filix-femina)
leather fern *(Polypodium scouleri)*
licorice fern *(Polypodium glycyrrhiza)*
littleleaf Oregon grape *(Berberis*
  nervosa)
liverwort (division Hepatophyta or
  Bryophyta)
(Pacific) madrone *(Arbutus menziesii)*
magnolia *(Magnolia* spp.)
manroot *(Marah fabaceus)*
maple *(Acer* spp.)
mitewort *(Mitella caulescens)*
Monterey pine *(Pinus radiata)*
New Zealand kauri *(Agathis aurtralis)*
oak *(Quercus* spp.)
Oregon grape *(Berberis* sp.)
Oregon white oak *(Quercus*
  garryana)
oxalis *(Oxalis oregana)*
pampas grass *(Cortaderia jubata)*
parasol pine *(Sciadopitys verticillata)*
pennyroyal *(Mentha pulgium)*
periwinkle *(Vinca major)*
piggyback plant *(Tolmiea menziesii)*
pine *(Pinus* spp.)
plantain *(Plantago lanceolata)*
Port Orford cedar *(Chamaecyparis*
  lawsoniana)
purple star-thistle *(Centaurea*
  calcitrapa)
pygmy cypress *(Cupressus pygmaea)*
red alder *(Alnus rubra)*
red huckleberry *(Vaccinium*
  parviflorum)
redwood *(Sequoia sempervirens)*
redwood ivy *(Vancouveria hexandra,*
  V. planipetala)

rhododendron *(Rhododendron macrophyllum)*
round-fruited sedge *(Carex globosa)*
salal *(Gaultheria shallon)*
salmonberry *(Rubus spectabilis)*
sassafras *(Sassafras* spp.)
Scotch broom *(Cytisus scoparius)*
sheep sorrel (aliens *Rumex acetosella)*
Sitka spruce *(Picea sitchensis)*
slender false Solomon's seal *(Smilacina stellata)*
slink pod *(Scoliopus bigelovii)*
Spanish broom (heath) *(Spartium junceum)*
spruce *(Picea* spp.)
sugar pine *(Pinus lambertiana)*
sugar scoop *(Tiarella trifoliata)*
swamp harebell *(Campanula californica)*

sword fern *(Polystichum miniatum)*
(western) sycamore *(Platanus racemosa)*
Taiwania *(Taiwania cryptomerioides)*
tanoak *(Lithocarpus densiflorus)*
tansy ragwort *(Senecio jacobaea)*
Tasmanian cedar *(Athrotaxis* spp.)
Trailing black currant *(Ribes laxifolium)*
trillium *(Trillium ovatum)*
twayblade *(Listera cordata)*
vanilla grass *(Hierochloe occidentalis)*
vetch *(Vicia angustifolia)*
wax myrtle *(Myrica californica)*
western hemlock *(Tsuga heterophylla)*
western redcedar *(Thuja plicata)*
willow *(Salix* sp.)
yellow star-thistle *(Centaurea solstitalis)*

Literature Cited

Abbott, L. L. 1987. The effect of fire on subsequent growth of surviving trees in an old growth redwood forest in Redwood National Park, California. Master's thesis, Humboldt State University, Arcata, CA.

Adam, D. P. 1975. A late Holocene pollen record from Pearson's Pond, Weeks Creek landslide, San Francisco Peninsula, California. *Journal of Research. U.S. Geological Survey* 3:721–731.

———. 1988. Palynology of two upper Quaternary cores from Clear Lake, Lake County, California. U.S. Geological Survey Professional Paper 1363. Washington, DC: U.S. Government Printing Office.

Adam, D. P., and G. J. West. 1983. Temperature and precipitation estimates through the last glacial cycle from Clear Lake, California, pollen data. *Science* 219:168–170.

Adam, D. P., R. Byrne, and L. Edgar. 1981. A late Pleistocene and Holocene pollen record from Laguna De Las Trancas, northern coastal Santa Cruz County, California. *Madroño* 28:255–272.

Adams, D. D. 1980. A case study of young-growth redwood management of small non-industrial private forest in Santa Cruz County. Professional Paper, Dept. of Forestry and Resource Management. Berkeley: University of California.

Adams, J., M. Cerney, and D. Thornburgh. 1996. Comparison between growth of Douglas-fir and redwood regeneration in uniform and group selection silvicultural systems. Pp. 72–73 in *Proceedings of the conference on coast redwood forest ecology and management*. June 18–20, 1996. Arcata, CA: Humboldt State University.

Afek U., L. A. Lippett, J. A. Menge, and E. Pond. 1994. Vesicular-arbuscular mycorrhizal colonization of redwood and incense-cedar following storage. *Horticultural Science* 29:1362.

Agee, J. K. 1993. *Fire ecology of Pacific Northwest forests*. Washington, DC: Island Press.

Airola, D. A. 1988. *Guide to the California wildlife habitat relationship system*. Sacramento, CA: Jones and Stokes.

Allan, J. D. 1995. *Stream ecology: Structure and function of running waters.* London: Chapman and Hall.

Allen, B. A. 1987. Ecological type classification for California: The Forest Service approach. General Technical Report PSW-98. Berkeley: USDA Forest Service, Pacific Southwest Forest and Range Experiment Station.

Allgood, T. L. 1996. Comparison of residual structure, recovery, and diversity in clearcut and "new forestry" silvicultural treatments at the Yurok Experimental Forest, a coast redwood type. Master's thesis, Humboldt State University, Arcata, CA.

Alverson, W. S., D. M. Waller, and S. L. Solheim. 1988. Forests too deer: Edge effects in northern Wisconsin. *Conservation Biology* 2:348–358.

Anderson, D. 1998. Learning, surviving the past, North Coast's timber history rich in lessons. *Eureka Times-Standard,* Feb. 27, 1998.

Andrus, E. W. 1988. Woody debris and its contribution to pool formation in a coastal stream fifty years after logging. *Canadian Journal of Fisheries and Aquatic Sciences* 45:2080–2086.

Anekonda, T. S. 1992. A genetic architecture study of coast redwood. Ph.D. diss., University of California, Berkeley.

Anekonda, T. S., and W. J. Libby. 1996. Effectiveness of nearest-neighbor adjustments in a clonal test of redwood. *Silvae Genetica* 45:46–51.

Anekonda, T. S., R. S. Criddle, and W. J. Libby. 1994. Calorimetric evidence for site-adapted biosynthetic metabolism in coast redwood *(Sequoia sempervirens). Canadian Journal of Forest Research* 24:380–389.

Asquith, A., J. D. Lattin, and A. R. Moldenke. 1990. Arthropods: The invisible diversity. *Northwest Environmental Journal* 6:404–405.

Atwater, B. F., C. W. Hedel, and E. J. Helley. 1977. *Late Quaternary depositional history, Holocene sea-level changes and vertical crustal movement, southern San Francisco Bay, California.* U.S. Geological Survey Professional Paper 1014. Washington, DC: U.S. Government Printing Office.

Aulenback, K. R., and B. A. Le Page. 1998. *Taxodium wallisii* sp. nov.: First occurrence of Taxodium from the Upper Cretaceous. *International Journal of Plant Science* 159:267–390.

Axelrod, D. I. 1944a. The Sonoma flora. Pp. 167–206 in R. W. Chaney, ed., *Pliocene floras of California and Oregon.* Carnegie Institution of Washington Publication 553, Contributions to Paleontology. Washington, DC.

———. 1944b. The Mulholland flora. Pp. 103–146 in R. W. Chaney, ed., *Pliocene floras of California and Oregon.* Carnegie Institution of Washington Publication 553, Contributions to Paleontology. Washington, DC.

———. 1950. A Sonoma florule from Napa, California. Pp. 23–71 in *Studies in late Tertiary paleobotany.* Carnegie Institution of Washington Publication 590, Contributions to Paleontology. Washington, DC.

———. 1962. A Pliocene *Sequoiadendron* forest from western Nevada. *University of California Publications in Geological Sciences* 39:195–268. Berkeley: University of California Press.

———. 1966. The Eocene Copper Basin flora of northeastern Nevada. University of California Publications in Geological Sciences, vol. 59. Berkeley: University of California Press.

————. 1967. Geologic history of the Californian insular flora. Pp. 267–315 in R. N. Philbrick, ed., *Proceedings of the symposium on the biology of the California Islands*. Santa Barbara, CA: Santa Barbara Botanic Garden.

————. 1973. History of the Mediterranean ecosystem in California. Pp. 225–277 in F. diCastri and H. A. Mooney, eds., *Mediterranean type ecosystems: Origin and structure*. New York, NY: Springer-Verlag.

————. 1976. *History of the coniferous forests, California and Nevada*. University of California Publications in Botany, vol. 70. Berkeley: University of California Press.

————. 1977. Outline history of California vegetation. Pp. 139–193 in M. G. Barbour and J. Major, eds., *Terrestrial vegetation of California*. New York, NY: John Wiley and Sons.

————. 1979. Desert vegetation: Its age and origin. In J. R. Goodin and D. K. Northington, eds., *Arid land plant resources*. Proceedings of the International Arid Lands Conference on Plant Resources, Texas Tech University. Lubbock, TX: International Center Arid and Semi-Arid Land Studies, Texas Tech University.

————. 1981. Holocene climatic changes in relation to vegetation disjunction and speciation. *The American Naturalist* 117:847–870.

————. 1986. The Sierra redwood *(Sequoiadendron)* forest: End of a dynasty. *Geophytology* 16:25–36.

Axelrod, D. I., and T. A. Demere. 1984. A Pliocene flora from Chula Vista, San Diego County, California. *Transactions of the San Diego Society of Natural History,* 20:277–300.

Axelrod, D. I., and P. H. Raven. 1985. Origins of the Cordilleran flora. *Journal of Biogeography* 12:21–47.

Azevedo, J., and D. L. Morgan. 1974. Fog precipitation in coastal California forests. *Journal of Ecology* 55:135–1141.

Bailey, R. G. 1995. *Description of the ecosystems of the United States*. 2d ed. Miscellaneous Publication Number 1391 (rev.). Washington, DC: USDA Forest Service.

————. 1996. *Ecosystem geography.* New York, NY: Springer-Verlag.

Baker, F. S. 1949. A revised tolerance table. *Journal of Forestry* 47:179–181.

Bakun, A. 1990. Global climate change and intensification of coastal ocean upwelling. *Science* 247:198–201.

Barbour, M. G., and J. Major. 1988. *Terrestrial vegetation of California*. New York, NY: John Wiley and Sons.

Barbour, M., B. Pavlik, F. Drysdale, and S. Lindstrom. 1992. *California's changing landscapes*. Sacramento, CA: California Native Plant Society Press.

Baumhoff, M. A. 1978. Environmental background. Pp. 16–24 in R. F. Heizer, ed., *Handbook of North American Indians*. Vol. 8, *California*. Washington, DC: Smithsonian Institution.

Beardsley, D., and R. Warbington. 1996. Old growth in northwestern California national-al forests. PNW-RP-491. Portland, OR: USDA Forest Service.

Becker, H. F. 1969. Fossil plants of the Tertiary Beaverhead Basins in southwestern Montana. *Palaeontographica Abteilung B*, 127:1–142.

————. 1973. The York Ranch flora of the upper Ruby River Basin, southwestern Montana. *Palaeontographica Abteilug B*, 143:18–93.

Becking, R. W. 1968. The ecology of the coastal redwood forest and the impact of the 1964 floods upon vegetation. Unpublished Final Report, NSF Grant No. 3468.

————. 1970. Fasciation in coastal redwood. *Madroño* 20:382–383.

————. 1996. Seed germinative capacity and seedling survival of the coast redwood *(Sequoia sempervirens)*. Pp. 69–71 in J. LeBlanc, ed., *Proceedings of the conference on coast redwood forest ecology and management.* University of California Cooperative Extension Forestry Publication. Arcata, CA: Humboldt State University.

Begg, E. L., W. R. Allardice, S. S. Munn, and J. I. Mallory. 1984. *Laboratory data and descriptions for some typical pedons of California soils.* Vol. 2, *North Coast.* CA: Department of Land, Air, and Water Resources, Davis, University of California and California Department of Forestry Soil-Vegetation Survey.

Beier, P., and R. F. Noss. 1998. Do habitat corridors provide connectivity? *Conservation Biology* 12:1241–1252.

Beissinger, S. R. 1995. Population trends of the marbled murrelet projected from demographic analyses. Pp. 385–393 in C. J. Ralph, G. L. Hunt Jr., J. F. Piatt, and M. G. Raphael, eds., *Ecology and conservation of the marbled murrelet.* General Technical Report PSW-GTR-152. Albany, CA: USDA Forest Service.

Bella, D. A. 1987. Ethics and the credibility of applied science. Pp. 19–32 in G. H. Reeves, D. L. Bottom, and M. H. Brookes, coordinators, *Ethical questions for resource managers.* Portland, OR: USDA Forest Service General Technical Report PNW-GTR-288.

————. 1997. Organizational systems and the burden of proof. Pp. 617–638 in D. J. Stouder, P. A. Bisson, and R. J. Naiman, eds., *Pacific salmon and their ecosystems: Status and future options.* New York, NY: Chapman and Hall.

Bencala, K. E. 1984. Interactions of solutes and streambed sediments II: A dynamic analysis of coupled hydrologic and chemical process that determine solute transport. *Water Resources Research* 20:1804–1814.

Bergquist, J. R. 1977–1978. Depositional history and fault-related studies of Bolinas Lagoon, California. Ph.D. diss., Stanford University, Palo Alto, CA.

Beschta, R. L., R. E. Bilby, G. W. Brown, L. B. Holtby, and T. D. Hofstra. 1987. Stream temperature and aquatic habitat: Fisheries and forestry interactions. Pp. 199–232 in E. O. Salo and T. W. Cundy, eds., *Streamside management: Forestry and fishery interactions.* Contribution no. 57. Seattle, WA: Institute of Forest Resources, University of Washington.

Beyer, K. M., and R. T. Golightly. 1995. Distribution of Pacific fisher and other forest carnivores in coastal northwestern California. California Department of Fish and Game. Unpublished Report, Sacramento, CA.

Bickel, P. M. 1978. Changing sea levels along the California coast: Anthropological implications. *Journal of California Anthropology* 5:6–20.

Bilby, R. E., B. R. Fransen, and P. A. Bisson. 1996. Incorporation of nitrogen and carbon from spawning coho salmon into the trophic systems of small streams: Evidence from stable isotopes. *Canadian Journal of Fisheries and Aquatic Sciences* 53:164–173.

Binford, L. C., B. G. Elliott, and S. W. Singer. 1975. Discovery of a nest and the downy young of the marbled murrelet. *Wilson Bulletin* 87:303–319.

Bingham, B. B. 1984. Decaying logs as a substrate for conifer regeneration in an upland old-growth redwood forest. Master's thesis, Humboldt State University, Arcata, CA.

————. 1992. Unpublished data of the old-growth program. Arcata, CA: Redwood Sciences Laboratory, Pacific Southwest Experimental Station.

Bingham, B. B., and J. O. Sawyer. 1988. Volume and mass of decaying logs in an upland old-growth redwood forest. *Canadian Journal of Forest Research* 18:1649–1651.

———. 1991. Distinctive features and definitions of young, mature, and old-growth Douglas-fir/hardwood forests. Pp. 363–377 in L. F. Fuggiero, K. B. Aubry, A. B. Carey, and M. H. Huff, eds., *Wildlife and vegetation in unmanaged Douglas-fir forests.* Pacific Northwest Research Station General Technical Report PNW-GRT-285. Portland, OR: USDA Forest Service.

Blencowe, C. 1998. Forest Management keeps private forest land forested. *Journal of the Forest Steward Guild* 3 (Winter/Spring) 1998.

Bloom, A. L. 1983a. Sea level and coastal morphology through the late Wisconsin glacial maximum. Pp. 215–229 in S. C. Porter, ed., *Late quaternary environments of the United States.* Vol. 1, *The late Pleistocene.* Minneapolis: University of Minnesota Press.

———. 1983b. Sea level and coastal changes. Pp. 42–51 in H. E. Wright Jr., ed., *Late Quaternary environments of the United States.* Vol. 2, *The Holocene.* Minneapolis: University of Minnesota Press.

Boe, K. N. 1968. Cone production, seed dispersal, germination in old-growth redwood cut and uncut stands. USDA Forest, Research Note PSW-184. Berkeley, CA: Southwest Forest and Range Experiment Station.

Bonar, L. 1942. Studies on some California fungi II. *Mycologia* 34:180–192.

———. 1965. Studies on some California fungi IV. *Mycologia* 57:379–396.

Bonham C. D. 1989. *Measurements of terrestrial vegetation.* New York, NY: John Wiley and Sons.

Borchert, M. D. Segotta, M. D. Purser. 1988. Coast redwood ecological types of southern Monterey, California. USDA Forest Service Report PSW-107. Berkeley, CA: Southwest Forest and Range Experiment Station.

Bormann, F. H., and G. E. Likens. 1979. *Pattern and process in a forested ecosystem.* New York, NY: Springer-Verlag.

Boulton, A. J., and P. S. Lake. 1991. Benthic organic matter and detritivorous macroinvertebrates in two intermittent streams in south-eastern Australia. *Hydrobiologia* 241:107–118.

Bounty, D. A., and D. A. Rossman. 1995. Allocation and status of the garter snake names *Coluber infernalis* Blainville, *Eutaenia sirtalis tetrataenia* Cope, and *Eutaenia imperialis* Coues and Yarrow. *Copeia* 1995:236–240.

Braun-Blanquet, L. 1932/1951. *Plant sociology* (Eng. trans.). New York, NY: McGraw-Hill.

Bray, J. R., and E. Gorham. 1964. Litter production in forests of the world. *Advances in Ecological Research* 2:101–158.

Brean, R., and L. Svensgaard-Brean. 1998. Prescribed burning in state parks: A perspective. *California Parklands* (Spring).

Brode, J. 1995. Status review of the southern torrent salamander *(Rhyacotriton variegatus)* in California. Rancho Cordova, CA: California Department of Fish and Game Report.

Broecker, W. S., and G. H. Denton. 1990. What drives glacial cycles? *Scientific American* 269(1): 49–56.

Brosofske, K. B., J. Chen, R. J. Naiman, and J. F. Franklin. 1997. Harvesting effects on microclimatic gradients from small streams to uplands in western Washington. *Ecological Applications* 7:1188–1200.

Brown, J. H., and A. C. Gibson. 1983. *Biogeography.* St Louis, MO: C. V. Mosby.

Brown, P. M. 1991. Dendrochronology and fire history in a stand of northern California coast redwood. M.S. thesis, University of Arizona, Tucson.

Brown, L. R., and P. B. Moyle. 1997. Invading species in the Eel River, California: Successes, failures, and relationships with resident species. *Environmental Biology of Fishes* 49: 271–291.

Brown, L. R., P. B. Moyle, and R. M. Yoshiyama. 1994. Historical decline and current status of coho salmon in California. *North American Journal of Fisheries Management* 14(2): 237–261

Brown, P. M., and T. W. Swetnam. 1994. A cross-dated fire history from coast redwood near Redwood National Park, California. *Canadian Journal of Forest Research* 24:2131.

Bruce, D. 1923. Preliminary yield tables for second-growth redwood. Pp. 425–467 in *Calif. Univ. Agr. Expt. Sta. Bul.* 361, Berkeley, CA.

Bruijnzeel, L. A. 1991. Hydrological impacts of tropical forest conversion. *Nature and Resources* 27:36–46.

Brunsfeld, S. J., P. S. Soltis, D. E. Soltis, P. A. Gadek, C. J. Quinn, D. D. Strenge, and T. A. Ranker. 1994. Phylogenetic relationships among the genera of Taxodiaceae and Cupressaceae: Evidence from rbcL sequences. *Systematic Botany* 19:253–262

Bunnell, F. L., and A. C. Chan-McLeod. 1997. Terrestrial vertebrates. Pp. 103–130 in P. K. Schoonmaker, B. von Hagen, and E. C. Wolf, eds., *The rainforests of home: Profile of a North American bioregion.* Washington, DC: Island Press.

Burke, R. M., and P. W. Birkeland. 1983. Holocene glaciation in the mountain ranges of the western United States. Pp. 3–11 in H. E. Wright Jr., ed., *Late–Quaternary environments of the United States.* Vol. 2, *The Holocene.* Minneapolis: University of Minnesota Press.

Burns, J. W. 1972. Some effects of logging and associated road construction on northern California streams. *Trans. Amer. Fish. Soc.* 101:1–17.

Burstall, S. W., and E. V. Sale. 1984. *Great trees of New Zealand.* Wellington, New Zealand: A. H. and A. W. Reed Ltd.

Burton, T. M., and G. E. Likens. 1975. Salamander populations and biomass in the Hubbard Brook Experimental Forest, New Hampshire. *Copeia* 1975:541–5446.

Busby, P. J., T. C. Wainwright, and R. S. Waples. 1994. Status review for Klamath Mountain province steelhead. U.S. Dept. Commerce, NOAA Technical Memorandum NMFS-NWFSC-19.

Buskirk, S. W. 1993. The refugium concept and the conservation of forest carnivores. Pp. 242–245 in I. D. Thompson, ed., *Proceedings of the International Union of Game Biologists XXI Congress, Forest and wildlife towards the 21st century.* August 15–20, 1993. Halifax, Nova Scotia.

———. In press. Mesocarnivores of Yellowstone. In T. W. Clark, S. C. Minta, P. K. Karieva, and A. P. Curlee, eds., *Carnivores in ecosystems.*

Buskirk, S. W., and R. A. Powell. 1994. Habitat ecology of fishers and American marten. Pp. 283–296 in S. W. Buskirk, A. S. Harestad, M. G. Raphael, and R. A. Powell, eds., *Martens, sables and fishers: Biology and conservation.* Ithaca, NY: Cornell University Press.

Buskirk, S. W., and L. F. Ruggiero. 1994. American marten. Pp. 38–73 in L. F. Ruggiero, K. B. Aubry, S. W. Buskirk, L. J. Lyon, and W. J. Zielinski, tech. eds., *The scientific basis for conserving forest carnivores, American marten, fisher, lynx, and wolverine in the western United States*. General Technical Report RM-254, Rocky Mountain Forest and Range Experiment Station. Fort Collins, CO: USDA Forest Service.

Byers, H. R. 1953. Coastal redwoods and fog drip. *Ecology* 34:92–193.

California Fish and Game Commission. 1992. Animal of California declared to be endangered or threatened. Gov. Code Sec. 11346.2, Reg. 92:1–8.

Cannon, W. A. 1901. On the relation of redwoods and fog to the general precipitation in the redwood belt of California. *Torreya* 1:137–139.

Canny, M. J. 1997a. Vessel content of leaves after excision: A test of Scholander's assumption. *American Journal of Botany* 84:1217–1222.

———. 1997b. Vessel contents during transpiration: Embolisms and refilling. *American Journal of Botany* 84:1223–1230.

Carder, A. C. 1995. *Forest giants of the world, past and present*. Marklam, Ontario, Canada: Fitzherny and Whiteside.

Carey, A. B., and R. O. Curtis. 1996. Conservation of biodiversity: A useful paradigm for forest ecosystem management. *Wildlife Society Bulletin* 24:610–620.

Carr, S. B. 1958. Growth increase in young redwood stands after thinning. *Journal of Forestry* 56(7): 512.

Carraway, L. N. 1990. A morphologic and morphometric analysis of the "Sorex vagrans species complex" in the Pacific Coast region. Special Publications, The Museum, Texas Tech University 32:1–76.

Carroll, C. 1997. Predicting the distribution of the fisher *(Martes pennanti)* in northwestern California, U.S.A., using survey data and GIS modeling. M.S. thesis, Oregon State University, Corvallis.

Carroll, C., W. J. Zielinski, and R. F. Noss. 1999. Using presence-absence data to build and test spatial habitat models for the fisher in the Klamath region, USA. *Conservation Biology* 13: in press.

Cederholm, C. J., D. B. Houston, D. L. Cole, and W. J. Scarlett. 1989. Fate of coho salmon *(Oncorhynchus kisutch)* carcasses in spawning streams. *Canadian Journal of Fisheries and Aquatic Sciences* 46:1347–1355.

Chamberlin, T. W., R. D. Harr, and F. H. Everest. 1991. Timber harvesting, silviculture and watershed processes. *American Fisheries Society Special Publication* 19:181–205.

Chaney, R. W. 1944. The Troutdale flora. Pp. 323–351 in R. W. Chaney, ed., *Pliocene floras of California and Oregon*. Publication 553, Contributions to Paleontology. Washington, DC: Carnegie Institution of Washington.

———. 1948. The bearing of the living Metasequoia on problems of Tertiary paleobotany. *Proceeding of the National Academy of Sciences* 34:503–515.

———. 1951a. A revision of fossil *Sequoia* and *Taxodium* in western North America based on the recent discovery of *Metasequoia*. *Transactions of the American Philosophical Society*, n.s. (40:3): 171–263.

———. 1951b. Prehistoric forests of the San Francisco Bay area. Pp. 193–202 in *Geologic guidebook of the San Francisco Bay counties (history, landscape, geology, fossils, minerals, industry, and routes to travel)*. Bulletin 154. San Francisco, CA: Division of Mines.

———. 1959. *Miocene floras of the Columbia Plateau*, part 1, *Composition and interpretation*, pp. 1–134. Publication 617, Contributions to Paleontology. Washington, DC: Carnegie Institution of Washington.

———. 1967. Miocene forests of the Pacific Basin: Their ancestors and their descendants. Pp. 209–239 in Jubilee publication commemorating Professor Sasa's sixtieth Birthday (reprinted 1972 in *Tertiary floras of Japan*, vol. 2. Association of Paleobotanical Research in Japan).

Chaney, R. W., and D. I. Axelrod. 1959. *Miocene floras of the Columbia Plateau*, part 2, *Systematic considerations*, pp. 135–237. Publication 617, Contributions to Paleontology. Washington, DC: Carnegie Institution of Washington.

Chaney, R. W., and H. L. Mason. 1933. A Pleistocene flora from the asphalt deposits at Carpinteria, California. Pp. 45–79 in *Studies of the Pleistocene palaeobotany of California*. Publication 415, Contributions to Palaeontology. Washington, DC: Carnegie Institution of Washington.

Ching, K. K., and D. Bever. 1960. Provenance study of Douglas-fir in the Pacific Northwest region I: Nursery performance. *Silvae Genetica* 9:11–17.

Christensen, N. L., A. M. Bartuska, J. H. Brown, S. Carpenter, C. D'Antonio, R. Francis, J. F. Franklin, J. A. MacMahon, R. F. Noss, D. J. Parsons, C. H. Peterson, M. G. Turner, and R. G. Woodmansee. 1996. *The report of the Ecological Society of America Committee on the Scientific Basis for Ecosystem Management*. Washington DC: Ecological Society of America.

Clarke, S. H. Jr., and G. A. Carver. 1992. Late Holocene tectonics and paleoseismicity, southern Cascadia subduction zone. *Science* 255:188–192.

Clebsch, E. E. C., and R. T. Busing. 1989. Secondary succession, gap dynamics, and community structure in a southern Appalachian cove forest. *Ecology* 70:728–735.

COHMAP (Cooperative Holocene Mapping Project). 1988. Climatic changes of the last 18,000 years: Observations and model simulations. *Science* 241:1043–1052.

Cole, D. W. 1983. Redwood sprout growth three decades after thinning. *Journal of Forestry* 81:148–150, 157.

Connell, J. H. 1978. Diversity in tropical rain forests and coral reefs. *Science* 199:1302–1310.

Conness, John. 1864. *Congressional Globe*, p. 2301. Testimony of Senator John Conness to the 38th Cong., 1st sess., May 17, Washington, DC.

Cook, T. D. 1978. *Soil survey of Monterey County, California*. Washington, DC: USDA Soil Conservation Service.

Cooke, M. C., and H. W. Harkness. 1884–1886. Fungi from the Harkness Herbarium published during the past year, in *Grevillea*. *Bulletin of the California Academy of Science* 1:268–232.

———. 1886. California fungi. *Grevillea* 13:111–114.

Cooper, D. W. 1965. *The coast redwood and its ecology*. Eureka, CA: Humboldt County Agricultural Extension Service.

Cooper, W. S. 1917. Redwoods, rainfall and fog. *Plant World* 20:79–189.

Cooperrider, A. Y. 1986a. Habitat evaluation systems. Pp. 757–776 in A. Y. Cooperrider, R. J. Boyd, and H. R. Stuart, eds., *Inventory and monitoring of wildlife habitat.* Denver, CO: USDI Bureau of Land Management.

———. 1986b. Terrestrial physical features. Pp. 587–601 in A. Y. Cooperrider, R. J. Boyd, and H. R. Stuart, eds., *Inventory and monitoring of wildlife habitat.* Denver, CO: USDI Bureau of Land Management.

Cooperrider, A. Y., R. J. Boyd, and H. R. Stuart, eds. 1986. *Inventory and monitoring of wildlife habitat.* Denver, CO: USDI Bureau of Land Management.

Corn, P. S., and R. B. Bury. 1989. Logging in western Oregon: Responses of headwater habitats and stream amphibians. *Forest Ecology and Management* 29:39–57.

Coulter, M. W. 1960. The status and distribution of fisher in Maine. *Journal of Mammalogy* 41:1–9.

Coyle, F. A. 1971. Systematics and natural history of the megalomorph spider genus *Antrodiaetus* and related genera. *Bulletin of the Museum of Comparative Zoology* (Harvard) 141:269–402.

Crist, P. J., T. C. Edwards Jr., C. G. Homer, and S. D. Bassett. 1998. *Mapping and categorizing land stewardship.* U.S. Geological Survey, Gap Analysis Program. <http://www.gap.uidaho.edu/gap/AboutGAP/Handbook/SMC.htm>

Csuti, B. 1996. Mapping animal distribution areas for gap analysis. Pp. 135–145 in J. M. Scott, T. H. Tear, and F. W. Davis, eds., *Gap analysis: A landscape approach to biodiversity planning.* Bethesda, MD: American Society for Photogrammetry and Remote Sensing.

Cummins, K. W. 1974. Structure and function of stream ecosystems. *BioScience* 24:631–641.

Cummins, K. W., and M. J. Klug. 1979. Feeding ecology of stream invertebrates. *Annual Review of Ecology and Systematics* 10:147–172.

Dahm, C. N. 1981. Pathways and mechanisms for removal of dissolved organic carbon from leaf leachate in streams. *Canadian Journal of Fisheries and Aquatic Sciences* 38:68–76.

Dansereau, P. 1957. *Biogeography.* New York, NY: Ronald Press.

Dark, S. J. 1997. A landscape-scale analysis of mammalian carnivore distribution and habitat use by fisher. Master's thesis, Humboldt State University, Arcata, CA.

Darlington, P. J. Jr. 1957. *Zoogeography: The geographical distribution of animals.* New York, NY: John Wiley and Sons.

Daubenmire, R. F. 1968. *Plant communities.* New York, NY: Harper and Row.

Daubenmire, R. F., and J. Daubenmire. 1975. The status of the coastal redwood, *Sequoia sempervirens.* Unpublished report prepared for the USDI, National Park Service. Arcata, CA: Redwood National Park.

Daugherty, C. H., and A. L. Sheldon. 1982a. Age-determination, growth, and life-history of a Montana population of the tailed frog, *Ascaphus truei. Herpetologica* 38:461–468.

———. 1982b. Age-specific movement patterns of the tailed frog, *Ascaphus truei. Herpetologica* 38:468–474.

Davidson, J. G. N. 1970. Seed and cone mortality of coast redwood. *Phytopathology* 60:1533.

———. 1971. Pathological problems in redwood regeneration from seed. Ph.D. diss., University of California, Berkeley.

Davis, M. B. 1981. Palynology and environmental history during the Quaternary period: Climates past and present. Readings from *American Scientist,* pp. 71–78.

Davis, O. K. 1984. Multiple thermal maxima during the Holocene. *Science* 225: 617–619.

Dawson, T. E. 1993. Water sources of plants as determined from xylem-water isotopic composition: Perspectives on plant competition, distribution, and water relations. Pp. 465–496 in J. R. Ehleringer, A. E. Hall, and G. D. Farquhar, eds., *Stable isotopes and plant carbon-water relations.* San Diego, CA: Academic Press.

———. 1996. The use of fog precipitation by plants in coastal redwood forests. Pp. 90–93 in J. LeBlanc, ed., *Proceedings of a conference on coastal redwood ecology and management.* University of California Cooperative Extension Forestry Publication. Arcata, CA: Humboldt State University.

———. 1998. Fog in the California redwood forest: Ecosystem inputs and use by plants. *Oecologia* 117:476–485.

Deevey, E. S., and R. F. Flint. 1957. Postglacial Hypsithermal interval. *Science* 125: 182–125.

Del Tredici, P. 1998. Lignotuber formation in *Sequoia sempervirens:* Development and ecological significance. *Madroño,* 45:255–260.

———. 1999. Lignotuber formation in *Sequoia sempervirens:* Developmental morphology and ecological significance. In C. Edelin, ed., Arbre: Biologie et developpement, 3rd conference. *Naturalia Monspeliansia,* in press.

DellaSala, D. A., and D. M. Olson. 1996. Seeing the forests for more than just the trees. *Wildlife Society Bulletin* 24:770–776.

Denison, W. C. 1973. Life in tall trees. *Scientific American* 228:74–80.

Diamond, J. M. 1975. The island dilemma: Lessons of modern biogeographic studies for the design of natural preserves. *Biological Conservation* 7:129–146.

———. 1976. Island biogeography and conservation: Strategy and limitations. *Science* 193:1027–1029.

Diaz, N. M., and S. Bell. 1997. Landscape analysis and design. Pp. 255–269 in K. A. Kohm and J. F. Franklin, eds., *Creating a forestry for the twenty-first century.* Washington, DC: Island Press.

Dill, W. A., and A. J. Cordone. 1997. *History and status of introduced fishes in California, 1871–1996.* Fish Bulletin 178. Sacramento, CA: California Dept. of Fish and Game.

Diller, L. V., and R. L. Wallace. 1994. Distribution and habitat of *Plethodon elongatus* on managed, young growth forests in north coastal California. *Journal of Herpetology* 28:310–318.

———. 1996. Distribution and habitat of *Rhyacotriton variegatus* on managed, young growth forests in north coastal California. *Journal of Herpetology* 30:184–191.

Dilsaver, Lary M, and William C. Tweed. 1990. Challenge of the big trees: A resource history of Sequoia and Kings Canyon National Parks. Three Rivers, CA: Sequoia Natural History Association.

Dixon, H. H., and J. Joly. 1894. On the ascent of sap. *Philosophical Transactions of the Royal Society of London* 186:563–576.

Dixon, J. 1924. A closed season needed for fisher, marten, and wolverine in California. *California Fish and Game* 23–25. Sacramento, CA: California Dept. of Fish and Game.

Doppelt, B., M. Scurlock, C. Frissell, and J. Karr. 1993. *Entering the watershed: A new approach to save America's river systems.* Washington, DC: Island Press.

Drury, Newton B. 1972. *Newton Bishop Drury, parks and redwoods, 1917–1971.* University of California, Berkeley: Regional Oral History Office.

Duff, J. H., and F. J. Triska. 1990. Denitrification in sediments from the hyporheic zone adjacent to a small forested stream. *Canadian Journal of Fisheries and Aquatic Science* 47:1140–1147.

Edwards, T. C. Jr., C. G. Homer, S. D. Bassett, A. Falconer, R. D. Ramsey, and D. W. Wright. 1995. *Utah gap analysis: An environmental information system.* Final Project Report 95-1. Logan, UT: Utah Cooperative Fish and Wildlife Research Unit.

Edwards, T. C. Jr., E. T. Deshler, D. Foster, and G. G. Moisen. 1996. Adequacy of wildlife habitat relation models for estimating spatial distributions of terrestrial verte-brates. *Conservation Biology* 10:263–270.

Eisenberg, J. F. 1981. *The mammalian radiations: An analysis of trends in evolution, adaptation, and behavior.* Chicago, IL: University of Chicago Press.

Ellis, J. B., and H. W. Harkness. 1881. Some new species of North American fungi. *Bulletin of the Torrey Botanical Club* 8:51.

Engbeck, Joseph H. Jr. 1973. *The enduring giants: The epic story of giant sequoia and the big trees of Calaveras.* Berkeley: University Extension, University of California.

———. 1980. *State parks of California, from 1864 to the present.* Portland, OR: Graphic Arts Center Publishing Co.

———. 1995. *Redwoods, the United Nations, and world peace.* San Francisco, CA: Save-the-Redwoods League.

Espinosa-Garcia, F. J., and J. H. Langenheim. 1990. The endophytic fungal communi-ty in leaves of a coastal redwood population: Diversity and spatial patterns. *New Phytology* 116:89–97.

Espinosa-Garcia, F. J., J. L. Rollinger, and J. H. Langenheim. 1996. Coastal redwood leaf endophytes: Their occurrence, interactions, and response to host volatile terpenoids. Pp. 101–130 in S. C. Redlin and L. M. Carris eds., *Endophytic fungi in grasses and woody plants: Systematics, ecology, and evolution.* St. Paul, MN: American Phytopath-ology Society.

Farquhar, G. D. 1997. Carbon dioxide and vegetation. *Science* 278:1411.

Farr, D. F., G. F. Bills, G. F. Chamuris, and A. Y. Rossman. 1989. *Fungi on plants and plant products in the United States.* St. Paul, MN: American Phytopathology Society.

FEMAT (Forest Ecosystem Management Assessment Team). 1993. *Forest ecosystem management: An ecological, economic, and social assessment.* Report of the Forest Ecosystem Management Assessment Team. Portland, OR: USDA Forest Service.

Finney, M. A. 1991. Ecological effects of prescribed and simulated fire on coastal red-wood (*Sequoia sempervirens* [D. Don] Endl.). Ph.D. diss., University of California, Berkeley.

Finney, M. A., and R. E. Martin. 1992. Short fire intervals recorded by redwoods at Annadel State Park, California. *Madroño* 39:251–262.

Ford, L. S., and D. C. Canatella. 1993. The major clades of frogs. *Herpetological Monographs* 7:94–117.

Forman, R. T. T. 1995. *Land mosaics: The ecology of landscapes and regions.* Cambridge, U.K.: Cambridge University Press.

Forman, R. T. T., and M. Godron. 1986. *Landscape ecology.* New York, NY: John Wiley and Sons.

Frankel, O. H., and M. E. Soulé. 1981. *Conservation and evolution.* Cambridge, UK: Cambridge University Press.

Frankland, J. C. 1998. Fungal succession-unravelling the unpredictable. *Mycological Research* 102:1–15.

Franklin, J. F. 1982. Old-growth forests in the Pacific Northwest: An ecological view, in *Old growth: A balanced perspective.* Washington, DC: Bureau of Government Research and Service.

———. 1995. Sustainability of managed temperate forest ecosystems. Pp. 355–385 in M. Munasingha and W. Shearer, eds., *Defining and measuring sustainability: The biophysical foundations.* United Nations University.

Franklin, J. F., and C. T. Dyrness. 1988. *Natural vegetation of Oregon and Washington.* Corvallis, OR: Oregon State University Press.

Franklin, J. F., and R. H. Waring. 1980. Distinctive features of the northwestern coniferous forest: Development, structure, and function. Pp. 59–85 in R. H. Waring, ed., *Forests: Fresh perspectives from ecosystem analysis.* Annual Biology Colloquium, 1979. Corvallis, OR: Oregon State University.

Franklin, J. F., K. Cromack, W. Denison, A. McKee, C. Maser, J. Sedell, F. Swanson, and G. Juday. 1981. Ecological characteristics of old-growth Douglas-fir forests. Gen. Tech. Rep. PNW-118, Pacific Northwest Forest and Range Experiment Station. Portland, OR: USDA Forest Service.

Freeman, G. J. 1971. Summer fog-drip in the coastal redwood forest. Master's thesis, Humboldt State University, Arcata, CA.

Frest, T. J., and E. J. Johannes. 1993. *Mollusc species of special concern within the range of the northern spotted owl.* Final Report prepared for the Forest Ecosystem Management Working Group, Pacific Northwest Region. Portland, OR: USDA Forest Service.

———. 1996. *Additional information on certain mollusk species of special concern occurring within the range of the northern spotted owl.* Seattle, WA: Deixis Consultants.

Friedrich, W. 1966. *Zur Geologie von Brjanslaekur (Nordwest-Island) unter besonderer Berucksichtigung der fossilen Flora.* Sonderveroffentlichungen des Geologischen Institutes der Universitat Koln, vol. 10. 108 pp. Bonn: Kommissionsverlag und Auslieferung fur den Buchhandel Wilhelm Stollfuss Verlag.

Frissell, C. A. 1992. Cumulative effects of land use on salmon habitat in southwest Oregon coastal streams. Ph.D. diss., Oregon State University, Corvallis, OR.

Frissell, C. A., and D. Bayles. 1996. Ecosystem management and the conservation of aquatic biodiversity and ecological integrity. *Water Resources Bulletin* 32:229–240.

Frissell, C. A., W. J. Liss, R. E. Gresswell, R. K. Nawa, and J. L. Ebersole. 1997. A resource in crisis: Changing the measure of salmon management. Pp. 411–444 in D. J. Stouder, P. A. Bisson, and R. J. Naiman, eds., *Pacific salmon and their ecosystems: Status and future options.* New York, NY: Chapman and Hall.

Fritz, C. W., and L. Bonar. 1931. *Poria sequoiae,* n. sp. *Journal of Forestry* 29:377.

Fritz, E. 1931. The role of fire in the redwood region. *Journal of Forestry* 29:939–950.

———. 1957. The life and habits of redwood: The extraordinary are described by an authority. *Western Conservation Journal* 14:4–7, 38.

Fritz, E., and J. A. Redelius. 1966. *Redwood reforestation problems: An experimental approach to their solution.* San Francisco, CA: Foundation for American Resource Management.

Fritzke, S., and P. Moore. 1998. Exotic plant management in national parks of California. *Fremontia* 26(4):49–53.

Fujimori, T. 1977. Stem biomass and structure of a mature *Sequoia sempervirens* stand on the Pacific Coast of northern California. *Journal of the Japanese Forestry Society* 59:431–441.

Fuller, D. D., and A. J. Lind. 1992. Implications of fish habitat improvement structures for other stream vertebrates. Pp. 96–104 in R. R. Harris and D. E. Erman, tech. coordinators, and H. M. Kemer, ed., *Proceedings of the symposium on biodiversity of northwestern California.* October 28–30, 1991, Santa Rosa, California. Wildland Resource Center Report 29. Berkeley: University of California.

Furniss, M. J., T. D. Roelofs, and C. E. Yee. 1991. *Road construction and maintenance.* American Fisheries Society Special Publication 19.

Gardener, J. V., L. E. Heusser, P. J. Quinterno, S. M. Stone, J. A. Barron, and R. Z Poore. 1988. Clear Lake record vs. the adjacent marine record: A correlation of their past 20,000 years of paleoclimatic and paleoceanographic responses. Pp. 171–182 in J. D. Sims, ed., *Late Quaternary climate, tectonism, and sedimentation in Clear Lake, northern California coast ranges.* Geological Society of America Special Paper 214.

Garrison, B. A. 1992. Biodiversity of wildlife in a coast redwood forest: An analysis using the California wildlife habitat relationships system. Pp. 175–181 in R. R. Harris and D. E. Erman, tech. coordinators, and H. M. Kemer, ed., *Proceedings of the symposium on biodiversity of northwestern California.* October 28–30, 1991, Santa Rosa, California. Wildland Resources Center Report No. 29. Berkeley: University of California.

Gates, J. H. 1983. *Falk's claim: The life and death of a redwood lumber town.* Eureka, CA: Pioneer Graphics.

Gauch, H. G. 1982. *Multivariate analysis in community ecology.* Cambridge Studies in Ecology. New York, NY: Cambridge University Press.

Gellman, S. T., and W. J. Zielinski. 1996. Use by bats of old-growth redwood hollows on the north coast of California. *Journal of Mammalogy* 77:255–265.

Gerdemann, J. W. 1968. Vesicular-arbuscular mycorrhiza and plant growth. *Annual Review of Phytopathology* 6:397–418.

Gerstung, E. R. 1997. Status of coastal cutthroat trout in California. Pp. 43–56 in J. D. Hall, P. A. Bisson, and R. E. Gresswell, eds., *Sea-run cutthroat trout: Biology, management, and future conservation.* Corvallis, OR: Oregon Chapter, American Fisheries Society.

Giusti, G. A. 1990. Black bear feeding on second-growth redwoods: A critical assessment. Pp. 214–217 in L. Davis and R. Marsh, eds., *Proceedings of the fourteenth vertebrate pest conference.* March 6–8, 1990, Sacramento, CA.

———. 1999. Dusky-footed woodrats and northern spotted owls in the coastal redwoods. *Journal of Forestry,* in press.

Gleason's Pictorial Drawing-Room Companion, October 1853. Quoted in Joseph H. Engbeck Jr., *The enduring giants.* Berkeley: University Extension, University of California, 1973.

Good, D. A. 1989. Hybridization and cryptic species in *Dicamptodon* (Caudata: Dicamptodontidae). *Evolution* 43:728–744.

Good, D. A., and D. B. Wake. 1992. Geographic variation and speciation in the torrent salamanders of the genus *Rhyacotriton* (Caudata: Rhyacotritonidae). University of California Publications in Zoology 126. 91 pp.

Gorton-Linsley, E. 1963. Bering Arc relationships of Cerambycidae and their host plants. Pp. 159–178 in J. Linsley Gressitt, ed., *Pacific Basin biogeography: A symposium*. Honolulu, HI: Bishop Museum Press.

Graham, R. L., and K. Cromack Jr. 1982. Mass, nutrient content, and decay rate of dead boles in rain forests of Olympic National Park. *Canadian Journal of Forest Research* 12:511–521.

Grant, Madison. 1919. Saving the redwoods. *Bulletin of the New York Zoological Society* (September).

Gray, J. 1964. Northwest American Tertiary palynology: The emerging picture. Pp. 21–30 in L. Cranwell, ed., *Ancient Pacific floras*. Tenth Pacific Science Congress Series. Honolulu, HI: University of Hawai'i Press.

———. 1985. Interpretation of co-occurring megafossils and pollen: A comparative study with Clarkia as an example. Pp. 185–244 in C. J. Smiley, ed., *Late Cenozoic history of the Pacific Northwest: Interdisciplinary studies on the Clarkia fossil beds of northern Idaho*. San Francisco, CA: Pacific Division of the American Association for the Advancement of Science.

Greenlee, J. M. 1983. Vegetation, fire history, and fire potential of Big Basin Redwoods State Park, California. Ph.D. diss., University of California, Santa Cruz, CA.

Greenlee, J. M., and J. H. Langenheim. 1990. Historic fire regimes and their relation to vegetation patterns in the Monterey Bay area of California. *American Midland Naturalist* 124:239–253.

Gregory, S. V., G. A. Lamberti, D. C. Erman, K. V. Koski, M. L. Murphy, and J. R. Sedell. 1987. Influences of forest practices on aquatic production. Pp. 233–255 in E. O. Salo and T. W. Cundy, eds., *Streamside management: Forestry and fishery interactions*. Proc. Sympos., Feb. 12–14, 1986. Seattle, WA: University of Washington.

Gregory, S. V., G. A. Lamberti, and K. M. S. Moore. 1989. Influence of valley floor landforms on stream ecosystems. Pp. 3–9 in D. L. Abell, ed., *Proceedings of the California Riparian Systems Conference: Protection, management, and restoration for the 1990s*. September 22–24, 1988, Davis, CA. General Technical Report PSW-110, Pacific Southwest Forest and Range Experiment Station. Berkeley, CA: USDA Forest Service.

Gregory, S. V., F. J. Swanson, W. A. McKee, and K. W. Cummins. 1991. An ecosystem perspective of riparian zones. *Bioscience* 41(8): 540–551.

Griffin, J. R., and W. B. Critchfield. 1972. The distribution of forest trees in California. Research Paper PSW-82, Pacific Southwest Forest and Range Experiment Station. Berkeley, CA: USDA Forest Service.

Griffith, R. S. 1992. *Sequoia sempervirens*. In W. C. Fischer, comp., The fire effects information system. Database, Intermountain Fire Sciences Laboratory. Missoula, MT: USDA Forest Service. <http://www.fs.fed.us/database/feis>

Grimm, N. B., and S. G. Fisher. 1984. Exchange between interstitial and surface water:

Implications for stream metabolism and nutrient cycling. *Hydrobiologia* 111:219–228.

Grinnell, J., and J. S. Dixon. 1926. Two new races of the pine marten from the Pacific Coast of North America. *Zoology* 21:411–417

Grinnell, J., J. S. Dixon, and J. Linsdale. 1937. *Fur-bearing mammals of California: Their natural history, systematic status, and relations to man*, vols. 1 and 2. Berkeley: University of California Press.

Grossman D. K., K. L. Goodin, and C. L. Reuss. 1994. *Rare plant communities of the conterminous United States*. Arlington, VA: The Nature Conservancy.

Grumbine, R. E. 1992. *Ghost bears: Exploring the biodiversity crisis*. Washington, DC: Island Press.

———. 1994. What is ecosystem management? *Conservation Biology* 8:27–38.

Hagans, D. K., and W. E. Weaver. 1987. Magnitude, cause and basin response to fluvial erosion, Redwood Creek basin, northern California. Pp. 419–428 in R. L. Beschta, T. Blinn, G. E. Grant, G. G. Ice, and F. J. Swanson, eds., *Erosion and sedimentation in the Pacific Rim*. International Association of Hydrological Sciences Publication No. 165.

Hale, M. 1979. *How to know the lichens*. Dubuque, IA: William C. Brown Co.

Hale, M. E., and M. Cole. 1988. *Lichens of California*. Berkeley: University of California Press.

Hall, E. R. 1981. *The mammals of North America*. 2d ed. New York, NY: John Wiley and Sons.

Hall, G. D., and J. H. Langenheim. 1987. Geographic variation in leaf monoterpenes of *Sequoia sempervirens. Biochemical Systematics and Ecology* 15:31–43.

Hallam, A. 1994. *An outline of Phanerozoic biogeography*. Oxford, UK: Oxford University Press.

Hallin, W. 1934. Fast growing redwood. *Journal of Forestry* 32:612–613.

Hamer, T. E., and S. K. Nelson. 1995. Characteristics of marbled murrelet nest trees and nesting stand. Pp. 69–82 in C. J. Ralph, G. L. Hunt Jr., J. F. Piatt, and M. G. Raphael, eds., *Ecology and conservation of the marbled murrelet*. General Technical Report PSW-GTR-152. Albany, CA: USDA Forest Service.

Hansen, A. J., T. A. Spies, F. J. Swanson, and J. L. Ohmann. 1991. Conserving biodiversity in managed forests. *BioScience* 41:382–392.

Hanski, I., and M. Gilpin, eds. 1997. *Metapopulation biology: Ecology, genetics and evolution*. London: Academic Press.

Hansson, L., L. Fahrig, and G. Merriam. 1995. Mosaic landscapes and ecological processes. New York: Chapman and Hall.

Hargis, C. D., and J. A. Bissonette. 1997. Effects of forest fragmentation on populations of American marten in the intermountain west. Pp. 437–451 in G. Proulx, H. N. Bryant, and P. M. Woodard, eds., *Martes: Taxonomy, ecology, techniques, and management*. Edmonton, Alberta: Provincial Museum of Alberta.

Harkness, H. W. 1884–1886a. Fungi of the Pacific Coast I. *Bulletin of the California Academy of Science* 1:159–176.

———.1884–1886b. Fungi of the Pacific Coast IV. *Bulletin of the California Academy of Science* 1:256–268.

Harmon, M. E., J. F. Franklin, F. J. Swanson, P. Sollins, S. V. Gregory, J. D. Lattin, N. H. Anderson, S. P. Cline, N. G. Aumen, J. R. Sedell, G. W. Liekaemper, K. Cromack Jr., and K. W. Cummins. 1986. Ecology of coarse woody debris in temperate ecosystems. *Advances in Ecological Research* 15:133–302.

Harr, R. D., and R. A. Nichols. 1993. Stabilizing forest roads to help restore fish habitats: A northwest Washington example. *Fisheries* 18(4): 18–22.

Harrington, J. M. 1983. An evaluation of techniques for collection and analysis of benthic invertebrate communities in second order streams in Redwood National Park. Master's thesis, Humboldt State University, Arcata, CA.

Harris, D. 1995. *The last stand: The war between Wall Street and Main Street over California's ancient redwoods.* New York, NY: Random House/Times Books.

Harris, L. D. 1984. *The fragmented forest: Island biogeography theory and the preservation of biotic diversity.* Chicago, IL: University Chicago Press.

Harris, S. A. 1987. Relationship of convection fog to characteristics of the vegetation of Redwood National Park. Master's thesis, Humboldt State University, Arcata, CA.

Harrison, S. 1994. Metapopulations and conservation. Pp. 111–128 in P. J. Edwards, R. M. May, and N. R. Webb, eds., *Large-scale ecology and conservation biology.* Oxford, UK: Blackwell Scientific Publications.

Harrison S., and A. D. Taylor. 1997. Empirical evidence for metapopulation dynamics: A critical review, in I. Hanski and M. E. Gilpin, eds., *Metapopulation biology: Ecology, genetics and evolution.* New York, NY: Academic Press.

Hauxwell, D. H., S. P. Bulkin, and C. A. Hanson. 1981. The influence of elevation, vegetation type, and distance from the coast on soil temperature in Humboldt County, California. *Agronomy Abstracts,* American Society of Agronomy, p. 198.

Hawkins, C. P., K. L. Murphy, H. H. Anderson, and M. A. Wilzbach. 1983. Density of fish and salamanders in relation to riparian canopy and physical habitat in streams of the northwestern United States. *Canadian Journal of Fish and Aquatic Sciences* 40:1173–1185.

Hays, J. D., J. Imbrie, and N. J. Shackleton. 1976. Variations in the earth's orbit: Pacemaker of the ice ages. *Science* 194:1121–1132.

Healey, M. C. 1991. Life history of Chinook salmon *(Oncorhynchus tshawytscha).* Pp. 313–393 in C. Groot and L. Margolis, eds., *Pacific salmon life histories.* Vancouver, BC: University of British Columbia Press.

Hedberg, H. D, ed. 1976. *International stratigraphic guide: A guide to stratigraphic classification, terminology, and procedure.* New York, NY: John Wiley and Sons.

Heinemeyer, K., and J. L. Jones. 1994. *Fisher biology and management in the western United States: A literature review and adaptive management strategy.* Northern Region and Interagency Forest Carnivore Working Group. Missoula, MT: USDA Forest Service.

Hellmers, H. 1963. Effects of soil and air temperatures on growth of redwood seedlings. *Botanical Gazette* 124:172–177.

———. 1966. Growth response of redwood seedlings to thermoperiodism. *Forest Science* 12:276–283.

Helms, J. A. 1995. The California region. Pp. 441–498 in J. W. Barrett, ed., *Regional silviculture of the United States.* New York, NY: John Wiley and Sons.

Helms, J. A., and C. Hipkin. 1996. Growth of coast redwood under alternative silvicultural systems. Pp. 74–77 in *Proceedings of the conference on coast redwood forest ecology and management*. Arcata, CA: Humboldt State University.

Henson, P., and D. J. Usner. 1993. *The natural history of Big Sur*. Berkeley: University of California Press.

Hepting, G. H. 1971. *Diseases of forest and shade trees of the United States*. Agriculture Handbook 386. Washington, DC: U.S. Department of Agriculture.

Hermann, R. K., and D. P. Lavender. 1990. *Pseudotsuga menziesii* (Mirb) Franco. Pp. 527–540 in R. M. Barnes and B. H. Honkala, eds., *Silvics of North America*. Vol. 1, *Conifers*. Agriculture Handbook 654. Washington, DC: USDA Forest Service.

Heusser, C. J. 1960. Late-Pleistocene environments of North Pacific North America. American Geographical Society Special Publication No. 35.

Heusser, C. J., L. E. Heusser, and S. S. Streeter. 1980. Quaternary temperatures and precipitation for the north-west coast of North America. *Nature* 286:702–704.

Heusser, L. E. 1982. Quaternary paleoecology of northwest California and southwest Oregon. Abstracts, seventh biennial conference American Quaternary Association, June 28–30, 1982. Seattle: University of Washington.

———. 1983. Contemporary pollen distribution in coastal California and Oregon. *Palynology* 7:19–42.

Heusser, L. E., and W. L. Balsam. 1977 Pollen distribution in the northeast Pacific Ocean. *Quaternary Research* 7:45–62.

Hibbard, C. J. 1996. Tree retention for future northern spotted owl habitat in timber harvest plans: An HCP strategy. Poster in *Proceedings of the conference on coast redwood forest ecology and management*. June 18–20, 1996. Arcata, CA: Humboldt State University.

Hicks, B. J., J. D. Hall, P. A. Bisson, and J. R. Sedell. 1991. *Responses of salmonids to habitat changes*. American Fisheries Society Special Publication 19:483–5118.

Hildebrandt, W., and M. Hayes. 1993. Settlement pattern change in the mountains of northwest California: A view from Pilot Ridge. Pp. 107–120 in G. White, P. Mikkelsen, W. R. Hildebrandt, and M. E. Basgall, eds., *There grows a green tree: Papers in honor of David A. Fredrickson*. Publication no. 11. Davis, CA: Center for Archaeological Research at Davis.

Hill, Frank E., and Florence W. Hill. 1927. *The acquisition of California Redwood Park*. San Jose, CA: Florence W. Hill.

Hoekstra, J. M., R. T. Bell, A. E. Launer, and D. D. Murphy. 1995. Soil arthropod abundance in coast redwood forest: Effect of selective timber harvest. *Environmental Entomology* 24:246–252.

Holbrook, N. M., M. J. Burns, and C. B. Field. 1995. Negative xylem pressures in plants: A test of the balancing pressure technique. *Science* 270:1193–1194.

Holling, C. S., ed. 1978. *Adaptive environmental assessment and management*. New York, NY: John Wiley and Sons.

Holt, R. D., and M. S. Gaines. 1992. Analysis of adaptation in heterogeneous landscapes: The manifold role of habitat selection. *Evolution and Ecology* 1992:433–447.

Hough, J. 1988. Biosphere reserves: Myth and reality. *Endangered Species Update* 6(1 and 2): 1–4.

Hunter, J. C. 1997. Fourteen years of change in two old-growth *Pseudotsuga-Lithocarpus* forests in northern California. *Journal of the Torrey Botanical Society* 124:273–279.

Hunter, M. L. Jr. 1989. What constitutes an old-growth stand? *Journal of Forestry* 87(8): 33–35.

Huston, M. 1979. A general hypothesis of species diversity. *American Naturalist* 113:81–101.

Ingles, L. G. 1965. *Mammals of the Pacific states.* Stanford, CA: Stanford University Press.

Iwatsubo, R. T., and R. C. Averett. 1981. Aquatic biology of the Redwood Creek and Mill Creek drainage basins. Redwood National Park, Humboldt and Del Norte Counties, California. U.S. Geological Survey Open File Report 81-143.

Iwatsubo, R. T., K. M. Nolan, D. R. Harden, and G. D. Glysson. 1976. Redwood National Park studies, data release number 2, Redwood Creek, Humboldt County, and Mill Creek, Del Norte County, California, April 11, 1974–Sept. 30, 1975. U.S. Geological Survey Open-File Report 76–678.

Iwatsubo, R. T., K. M. Nolan, D. R. Harden, G. D. Glysson, and R. J. Janda. 1975. Redwood National Park studies, data release number 1, Redwood Creek, Humboldt County, and Mill Creek, Del Norte County, California, Sept. 1, 1973–April 10, 1974. U.S. Geological Survey Open File Report.

Jacobs, D. F. 1987. The ecology of redwood (*Sequoia sempervirens* [D. Don] Endl.) seedling establishment. Ph.D. diss., University of California, Berkeley.

Jacobs, D. F., D. W. Cole, J. R. McBride. 1985. Fire history and perpetuation of natural coast redwood ecosystems. *Journal of Forestry* 83:494–497.

Janda, R. J., K. M. Nolan, D. R. Harden, and S. M. Coleman. 1975. Watershed conditions in the drainage of Redwood Creek, Humboldt County, California, as of 1973. U.S. Geological Survey, Open File Report 75-568, Menlo Park, CA.

Jenney, H., R. J. Arkley, A. M. Schultz. 1969. The pygmy forests ecosystem and its dune associates of the Mendocino coast. *Madroño* 20:60–74.

Jepson, W. L. 1910. The silvae of California. *Memoirs of the University of California* 2:16–18, 128–139.

Jimerson, T. M., E. A. McGee, D. W. Jones, R. J. Svilich, E. Hotalen, G. DeNitto, T. Laurent, J. D. Tenpas, M. E. Smith, K. Hefner-McClelland, and J. Mattison. 1996. *A field guide to the tanoak and the Douglas-fir plant associations in northwestern California.* USDA Forest Service R5-ECOL-TP-009.

Johnson, D. L. 1977. The late Quaternary climate of coastal California: Evidence for an Ice Age refugium. *Quaternary Research* 8:154–179.

Johnson, M. L., and S. B. George. 1991. Species limits within the *Arborimus longicaudus* species-complex (Mammalia: Rodentia) with a description of a new species from California. *Natural History Museum of Los Angeles County Contributions in Science* 429:1–16.

Johnston, V. R. 1994. *California forests and woodlands.* Berkeley: University of California Press.

Jones, T. L. 1992. Settlement trends along the California coast. Pp. 1–37 in T. L. Jones, ed., *Essays on the prehistory of maritime California.* Publication No. 10. Davis, CA: Center for Archaeological Research at Davis.

Jouzel, J., N. I. Barkov, J. M. Barnola, M. Bender, J. Chappellaz, C. Genthon, V. M. Kotlyakov, V. Lipenkov, C. Lorius, J. R. Petit, D. Raynaud, G. Raisbeck, C. Ritz, T. Sowers, M. Stievenard, F. Yiou, and P. Yiou. 1993. Extending the Vostok ice-core record of palaeoclimate to the penultimate glacial period. *Nature* 364:407–412.

Judah, R. 1983. Old growth Douglas-fir redwood forest: Forty-sixth breeding bird census. *American Birds* 37(1):89–90.

Karraker, N. E., and G. S. Beyersdorf. 1997. A tailed frog (Ascaphus truei) nest site in northwestern California. *Northwestern Naturalist* 78:110–111.

Keller, E. A., A. MacDonald, T. Tally, and N. J. Merrit. 1995. Effects of large organic debris on channel morphology and sediment storage in selected tributaries of Redwood Creek, northwestern California. Pp. 1–29 in K. M. Nolan, H. M. Kelsey, and D. C. Marron, eds., *Geomorphic processes and aquatic habitats in the Redwood Creek Basin, northwestern California.* U.S. Geological Survey Professional Paper 1454-P. Washington, DC: U.S. Government Printing Office.

Kelsey, H. M. 1980. A sediment budget and an analysis of geomorphic process in the Van Duzen River basin, north coastal California, 1941–1975. *Geological Society of America Bulletin* 91:119–1216.

Kennedy, C. E. 1983. A study of uneven-aged management strategies in young-growth redwood at Jackson Demonstration State Forest. Professional Paper, Dept. of Forestry and Resource Management Berkeley: University of California.

Kennett, J. P., and B. L. Ingram. 1995. A 20,000-year record of ocean circulation and climate change from the Santa Barbara basin. *Nature* 377:510–514.

Kent, M., and P. Coker. 1992. *Vegetation description and analysis.* Boca Raton, FL: CRC Press.

Keppeler, E. T. 1998. The summer flow and water yield response to timber harvest. Pp. 37–46 in R. Ziemer, ed., *Proceedings of the conference on coastal watersheds: The Caspar Creek story.* USDA FS PS-GTR 165.

Keppeler, E. T., and D. Brown. 1998. Subsurface drainage processes and management impacts. Pp. 25–35 in R. Ziemer, ed., *Proceedings of the conference on coastal watersheds: The Caspar Creek story.* USDA FS PS-GTR 165.

Kerfoot, O. 1968. Mist precipitation on vegetation. *Forest Abstracts* 29:8–20.

Kerns, S. J., and R. A. Miller. 1995. Two marbled murrelet nest sites on private commercial forest lands in northern California. Pp. 40–42 in S. K. Nelson and S. G. Sealy, eds., Biology of the marbled murrelet: Inland and at sea. *Northwestern Naturalist* 76.

Kirchgessner, K. A., and W. J. Libby. 1985. Inbreeding depression in selfs of redwood: Rooting. California. *Forestry and Forest Products* 60.

Klug, R. R. 1996. Occurrence of Pacific fisher in the redwood zone of northern California and the habitat attributes associated with their detection. Master's thesis, Humboldt State University, Arcata, CA.

Kohm, K. A., and J. Franklin, eds. 1997. *Creating a forestry for the twenty-first century.* Washington, DC: Island Press.

Kough, J. L., R. Molina, and R. G. Linderman. 1985. Mycorrhizal responsiveness of four cedar and redwood species of western North America. P. 259 in R. Molina, ed., *Proceedings of the sixth North American conference on Mycorrhiza.* June 25–29, 1984. Bend, Oregon. Corvallis, OR: Forest Research Laboratory.

Krohn, W. B., K. D. Elowe, and R. B. Boone. 1995. Relations among fishers, snow, and martens: Development and evaluation of two hypotheses. *Forestry Chronicle* 71:97–105.

Kucera, T. E., A. M. Soukkala, and W. J. Zielinski. 1995. Photographic bait stations. Pp. 25–61 in W. J. Zielinski and T. E. Kucera, tech. eds., *American marten, fisher, lynx and wolverine: Survey methods for their detection*. General Technical Report PSW-GTR-157, Pacific Southwest Research Station. Albany, CA: USDA Forest Service.

Küchler, A. W. 1967. *Vegetation mapping*. New York, NY: Ronald Press.

———. 1975. Map of potential natural vegetation of the conterminous United States. Scale 1:3,168,000. Special Publication No. 36. New York, NY: American Geographic Society.

Kuperburg, S. J. 1994. Exotic larval bullfrogs *(Rana catesbeiana)* as prey for native garter snakes: Functional and conservation implications. *Herpetological Review* 25(3): 95–97.

———. 1996. Hydrologic and geomorphic factors affecting conservation of a river-breeding frog *(Rana boylii)*. *Ecological Applications* 6:1332–1344.

Kuser, J. E., and K. K. Ching. 1980. Provenance variation in phenology and cold hardiness of western hemlock seedlings. *Forest Science* 26:463–470.

Kuser, J. E., W. J. Libby, J. A. Martin, and J. A. Rydelius. 1984. International rangewide provenance test of redwood, *Sequoia sempervirens* ([D. Don] Endl.). Unpublished report.

Kuser, J. E., A. Bailly, A. Franclet, W. J. Libby, J. Martin, J. Rydelius, R. Scheonike, and N. Vagel. 1997. Early results of a rangewide provenance test of *Sequoia sempervirens*. Forest Genetic Resources No. 23. <http://www.fao.org/WAICENT/faoinfo>

Kutzbach, J. E., and P. J. Guetter. 1986. The influence of changing orbital patterns and surface boundary conditions on climate simulations for the past 18,000 years. *Journal of Atmospheric Sciences* 43:1726–1759.

Lakhanpal, R. N. 1958. *The Rujada flora of west central Oregon*. University of California Publications in Geological Sciences, vol. 35, no. 1, pp. 1–66. Berkeley: University of California Press.

Largent, D. L. 1985. *The Agaricales (gilled fungi) of California*. Vol. 5, *Hygrophoraceae*. Eureka, CA: Mad River Press.

———. 1994. *Entolomatoid fungi of western North American and Alaska*. Eureka, CA: Mad River Press.

———. 1998. Species of mushrooms, boletes, chanterelles, coral fungi, and tooth fungi found in the Redwood National and States Parks between 1993 and 1997. Unpublished report. Arcata, CA: Redwood National Park.

Larson, C. J. 1991. A status review of the marbled murrelet *(Brachyramphus marmoratus)* in California. Report to the Fish and Game Commission, California Department of Fish and Game, Wildlife Management Division, Sacramento.

Lattin, J. 1990. Arthropod diversity in northwest old-growth forests. *Wings* 15:7–10.

———. 1993. Arthropod diversity and conservation in old-growth Northwest forests. *American Zoologist* 33:578–587.

Lattin, J. D., and A. R. Moldenke. 1992. Ecologically sensitive invertebrate taxa of Pacific Northwest old-growth conifer forests. Report prepared for the Northern Spotted Owl Recovery Team, Other Species, and Ecosystems Committee.

Lauck, D. R. 1964. Miscellaneous groups of animals occurring in the redwood region. Pp. 107–202 in *Ecology of the coast redwood region*. Arcata, CA: Humboldt State University.

Laurance, W. F., and E. Yensen. 1990. Predicting the impacts of edge effects in fragmented habitats. *Biological Conservation* 55:77–92.

Ledwith, T. 1996. The effects of buffer strip width on air temperature and relative humidity on a stream riparian zone. *Watershed Management Council Networker* 6(4): 6–7.

Lenihan, J. M. 1986. The forest associations of the Little Lost Man Creek Research Natural Area, Redwood National Park, California. Master's thesis, Humboldt State University, Arcata, CA.

Lenihan J. M., W. S. Lennox, E. H. Muldavin, and S. D. Veirs. 1983. A handbook for classifying early post-logging vegetation in the lower Redwood Creek basin. Redwood National Park Technical Report Number 7. Arcata, CA: USDI, National Park Service, Redwood National Park.

Leonard, W. P, H. A. Brown, L. L. C. Jones, K. R. McAllister, and R. M. Storm. 1993. *Amphibians of Washington and Oregon.* Seattle, WA: Seattle Audubon Society.

Leopold, A. 1933. *Game management.* New York, NY: Charles Scribners Sons.

Lewis, J. 1982. Soil moisture and temperature regimes in the Mad River District of Six Rivers National Forest. Master's thesis, Humboldt State University, Arcata, CA.

———. 1998. Evaluating the impacts of logging activities on erosion and suspended sediment transport in the Caspar Creek Watersheds. Pp. 59–74 in R. Ziemer, ed., *Proceedings of the conference on coastal watersheds: The Caspar Creek story.* USDA FS PS-GTR 165.

Lewis, J. C., and W. J. Zielinski. 1996. Historical harvest and incidental capture of fishers in California. *Northwest Science* 70:291–297.

Libby, W. J. 1996. Ecology and management of coast redwood. Pp. 1–3 in J. LeBlanc, ed., *Proceedings of the conference on coast redwood forest ecology and management.* Arcata, CA: University of California Cooperative Extension Forestry Publication, Humboldt State University.

Libby, W. J., and W. B. Critchfield. 1987. Patterns of genetic architecture. *Annales Forestales* 13:77–92.

Libby, W. J., and B. G. McCutchan. 1978. Taming the Redwood. *American Forests* 84:18–23, 37–39.

Libby, W. J., B. G. McCutchan, and C. I. Millar. 1981. Inbreeding depression in selfs of redwood. *Silvae Genetica* 30:15–25.

Libby, W. J., T. S. Anekonda, and J. E. Kuser. 1996. The genetic architecture of coast redwood. Extended abstract. Pp. 147–149 in J. LeBlanc, ed., *Proceedings of the conference on coast redwood forest ecology and management.* Arcata, CA: University of California Cooperative Extension Forestry Publication, Humboldt State University.

Lind, A. J., and H. H. Welsh Jr. 1990. Predation of *Thamnophis couchii* on *Dicamptodon ensatus. Journal of Herpetology* 24:104–106.

———. 1994. Ontogenetic changes in the foraging behavior and habitat use of the Oregon garter snake, *Thamnophis attratus hydrophilus. Animal Behavior* 48:1261–1273.

Lind, A. J., H. H. Welsh Jr., and R. A. Wilson. 1996. The effects of a dam on breeding habitat and egg survival of the foothill yellow-legged frog *(Rana boylii). Herpetological Review* 27:62–67.

Lindquist, J. L. 1974. *Redwood, an American wood.* USDA Forest Service Report FS-264. Washington, DC: USDA Forest Service.

———. 1979. Sprout regeneration of young-growth redwood. USDA, Forest Service

Research Note PSW-337. Berkeley: Pacific Southwest Forest and Range Experiment Station.

Lindquist, J. L., and M. N. Palley. 1963. Empirical yield tables for young-growth redwood. California Agricultural Experiment Station Bulletin 796, Berkeley, CA.

Lisle, T. E. 1982. Effects of aggradation and degradation on riffle-pool morphology in natural gravel channels, northwestern California. *Water Resources Research* 18:1643–1651.

―――. 1989. Channel-dynamic control on the establishment of riparian trees after large floods in northwestern California. Pp. 9–13 in D. L. Abell, ed., *Proceedings of the California riparian systems conference.* September 9–13, 1988. Berkeley, CA. USDA Forest Service General Technical Report PSW-110.

Lisle, T. E., and M. B. Napolitano. 1998. Effects of recent logging on the main channel of North Fork Caspar Creek. Pp. 87–93 in R. Ziemer, ed., *Proceedings of the conference on coastal watersheds: The Caspar Creek story.* USDA FS PS-GTR 165.

Little, B., and D. L. Largent. 1993. Epigeous fungi (Agaricales and Boletales) inventory and monitoring handbook for Redwood National and State Parks, California. Unpublished report. Arcata, CA: Redwood National Park.

Lucus, Ronda. 1998. Spotted owl illustrates weakness of ESA. *Forest Landowner,* Spring 1998.

MacArthur, R. H., J. W. MacArthur, and J. Preer. 1962. On bird species diversity: Prediction of bird censuses from habitat measurements. *American Naturalist* 96:167–174.

MacGinitie, H. D. 1953. Fossil plants of the Florissant beds, Colorado. Carnegie Institution of Washington Publication 599, Contributions to Paleontology. Washington, DC.

Madej, M. A. 1987. Residence times of channel-stored sediment in Redwood Creek, northwestern California. Pp. 429–438 in R. L. Beschta, T. Blinn, G. E. Grant, G. G. Ice, and F. J. Swanson, eds., *Erosion and sedimentation in the Pacific Rim.* International Association of Hydrological Sciences Publication No. 165, Wallingford, UK.

Madej, M. A., and V. Ozaki. 1996. Channel response to sediment wave propagation and movement, Redwood Creek, California, USA. *Earth Surface Processes and Landforms* 21:911–927.

Main, B. Y. 1987. Persistence of invertebrates in small areas: Case studies of trapdoor spiders in western Australia. Pp. 29–39 in D. A. Saunders, G. W. Arnold, A. A. Burbidge, and A. J. M. Hopkins, eds., *Nature conservation: The role of remnants of native vegetation.* Chipping Norton, NSW, Australia: Surrey Beatty and Sons.

Major, J. 1988. California climate in relation to vegetation. Pp. 11–74 in M. G. Barbour and J. Major, eds., *Terrestrial vegetation of California.* New York, NY: John Wiley and Sons.

Marden, M. 1993. The tolerance of *Sequoia sempervirens* to sedimentation, East Coast region, New Zealand. *New Zealand Forestry* 38(3): 22–24.

Martin, L. D. 1983. The origin and early radiation of birds. Pp. 291–338 in A. H. Brush and G. A. Clark Jr., eds., *Perspectives in ornithology.* Cambridge, UK: Cambridge University Press.

Martin, P. S. 1984. Prehistoric overkill: The global model. Pp. 354–403 in P. S. Martin and R. G. Klein, eds., *Quaternary extinctions: A prehistoric revolution.* Tucson, AZ: University of Arizona Press.

Maser, C. 1988. *The redesigned forest.* San Pedro, CA: R. and E. Miles.

Maser, C., B. R. Mate, J. F. Franklin, and C. T. Dyrness. 1981. Natural history of Oregon coast mammals. General Technical Report PNW-133. Portland OR.

Maser, C., R. F. Tarrant, J. M. Trappe, and J. F. Franklin, eds. 1988. *From the forest to the sea: A story of fallen trees.* U.S. Forest Service, Pacific Northwest Research Station PNW-GTR-229. Portland, OR: USDA Forest Service.

Mastrogiuseppe, R., and J. Lindquist. 1996. Dynamics of second growth redwood: *Sequoia sempervirens* forest establishment in Redwood National Park, California. In *Proceedings of the conference on coast redwood forest ecology and management.* June 18–20, 1996. Arcata, CA: Humboldt State University.

Matthews, S. C. 1986. Old-growth forest associations of the Bull Creek watershed, Humboldt Redwoods State Park, CA. Master's thesis, Humboldt State University, Arcata, CA.

Mattole Sensitive Watershed Group. 1996. A nomination from concerned citizens of the Mattole River Watershed of the State of California proposing the Mattole River as a sensitive watershed. Prepared for the California Board of Forestry and the California Department of Forestry, Sacramento, CA.

Mayer, K. E., and W. F. Laudenslayer. 1988. *A guide to wildlife habitats of California.* Sacramento, CA: California Department of Forestry and Fire Protection.

McCune, B. 1993. Gradients in epiphyte biomass in three *Pseudotsuga-Tsuga* forests of different ages in western Oregon and Washington. *Bryologist* 96:405–411.

McGinnis, S. M. 1984. *Freshwater fishes of California.* Berkeley: University of California Press.

McGurk, B. J., and D. F. Fong. 1995. Equivalent roaded area as a measure of cumulative effect of logging. *Environmental Management* 19:609–621.

McIver, J. D., A. R. Moldenke, and G. L. Parsons. 1990. Litter spiders as bio-indicators of recovery after clearcutting in a western coniferous forest. *Northwest Environmental Journal* 6:410–412.

McKelvey, K. S., B. R. Noon, and R. H. Lamberson. 1993. Conservation planning for species occupying fragmented landscapes: The case of the northern spotted owl. Pp. 424–450 in P. M. Karieva, J. G. Kingsolver, and R. B. Huey, comps. eds., *Biotic interactions and global change.* Sunderland, MA: Sinauer.

McKenzie, M., S. T. Gellman, and W. J. Zielinski. 1994. Interim biological assessment of the proposed Wilson Creek realignment project route 101, Del Norte County, California. Report 01-DN-101-12.5/16.3. Eureka, CA: California Dept. Transportation.

Meehan, W. R., ed. 1991. *Influences of forest and rangeland management on salmonid fishes and their habitats.* American Fisheries Society Special Publication 19.

Meffe, G. K., and C. R. Carroll, eds. 1997. *Principles of conservation biology.* Second edition. Sunderland, MA: Sinauer.

Menges, K. M. 1994. Bird abundance in old-growth, thinned and unthinned redwood Forests. Unpublished report, Forestry. Arcata, CA: Humboldt State University.

Merritt, R. W., and K. W. Cummins. 1978. *An introduction to the aquatic insects of North America.* Dubuque, IA: Kendall/Hunt.

Miles, S. R., and C. B. Goudey. 1997. Ecological subsections of California. USDA Forest Service, Pacific Southwest Region R5-EM-TP.005 Report. San Francisco, CA.

Millar, C. I., and W. J. Libby. 1989. Disneyland or native ecosystem: Genetics and the restorationist. *Restoration and Management Notes* 7:18–24.

———. 1991 Strategies for conserving clinal, ecotypic, and disjunct population diversity in widespread species. Pp. 149–170 in D. A. Falk and K. E. Holsinger, eds., *Genetics and conservation of rare plants.* New York, NY: Oxford University Press.

Millar, C. I., J. M. Dunlap, and N. K. Walker. 1985. Analysis of growth and specific gravity in a twenty-year-old provenance test of *Sequoia sempervirens.* California Forestry and Forest Products 59. CA.

Miller, C. N. Jr. 1977. Mesozoic conifers. *Botanical Review* 43:217–280.

———. 1988. The origin of modern conifer families. Pp. 448–486 in C. B. Beck, ed., *Origin and evolution of gymnosperms.* New York: Columbia University Press.

Miller, C. N. Jr., and D. R. Crabtree. 1989. A new taxodiacesous seed cone from the Oligocene of Washington. *American Journal of Botany* 76:133–142.

Miller, C. N. Jr., and C. A. LaPasha. 1983. Structure and affinities of Athrotaxites berryi Bell, an early Cretaceous conifer. *American Journal of Botany* 70:772–779.

Miller, S. L., and C. J. Ralph. 1995. Relationships of marbled murrelets with habitat characteristics at inland sites in California. Pp. 205–215 in C. J. Ralph, G. L. Hunt Jr., J. F. Piatt, and M. G. Raphael, eds., *Ecology and conservation of the marbled murrelet.* General Technical Report PSW-GTR-152. Albany, CA: USDA Forest Service.

Mills, L. S. 1995. Edge effects and isolation: Red-backed voles on forest remnants. *Conservation Biology* 9:395–403.

Minshall, G. W., R. C. Peterson, K. W. Cummins, T. L. Bott, J. R. Sedell, C. E. Cushing, and R. L. Vannote. 1983. Interbiome comparison of ecosystem dynamics. *Ecological Monographs* 51:1–25.

Minshall, G. W., K. W. Cummins, R. C. Peterson, C. E. Cushing, D. A. Bruns, J. R. Sedell, and R. L. Vannote. 1985. Developments in stream ecosystem theory. *Canadian Journal of Fisheries and Aquatic Sciences* 42:1045–1055.

Minshall, G. W., R. C. Peterson, T. L. Bott, C. E. Cushing, K. W. Cummins, R. L. Vannote, and J. R. Sedell. 1992. Stream ecosystem dynamics of the Salmon River, Idaho: An eighth order system. *Journal of the North American Benthological Society* 11(2): 111–137.

Mitchell, W. T. 1988. Microhabitat utilization and spatial segregation of juvenile coastal cutthroat and steelhead trout in the Smith River drainage, California. Master's thesis, Humboldt State University, Arcata, CA.

Mladenoff, D. J., T. A. Sickley, R. G. Haight, and A. P. Wydeven. 1995. A regional landscape analysis and prediction of favorable gray wolf habitat in the northern Great Lakes region. *Conservation Biology* 9:279–294.

Moldenke, A. R. 1990. One hundred twenty thousand little legs. *Wings* 15:11–14.

Moldenke, A. R., and J. D. Lattin. 1990a. Density and diversity of soil arthropods as biological probes of complex soil phenomena. *Northwest Environmental Journal* 6:409–410.

———. 1990b. Dispersal characteristics of old-growth soil arthropods: The potential for loss of diversity and biological function. *Northwest Environmental Journal* 6:408–409.

Monteith, J. L. 1963. Dew: Facts and fallacies. Pp. 37–56 in A. V. Rutter and F. H.

Whitehead, eds., *The water relations of plants.* Oxford, UK: Blackwell Scientific Publications.

Morrison, P. H. 1988. *Old growth in the Pacific Northwest: A status report.* Seattle, WA: The Wilderness Society.

Moss, M. L., and J. M. Erlandson. 1995. Reflections on North American Pacific Coast prehistory. *Journal of World Prehistory* 9:1–45.

Moyle, P. B. 1976. *Inland fishes of California.* Berkeley: University of California Press.

Moyle, P. B., and B. Herbold. 1987. Life-history patterns and community structure in stream fishes of western North American: Comparisons with eastern North America and Europe. Pp. 25–32 in W. J. Matthews and D. C. Heins, eds., *Community and evolutionary ecology of North American stream fishes.* Norman: University of Oklahoma Press.

Moyle, P. B., and G. M. Sato. 1991. On the design of preserves to protect native fishes. Pp. 155–169 in W. L. Minckley and J. E. Deacon, eds., *Battle against extinction: Native fish management in the American West.* Tucson, AZ: University of Arizona Press.

Moyle, P. B., and R. M. Yoshiyama. 1994. Protection of biodiversity in California: A five-tiered approach. *Fisheries* 19(2): 6–18.

Moyle, P. B., R. M. Yoshiyama, E. D. Wikramanayake, and J. W. Williams. 1995. *Fish of special concern.* 2d ed. Report to the California Department of Fish and Game. Sacramento, CA.

MRC (Mattole Restoration Council). 1989. Elements of recovery: An inventory of upslope sources of sedimentation in the Mattole River watershed. Report prepared by the MRC for the California Department of Fish and Game, Sacramento, CA.

Mueller-Dombois, D., and H. Ellenberg. 1974. *Aims and methods of vegetation ecology.* New York, NY: John Wiley and Sons.

Muldavin, E. H., J. M. Lenihan, W. S. Lennox, and S. D. Veirs. 1981. *Vegetation succession in the first ten years following logging of the coast redwood forests.* Redwood National Park Technical Report No. 6. Arcata, CA: USDI National Park Service.

Mulder, A. J., and C. J. M. de Waart. 1984. Preliminary architectural study of coastal redwood *(Sequoia sempervirens).* Paper of the Department of Silviculture, Agricultural University, Wageningen, The Netherlands.

Mulholland, P. J. 1992. Regulation of nutrient concentrations in a temperature forest stream: Roles of upland, riparian, and instream processes. *Limnology and Oceanography* 37:1512–1526.

Murphy, M. L., and J. D. Hall. 1981. Varied effects of clear-cut logging on predators and their habitats in small streams of the Cascade Mountains, Oregon. *Canadian Journal of Fisheries and Aquatic Sciences* 38:137–145.

Murphy, M. L., and W. R. Meehan. 1991. Stream ecosystems. *American Fisheries Society Special Publication* 19:17–46.

Murphy, M. L., C. P. Hawkins, and N. H. Anderson. 1981. Effects of canopy modification and accumulated sediment on stream communities. *American Fisheries Society* 110:469–478.

Myers, J. N. 1968. Fog. *Scientific American* (December): 37–44.

Myers, N. 1988. Threatened biotas: "Hot spots" in tropical forests. *Environmentalist* 8:187–208.

Naiman, R. J., and R. E. Bilby. 1998. *River ecology and management: Lessons from the Pacific coastal ecoregion.* New York, NY: Springer Verlag.

Naiman, R. J., and H. Decamps. 1997. The ecology of interfaces: Riparian zones. *Annual Review of Ecology and Systematics* 28:621–658.

Naiman, R. J., T. J. Beechie, L. E. Benda, D. R. Berg, P. A. Bisson, L. H. McDonald, M. D. O'Connor, P. L. Olson, and E. A. Steel. 1992. Fundamental elements of ecologically healthy watersheds in the Pacific Northwest coastal ecoregion. Pp. 127–188 in R. J. Naiman, ed., *Watershed management: Balancing sustainability and environmental change.* New York, NY: Springer-Verlag.

Naiman, R. J., K. L. Fetherston, S. J. McKay, and J. Chen. 1998. Riparian forests. Pp. 289–323 in R. J. Naiman and R. E. Bilby, eds., *River ecology and management: Lessons from the Pacific coastal ecoregion.* New York, NY: Springer-Verlag.

Nakamoto, R. J. 1998. Effects of timber harvest on aquatic vertebrates and habitat in the North Fork Caspar Creek. Pp. 95–104 in R. Ziemer, ed., *Proceedings of the conference on coastal watersheds: The Caspar Creek story.* USDA FS PS-GTR 165.

Napolitano, M. 1998. Persistence of historical logging impacts on channel form in mainstem North Fork Caspar Creek. Pp. 105–109 in R. Ziemer, ed., *Proceedings of the conference on coastal watersheds: The Caspar Creek story.* USDA FS PS-GTR 165.

Naveh, Z., and A. S. Lieberman. 1990. *Landscape ecology: Theory and application.* New York, NY: Springer-Verlag.

Neal, R. L. 1967. *Sprouting of old-growth redwood stumps, first year after logging.* USDA Forest Service, Research Note PSW-137. Southwest Forest and Range Experiment Station. Berkeley, CA: USDA Forest Service.

Nelson, S. K. 1997. Marbled murrelet *(Brachyramphus marmoratus).* Pp. 1–32 in A. Poole and F. Gill, eds., *Birds of North America.* No. 276, The Academy of Natural Sciences, Philadelphia, PA; Washington, DC: The American Ornithologists Union.

Nelson, S. K., and T. E. Hamer. 1995a. Nesting biology and behavior of the marbled murrelet. Pp. 57–68 in C. J. Ralph, G. L. Hunt Jr., J. F. Piatt, and M. G. Raphael, eds., *Ecology and conservation of the marbled murrelet.* General Technical Report PSW-GTR-152. Albany, CA: USDA Forest Service.

———. 1995b. Nest success and the effects of predation on marbled murrelets. Pp. 89–98 in C. J. Ralph, G. L. Hunt Jr., J. F. Piatt, and M. G. Raphael, eds., *Ecology and conservation of the marbled murrelet.* General Technical Report PSW-GTR-152, Albany, CA: USDA Forest Service.

Newbold, J. D., P. J. Mullholland, J. W. Elwood, and R. V. O'Neill. 1982. Organic carbon spiralling in stream ecosystems. *Oikos* 38:266–272.

Newman, E. I., and P. Reddell. 1987. The distribution of mycorrhizas among families of vascular plants. *New Phytologist* 106:745–751.

Nives, S. L. 1989. Fire behavior on the forest floor in coastal redwood forests, Redwood National Park. Master's thesis, Humboldt State University, Arcata, CA.

NMFS (National Marine Fisheries Service). 1998. See listing at <http://www.nwr.noaa.gov>.

NOAA. 1997. National Weather Service. Climatic data, thirty-year normals. See listing at <http://www.wrcc.dri.edu/summary>.

Nolan, K. M., and D. C. Marron. 1995. History, causes and significance of changes in the channel geometry of Redwood Creek, northwestern California. Pp. N1–N22 in

K. M. Nolan, H. M. Kelsey, and D. C. Marron, eds., *Geomorphic processes and aquatic habitat in the Redwood Creek Basin, northwestern California.* U.S. Geological Survey Professional Paper 1454. Washington, DC: U.S. Government Printing Office.

Noon, B. R., and K. S. McKelvey. 1996. Management of the spotted owl: A case history in conservation biology. *Annual Review of Ecological Systematics* 27:135–162.

Noon, B. R., and D. D. Murphy. 1997. Management of the spotted owl: The interaction of science, policy, politics, and litigation. Pp. 432–441 in G. K. Meffe and C. R. Carroll, eds., *Principles of conservation biology.* 2d ed. Sunderland, MA: Sinauer.

Norse, E. A. 1990. *Ancient forests of the Pacific Northwest.* Washington, DC: Island Press.

Noss, R. F. 1983. A regional landscape approach to maintain diversity. *BioScience* 33:700–706.

———. 1990. Indicators for monitoring biodiversity: A hierarchical approach. *Conservation Biology* 4:355–364.

———. 1992. The Wildlands Project: Land conservation strategy. *Wild Earth* (Special Issue):10–25.

———. 1993. Sustainable forestry or sustainable forests? Pp. 17–43 in G. H. Aplet, N. Johnson, J. T. Olson, and V. A. Sample, eds., *Defining sustainable forestry.* Washington, DC: Island Press.

———. 1998. A big-picture approach to forest certification: A report for World Wildlife Fund's Forest for Life Campaign in North America. In K. Kessler, D. A. DellaSala, and A. Hackman, eds., *Defining a forest vision: World Wildlife Fund's North American Forests for Life Conference.* Washington, DC: World Wildlife Fund.

———. 1999. Assessing and monitoring forest biodiversity: A suggested framework and indicators. *Forest Ecology and Management,* 115:135–146.

Noss, R. F., and A. Cooperrider. 1994. *Saving nature's legacy: Protecting and restoring biodiversity.* Washington, DC: Island Press.

Noss, R. F., and B. Csuti. 1997. Habitat fragmentation. Pp. 269–304 in G. K. Meffe and R. C. Carroll, eds., *Principles of conservation biology.* 2d ed. Sunderland, MA: Sinauer Associates.

Noss, R., E. Dinerstein, B. Gilbert, M. Gilpin, B. Miller, J. Terborgh, and S. Trombulak. 1999. Core areas: Where nature reigns. Pp. 99–128 n M. E. Soulé and J. Terborgh, eds., *Continental conservation: Scientific foundations of regional reserve networks.* Washington, DC: Island Press.

Noss, R. F., and L. D. Harris. 1986. Nodes, networks, and MUM's: Preserving diversity at all scales. *Environmental Management* 10:299–309.

Noss, R. F., E. T. LaRoe, and J. M. Scott. 1995. *Endangered ecosystems of the United States: A preliminary assessment of loss and degradation.* Biological Report 28. Washington, DC: USDI National Biological Service.

Noss, R. F., M. A. O'Connell, and D. D. Murphy. 1997. *The science of conservation planning: Habitat conservation under the Endangered Species Act.* Washington, DC: Island Press.

Noss, R. F., and R. L. Peters. 1995. *Endangered ecosystems of the United States: A status report and plan for action.* Washington, DC: Defenders of Wildlife.

Noss, R. F., H. B. Quigley, M. G. Hornocker, T. Merrill, and P. C. Paquet. 1996. Conservation biology and carnivore conservation in the Rocky Mountains. *Conservation Biology* 10:949–963.

Noss, R. F., J. R. Strittholt, K. Vance Borland, C. Carroll, and P. Frost. 1999. A conservation plan for the Klamath-Siskiyou ecoregion. *Natural Areas Journal.* 19: in press.

NRC (National Research Council). 1996. *Upstream: Salmon and society in the Pacific Northwest.* Washington DC: National Academy Press.

NRCS (Natural Resources Conservation Service). 1998a. *Official soil series descriptions.* USDA RCS, Soil Survey Division. Ames, IA: Iowa State University. <http://www.statlab.iastate.edu/cgi-bin/osd/osdname.cgi>

————. 1998b. Research data, USDA Natural Resources Conservation Service, Soil Survey Division, National Soil Survey Center. Lincoln, NE. <http://vmhost.cdp.state.ne.us/~nslsoil/soil.html>

Nussbaum, R. A. 1969. The nest of the Olympic salamander, *Rhyacotriton olympicus.* *Herpetologica* 25:277–278.

Oberlander, G. T. 1956. Summer fog precipitation on the San Francisco peninsula. *Ecology* 37:851–852.

O'Dell, T. E. 1996. Silviculture in the redwood region: An historical perspective. Pp. 15–17 in *Proceedings of the conference on coast redwood forest ecology and management.* June 18–20, 1996. Arcata, CA: Humboldt State University.

O'Dell, T. E., J. E. Smith, M. Castellano, and D. Luoma. 1996. Diversity and conservation of forest fungi. Pp. 19–23 in D. Pilz and R. Molina, eds., *Managing forest ecosystems to conserve fungus diversity and sustain wild mushroom harvests.* Pacific Northwest Station Technical Report PNW-GTR-371. Portland, OR: USDA Forest Service.

Old-Growth Definition Task Group. 1986. *Interim definitions for old-growth Douglas-fir and mixed conifer forests in the Pacific Northwest and California.* PNW-447. Portland, OR: USDA Forest Service.

Olive, S. P., F. S. Bowcutt, S. Bakken, W. J. Barry. 1982. Plant life. Pp. 1–71 in *Inventory of features: Jedediah Smith Redwoods State Park, Del Norte Redwoods State Park and Prairie Creek Redwoods State Park.* California State Redwood Parks General Plan. Resource Protection Division, Natural Heritage Section, Cultural Heritage Planning Unit, Sacramento, CA.

Oliver, W. W., J. L. Lindquist, and R. O. Strothmann. 1996. Response to thinning young-growth coast redwood stands. In *Proceedings of the conference on coast redwood forest ecology and management.* June 18–20, 1996. Arcata, CA: Humboldt State University.

Olmsted, Frederick Law. 1865. *Yosemite and the Mariposa Grove: A Preliminary Report, 1865.* Reprinted by The Yosemite Association, Yosemite National Park, 1995.

Olson, D. F., D. F. Roy, and G. A. Walters. 1990. *Sequoia sempervirens* (D. Don) Endl. Pp. 541–551 in R. M. Barnes and B. H. Honkala, eds., *Silvics of North America.* Vol. 1, *Conifers.* Agriculture Handbook 654. Washington, DC: USDA Forest Service.

Olson, D. M. 1992. The northern spotted owl conservation strategy: Implications for Pacific Northwest invertebrates and associated ecosystem processes. Final Report prepared for the Northern Spotted Owl EIS Team. USDA Forest Service, Order number 40-04HI-2-1650.

Omernik, J. M., and A. L. Gallant, eds. 1986. *Ecoregions of the Pacific Northwest.* EPA/600/3-86/033. Corvallis, OR: U.S. Environmental Protection Agency.

Ornduff, R. 1998. The *Sequoia sempervirens* (coast redwood) forest of the Pacific Coast, USA. Pp. 221–236 in A. D. Laderman ed., *Coastally restricted forests*. New York, NY: Oxford University Press.

Pacific Lumber Company. 1998. Sustained Yield Plan/Habitat Conservation Plan. Unpublished draft.

Pacific Watershed Associates. 1998. Sediment source investigation and sediment reduction plan for the Bear Creek watershed, Humboldt County, California. Report Prepared for the Pacific Lumber Company, Scotia, CA.

Packee, E. C. 1990. *Tsuga heterophylla* (Raf.) Sarg. Pp. 613–622 in R. M. Barnes and B. H. Honkala, eds., *Silvics of North America*. Vol. 1, *Conifers*. Agriculture Handbook 654. Washington, DC: USDA Forest Service.

Packer, W. C. 1960. Bioclimatic influences on the breeding of *Taricha rivularis*. *Ecology* 41:509–517.

Page, W. D., D. R. Packer, and T. Stephens. 1982. Speculations on the ages of marine terraces and Quaternary marine deposits, Trinidad, California. P. 26 in D. R. Harden, D. C. Marron, and A. MacDonald, eds., *Late Cenozoic history and forest geomorphology of Humboldt County, California: Friends of the Pleistocene Pacific Cell guidebook*, guidebook supplement.

Parker, M. S. 1994. Feeding ecology of stream-dwelling Pacific giant salamander larvae *(Dicamptodon tenebrosus)*. *Copeia* 1994:705–718.

Parsons, J. J. 1960. "Fog-drip" from coastal stratus, with special reference to California. *Weather* 15:58–62.

Peters, M. D., and D. C. Christophel. 1978. *Austrosequoia wintonensis*, a new taxodiaceous cone from Queensland, Australia. *Canadian Journal of Botany* 56:3119-3128.

Peterson, R. T. 1961. *A field guide to western birds*. Boston, MA: Houghton Mifflin.

Petranka, J. W. 1998. *Salamanders of the United States and Canada*. Washington, DC: Smithsonian Institution Press.

Phillips, W., and H. W. Harkness. 1884–1886. Fungi of California. *Bulletin of the California Academy of Science* 1:21–26.

Pickett, S. T. A., and J. Kolasa. 1989. Structure of theory in vegetation science. *Vegetatio* 83:7–15.

Pickett, S. T. A., and P. S. White, eds. 1985. *The ecology of natural disturbance and patch dynamics*. Orlando, FL: Academic Press.

Piirto, D., R. P. Thompson, and K. L. Piper. 1996. Implementing uneven-aged redwood management at Cal Poly's school forest. Pp. 78–82 in *Proceedings of the conference on coast redwood forest ecology and management*. June 18–20, 1996. Arcata, CA: Humboldt State University.

Pike, L. H., W. C. Denison, D. Tracy, M. Sherwood, and F. Rhoades. 1975. Floristic survey of epiphytic lichens and bryophytes growing on old-growth conifers in western Oregon. *Bryologist* 78:389–402.

Pike, L. H., R. A. Rydell, and W. C. Denison. 1977. A 400-year-old Douglas-fir tree and its epiphytes: Biomass, surface area, and their distributions. *Canadian Journal of Forest Research* 7:680–699.

Pillars, M. D., and J. D. Stuart. 1993. Leaf-little accretion and decomposition in interior and coastal old-growth redwood stands. *Canadian Journal of Forest Research* 23:552–557.

Pockman, W. T., J. S. Sperry, and J. W. O'Leary. 1995. Sustained and significant negative water pressure in xylem. *Nature* 378:715–716.

Popenoe, J. H. 1987. *Soil series descriptions and laboratory data from Redwood National Park.* Technical Report 20. Orick, CA: Redwood National Park.

Popenoe, J. H., K. A. Bevis, B. R. Gordon, N. K. Sturhan, and D. L. Hauxwell. 1992. Soil-vegetation relationships in Franciscan terrain of northwestern California. *Journal of the Soil Science Society of America* 56:1951–1959.

Porter, S. C. 1983. Introduction. Pp. xi–xiv in H. E. Wright Jr., ed., *Late Quaternary environments of the United States.* Vol. 1, *The late Pleistocene.* Minneapolis: University of Minnesota Press.

Powell, J. A., and C. L. Hogue. 1979. *California insects.* Berkeley: University of California Press.

Powell, R. A., and W. J. Zielinski. 1994. The fisher. Pp. 38–73 in L. F. Ruggiero, K. B. Aubry, S. W. Buskirk, L. J. Lyon, and W. J. Zielinski, tech. eds., *The scientific basis for conserving forest carnivores: American marten, fisher, lynx and wolverine in the western United States.* General Technical Report RM-254, Rocky Mountain Forest and Range Experiment Station. Fort Collins, CO: USDA Forest Service.

Power, M. E. 1991. Shifts in the effects of tuft-weaving midges on filamentous algae. *American Midland Naturalist* 125:275–285.

———. 1992. Hydrologic and trophic controls of seasonal blooms in northern California rivers. *Archive fur Hydrobiologie* 125:385–410.

Power, M. E., D. Tilman, J. A. Estes, B. A. Menge, W. J. Bond, L. S. Mills, G. Daily, J. C. Castilla, J. Lubchenco, and R. T. Paine. 1996. Challenges in the quest for keystones. *Conservation Biology* 46:609–620.

Powers, R. F., and H. V. Wiant. 1970. Sprouting of old-growth coastal redwood stumps. *Forest Science* 16:339–341.

Pressey, R. L., C. J. Humphries, C. R. Margules, R. I. Vane-Wright, and P. H. Williams. 1993. Beyond opportunism: Key principles for systematic reserve selection. *Trends in Ecology and Evolution* 8:124–128.

Pritchett, W. L. 1979. *Properties and management of forest soils.* New York, NY: John Wiley and Sons.

Ralph, C. J., and S. L. Miller. 1995. Offshore population estimates of marbled murrelets in California. Pp. 353–360 in C. J. Ralph, G. L. Hunt Jr., J. F. Piatt, and M. G. Raphael, eds., *Ecology and conservation of the marbled murrelet.* General Technical Report PSW-GTR-152, Albany, CA: USDA Forest Service.

Randall, J., M. Rejmanek, and J. Hunter. 1998. Characteristics of the exotic flora of California. *Fremontia* 26(4):

Raphael, M. 1988. Long-term trends in abundance of amphibians, reptiles, and mammals in Douglas-fir forests of northwest California. Pp. 23–31 in R. C. Szaro, K. E. Severson, and D. R. Patton, eds., *Management of amphibians, reptiles, and small mammals in North America.* General Technical Report RM-166. Rocky Mountain Forest and Range Experiment Station. Flagstaff, AZ: USDA Forest Service.

Raphael, M. G., and B. G. Marcot. 1986. Validation of a wildlife-habitat-relationship model: Vertebrates in a Douglas-fir sere. Pp. 129–138 in J. Verner, M. L. Morrison, and C. J. Ralph, eds., *Wildlife 2000: Modeling habitat relationships of terrestrial vertebrates*. Madison, WI: University of Wisconsin Press.

Raven, P. H., and D. I. Axelrod. 1978. *Origin and relationships of the California flora*. University of California Publications in Botany, vol. 72. Berkeley: University of California Press.

Raymo, M. E., and W. F. Ruddiman. 1992. Tectonic forcing of late Cenozoic climate. *Nature* 359:117–122.

Reese, D. A., and H. H. Welsh Jr. 1997. Utilization of terrestrial habitat by western pond turtles *(Clemmys marmorata)*: Implications for management. Pp. 352–357 in J. Van Abbema, ed., *Proceedings: Conservation, restoration, and management of tortoises and turtles: An international conference*. New York, NY: New York Turtle and Tortoise Society.

———. 1998. Habitat use by western pond turtles in the Trinity River, California. *Journal of Wildlife Management* 62:842–853.

Reeves, G. H., L. E. Benda, K. M. Burnett, P. A. Bisson, and J. R. Sedell. 1995. A disturbance-based ecosystem approach to maintaining and restoring fresh-water habitats of evolutionarily significant units of anadromous salmonids in the Pacific Northwest. *American Fisheries Society Symposium* 17:334–349.

Reeves, G. H., P. A. Bisson, and J. M. Dambacher. 1998. Fish communities. Chap. 9 in R. J. Naiman and R. E. Bilby, eds., *River Ecology and management: Lesson from the Pacific coastal ecoregion*. New York: Springer Verlag.

Regier, H. A. 1978. *A balanced science of renewable resources with particular reference to fisheries*. Seattle: University of Washington Press.

Reid, L. M., and S. Hilton. 1998. Buffering the buffer. Pp. 75–85 in R. Ziemer, ed., *Proceedings of the conference on coastal watersheds: The Caspar Creek story*. USDA FS PS-GTR 165.

Resh, V. H., et al. 1988. The role of disturbance in stream ecology. *Journal of the North American Benthological Society* 7:433–455.

Ricketts, T. H., E. Dinerstein, D. M. Olson, C. J. Loucks, W. M. Eichbaum, D. A. DellaSala, K. C. Kavanagh, P. Hedao, P. T. Hurley, K. M. Carney, R. A. Abell, and S. Walters. 1999. *A conservation assessment of the terrestrial ecoregions of North America*. Vol. 1, *The United States and Canada*. Washington, DC: Island Press.

Roelofs, T. D. 1983. Current status of California summer steelhead *(Salmo gairderni)* stocks and habitat, and recommendations for their management. Report to USDA Forest Service Region 5.

Rogers, D. L. 1994. Spatial patterns of allozyme variation and clonal structure in coast redwood *(Sequoia sempervirens)*. Ph.D. diss., University of California, Berkeley.

Rollinger, J. L., and J. H. Langenheim. 1993. Geographic survey of fungal endophyte community composition in leaves of coast redwood. *Mycologia* 85:149–156.

Rosenberg, K. V., and R. G. Raphael. 1986. Effects of forest fragmentation on vertebrates in Douglas-fir forests. Pp. 263–272 in J. Verner, M. L. Morrison, C. J. Ralph, eds., *Wildlife 2000: Modeling habitat relationships of terrestrial vertebrates*. Madison: University of Wisconsin Press.

Rossman, A. Y. 1994. A strategy for an all-taxa inventory of fungal biodiversity. Pp. 169–194 in C.-I. Peng and C. H. Chou eds., *Biodiversity and terrestrial ecosystems*. Institute of Botany, Academia Sinica Monograph Series No. 14. Taipei.

Rossman, D. A., N. B. Ford, and R. A. Seigel. 1996. *The garter snakes: Evolution and ecology.* Norman: University of Oklahoma Press.

Roth, R. R. 1976. Spatial heterogeneity and bird species diversity. *Ecology* 57:773–782.

Rowe, R. E., and K. K. Ching. 1973. Provenance study of Douglas-fir in the Pacific Northwest Region II. Field performance at age nine. *Silvae Genetica* 22:115–119.

Roy, D. F., 1966. *Silvical characteristics of redwood.* Research Paper PSW-28, Pacific Southwest Forest and Range Experiment Station. Berkeley, CA: USDA Forest Service.

———. 1972. Fasciation in redwood. *Madroño* 21:7.

Ruddiman, W. F., M. E. Raymo, W. L. Prell, and J. E. Kutzbach. 1997. The uplift-climate connection: A synthesis. Pp. 471–515 in W. F. Ruddiman, ed., *Tectonic uplift and climate change.* New York, NY: Plenum Press.

Ruggiero, L. F., K. B. Aubry, A. B. Carey, and M. H. Huff, tech. coordinators. 1991. *Wildlife and vegetation of unmanaged Douglas-fir forests.* General Technical Report PNW-GTR-285. Portland, OR: USDA Forest Service.

Ruggiero, L. F., K. B. Aubry, S. W. Buskirk, L. J. Lyon, and W. J. Zielinski. 1994. *American marten, fisher, lynx, and wolverine in the western United States.* General Technical Report RM-254. Fort Collins, CO: USDA Forest Service.

Russell, E. W. 1983. Pollen analysis of past vegetation at Point Reyes National Seashore, California. *Madroño* 30:1–11.

Russell, W. H., J. R. McBride, and K. Carnell. In press. Edge effects and the effective size of old-growth coast redwood preserves. In D. N. Cole and S. F. McCool, eds., *Proceedings: Wilderness science in a time of change.* Ogden, UT: USDA Forest Service.

Ryan, R. G., and B. J. Yoder. 1997. Hydraulic limits to tree height and tree growth. *BioScience* 47:235–242.

Rydelius, J. A., and W. J. Libby. 1993. Arguments for redwood clonal forestry. Pp. 159–168 in M. R. Ahuja and W. J. Libby, eds., *Clonal forestry II. Conservation and application.* Heidelberg: Springer Verlag.

Rypins, S., S. L. Reneau, Steven R. Byrne, and D. R. Montgomery. 1989. Palynologic and geomorphic evidence for environmental changes during the Pleistocene-Holocene transition at Point Reyes Peninsula, central coastal California. *Quaternary Research* 32:72–87.

Sakai, H. P., and B. R. Noon. 1993. Dusky-footed woodrat abundance in different-aged forests in northwestern California. *Journal of Wildlife Management* 373–382.

Sampson, N. 1998. Private forests: More owners, fewer acres. *American Forests* (Autumn).

Samways, M. J. 1994. *Insect conservation biology.* London, UK: Chapman and Hall.

Sandek, D. 1984. The effects of brush control on soil moisture and soil temperature in coastal Douglas-fir plantations. Master's thesis, Humboldt State University, Arcata, CA.

Sandercock, R. K. 1991. Life history of coho salmon *(Oncorhynchus kisutch).* Pp. 395–445 in C. Groot and L. Margolis, eds., *Pacific salmon life histories.* Vancouver: University of British Columbia Press.

Saunders, D. A., R. J. Hobbs, and C. R. Margules. 1991. Biological consequences of ecosystem fragmentation: A review. *Conservation Biology* 5:18–32.

Savage, J. M. 1960. Evolution of a peninsular herpetofauna. *Systematic Zoology* 9(3–4): 184–212.

Save-the-Redwoods League. Summary of Acquisitions, 1918–1996. Unpublished report.

Sawyer, J. O. 1996. Northern California. Pp. 20–42 in R. Kirk, ed., *The enduring forests.* Seattle, WA: The Mountaineers.

Sawyer, J. O., and T. Keeler-Wolf. 1995. *A manual of California vegetation.* Sacramento, CA: California Native Plant Society Press.

Saylor, L. C., and H. A. Simons. 1970. Karyology of Sequoia sempervirens: Karyotype and accessory chromosomes. *Cytologia* 35:294–303.

Schloemer-Jager, A. 1958. Altertertiare Pflanzen aus Flozen der Brogger-Hulbinsel Spitzbergen. *Palaeontographica Abteilung B* 104:39–103.

Schlosser, I. J. 1991. Stream fish ecology: A landscape perspective. *BioScience* 41:704–712.

Schowalter, T. D. 1989. Canopy arthropod community structure and herbivory in old-growth and regenerating forests in western Oregon. *Canadian Journal of Forest Research* 19:318–322.

———. 1990. Invertebrate diversity in old-growth versus regenerating forest. *Northwest Environmental Journal* 6:403–404.

Schowalter, T. D., and D. A. Crossley Jr. 1987. Canopy arthropods and their response to forest disturbance. Pp. 201–218 in D. A. Crossley Jr., and W. T. Swank, eds., *Forest hydrology and ecology at Coweeta.* New York, NY: Springer-Verlag.

Schrepfer, S. R. 1983. *The fight to save the redwoods: A history of environmental reform, 1917–1978.* Madison: University of Wisconsin Press.

Schumaker, N. 1996. Using landscape indices to predict habitat connectivity. *Ecology* 77:1210–1225.

Schwartz, D. L., H. T. Mullins, and D. F. Belknap. 1986. Holocene geologic history of a transform margin estuary: Elkhorn Slough, central California. *Estuarine, Coastal, and Shelf Science* 22:285–302.

Schweitzer, H. J. 1974. Die "Tertiaren" koniferen Spitzbergens. *Palaeontographica Abt. B* 149, Lfg. 1-4:1–89.

———. 1980. Environment and climate in the early Tertiary of Spitsbergen. *Palaeogeography, Palaeoclimatology, and Palaeoecology* 30:297–311.

Scott, J. M., F. Davis, B. Csuti, R. Noss, B. Butterfield, C. Groves, J. Anderson, S. Caicco, F. D'Erchia, T. C. Edwards, J. Ulliman, and R. G. Wright. 1993. Gap analysis: A geographical approach to protection of biological diversity. *Wildlife Monographs* 123:1–41.

Scott, R., and A. L. Kroeber. 1942. Yurok narratives. *American Archaeology and Ethnology* 35:143–256.

Sedell, J. R., P. A. Bisson, F. J. Swanson, and S. V. Gregory. 1988. What we know about large trees that fall into streams and rivers. Pp. 47–81 in C. Maser, R. F. Tarrant, J. M. Trappe, and J. F. Franklin, eds., *From the forest to the sea: A story of fallen trees.* Pacific Northwest Research Station PNW-GTR-229. Portland, OR: USDA Forest Service.

Sedell, J. R., G. H. Reeves, and P. A. Bisson. 1997. Habitat policy for salmon in the Pacific Northwest. Pp. 375–387 in D. J. Stouder, P. A. Bisson, and R. J. Naiman, eds., *Pacific salmon and their ecosystems: Status and future options.* New York, NY: Chapman and Hall.

Sedell, J. R., G. H. Reeves, and K. M. Burnett. 1994. Development and evaluation of aquatic conservation strategies. *Journal of Forestry* 92(4):28–31.

Semlitsch, R. D., D. E. Scott, J. H. K. Pechmann, and J. W. Gibbons. 1993. Phenotypic variation in the arrival time of breeding salamanders: Individual repeatibility and environmental influences. *Journal of Animal Ecology* 62:334–340.

———. 1996. Structure and dynamics of an amphibian community: Evidence from a sixteen-year study of a natural pond. Pp. 217–248 in M. L. Cody and J. A. Smallwood, eds., *Long-term studies of vertebrate communities.* San Diego, CA: Academic Press.

Shaffer, M. L. 1981. Minimum population sizes for species conservation. *BioScience* 31:131–134.

Shantz, H. L., and R. Zon. 1924. *Natural vegetation.* Atlas of American Agriculture, Pt. 1, Section E. Washington, DC: U.S. Department of Agriculture.

Shapovalov, L., and A. C. Taft. 1954. The life histories of the steelhead rainbow trout *(Salmo gairdneri gairdneri)* and silver salmon *(Oncorhynchus kisutch)* with special reference to Waddell Creek, California, and recommendations regarding their management. California Department of Fish and Game Fish Bulletin No. 98.

Sholars, T. 1998. Checklist of redwood forest lichens. Vouchered at the College of the Redwoods Herbarium, Fort Bragg, CA.

Sidle, R. C., A. J. Pearce, and C. L. O'Loughlin. 1985. *Hillslope stability and land use.* Water Resources Monograph Series, vol. 11. Washington, DC: American Geophysical Union.

Sillett, S. C. 1995. Branch epiphyte assemblages in the forest interior and on the clearcut edge of a 700-year-old Douglas fir canopy in western Oregon. *Bryologist* 98:301–312.

———. 1999. Tree crown structure and vascular epiphyte distribution in Sequoia sempervirens rain forest canopies. *Selbyana,* in press.

Silverton, J. W., and J. Lovett Doust. 1993. *Introduction to plant population ecology.* Oxford, UK: Blackwell Scientific Publications.

Simmons, N. 1973. Description of bud collars on redwood seedlings. Master's thesis, Humboldt State University, Arcata, CA.

Singer, S. W., N. L. Naslund, S. A. Singer, and C. J. Ralph. 1991. Discovery and observations of two tree nests of the marbled murrelet. *Condor* 93:330–339.

Singer, S. W., D. L. Suddjian, and S. A. Singer. 1995. Fledging behavior, flight patterns and habitat characteristics of marbled murrelet tree nests in California. Pp. 54–62 in S. K. Nelson and S. G. Sealy, eds., Biology of the marbled murrelet: Inland and at sea. *Northwestern Naturalist* 76.

Skinner, M. W., and B. M. Pavlik. 1994. *California Native Plant Society inventory of rare and endangered vascular plants of California.* 5th edition. Sacramento, CA: California Native Plant Society.

Small, A. 1974. *The birds of California.* New York, NY: Collier Books.

Smiley, C. J. 1963. The Ellensburg flora of Washington. *University of California Publications in Geological Sciences* 35(3): 159–276.

———. 1966. Cretaceous floras from Kuk River area, Alaska: Stratigraphic and climatic interpretations. *Geological Society of America Bulletin* 77:1–14.

——— 1969. Cretaceous floras of Chandler-Colville region, Alaska: Stratigraphy and preliminary floristics. *American Association of Petroleum Geologists Bulletin* 53:482–502.

Smiley, C. J., and W. C. Rember. 1985. Composition of the Miocene Clarkia flora. Pp.

95–112 in C. J. Smiley, ed., *Late Cenozoic history of the Pacific Northwest: Interdisciplinary studies on the Clarkia fossil beds of northern Idaho*. San Francisco, CA: Pacific Division of the American Association for the Advancement of Science.

Smith, A. H. 1947. North American species of *Mycena. University of Michigan Studies, Sciences Series* 17:1–521.

———. 1951. The North American species of *Naemataloma. Mycologia* 43:467–521.

———. 1952. New and rare agarics from the Douglas Lake Region and Tahquamenon Falls State Park, Michigan, and an account of the North American species of *Xeromphalina. Papers of the Michigan Academy of Science* 38:53–87.

Smith, H. M. 1978. *Amphibians of North America*. New York: Golden Press.

Smith, K. G., and D. G. Catanzaro. 1996. Predicting vertebrate distributions for gap analysis: Potential problems in constructing the models. Pp. 163–169 in J. M. Scott, T. H. Tear, and F. W. Davis, eds., *Gap analysis: A landscape approach to biodiversity planning*. Bethesda, MD: American Society for Photogrammetry and Remote Sensing.

Smith, S. E., and D. J. Read. 1997. *Mycorrhizal symbiosis*. 2d ed. San Diego, CA: Academic Press.

Snyder, J. 1925. The half-pounder of Eel River, a steelhead trout. *California Fish and Game* 11(2):49–55.

Snyder, J. A. 1992. The ecology of *Sequoia sempervirens:* An addendum to *On the edge: Nature's last stand for coast redwoods*. Master's thesis, San Jose State University, San Jose, CA.

Soil Survey Staff. 1998. *Keys to soil taxonomy*. 8th ed. Washington, DC: U.S. Government Printing Office.

Sokal, R. R. 1974. Classification: Processes, principles, progress, prospects. *Science* 185:1115–1123.

Soulé, M. E., ed. 1987. *Viable populations for conservation*. Cambridge, UK: Cambridge University Press.

Southwood, T. R. E. 1977. Habitat, the templet for ecological strategies? *Journal of Animal Ecology* 46:337–365.

Sowers, T., and M. Bender. 1995. Climate records covering the last deglaciation. *Science* 269:210–214.

Spence, B. C., G. A. Lomnicky, R. M. Hughes, and R. P. Novitzki. 1996. *An ecosystem approach to salmonid conservation*. TR-4501-96-6057. Corvallis, Oregon: ManTech Environmental Research Services Corp.

Sperry, J. S., N. Z. Saliendra, W. T. Pockman, H. Cochard, P. Cruziat, S. D. Davis, F. W. Ewers, and M. T. Tyree. 1996. New evidence for large negative xylem pressures and their measurement by the pressure chamber method. *Plant, Cell and Environment* 19:427–436.

Spies, T. 1997. Forest stand structure, composition, and function. Pp. 11–30 in K. A. Kohm and J. F. Franklin, eds., *Creating a forestry for the twenty-first century*. Washington, DC: Island Press.

Srinivasan, V., and E. M. Friis. 1989. Taxodiaceous conifers from the Upper Cretaceous of Sweden. Det Kongelige Danske Videnskabernes Selskab (The Royal Danish Academy of Sciences and Letters), *Biologiske Skrifter* 35:1–57.

Stanford, J. A., and J. V. Ward. 1988. The hyporheic habitat of river ecosystems. *Nature* 335:64–66.

————. 1992. Management of aquatic resources in large catchments: Recognizing interactions between ecosystem connectivity and environmental disturbance. Pp. 91–124 in R. J. Naiman, ed., *Watershed management: Balancing sustainability and environmental change*. New York, NY: Springer-Verlag.

Stebbins, G. L., and J. Major. 1965. Endemism and speciation in the California flora. *Ecological Monographs* 35:1–35.

Stebbins, R. C. 1985. *A field guide to western reptiles and amphibians*. Boston, MA: Houghton Mifflin.

Steele, D. T. 1986. *A review of the population status of the Point Arena mountain beaver* (Aplodontia rufa nigra). Final report 10188-5671-5, Endangered Species Office. Sacramento, CA: U.S. Fish and Wildlife Service.

————. 1989. *An ecological survey of endemic mountain beavers* (Aplodontia rufa) *in California, 1979–1983*. Administrative Report 89–1. Sacramento, CA: California Department of Fish and Game.

Stein, W. I. 1990. *Umbellularia californica* (Hook and Arn.) Nutt. Pp. 826–834 in R. M. Barnes and B. H. Honkala, eds., *Silvics of North America*. Vol. 2, *Hardwoods*. Agriculture Handbook 654. Washington, DC: USDA Forest Service.

Stephens, F. 1906. *California mammals*. San Diego, CA: West Coast Publishing Co.

Stewart, W. N. 1983. *Paleobotany and the evolution of plants*. Cambridge, UK: Cambridge University Press.

Stine, S. 1994. Extreme and persistent drought in California and Patagonia during mediaeval time. *Nature* 369:546–549.

Stone, E. C. 1957. Dew as an ecological factor. I. A review of the literature. *Ecology* 38:407–413.

Stone, E. C., and R. B. Vasey. 1968. Preservation of coast redwood on alluvial flats. *Science* 159:157–161.

Stouder, D. J., P. A. Bisson, and R. J. Naiman, eds. 1997. *Pacific salmon and their ecosystems*. New York, NY: Chapman and Hall.

Strahler, A. N. 1957. Quantitative analysis of watershed geomorphology. *Transactions of the American Geophysical Union* 38(6):913–920.

Strittholt, J. R., G. E. Heilman, and R. F. Noss. 1999. *A GIS-based model for assessing conservation focal areas for the redwood ecosystem*. Corvallis, OR: Conservation Biology Institute.

Stuart, J. D. 1987. Fire history of an old-growth forest of *Sequoia sempervirens* (Taxodiaceae) forest in Humboldt Redwoods State Park, CA. *Madroño* 34:128–141.

Sturtevant, B. R., and J. A. Bissonette. 1997. Stand structure and microtine abundance in Newfoundland: Implications for marten. Pp.182–198 in G. Proulx, H. N. Bryant, and P. M. Woodard, eds., *Martes: Taxonomy, ecology, techniques, and management*. Edmonton, Alberta: Provincial Museum of Alberta.

Sugihara, N. G. 1992. The role of treefall gaps and fallen trees in the dynamics of old growth coast redwood (*Sequoia sempervirens* [D. Don] Endl.) forests. Ph.D. diss. Berkeley: University of California.

————. 1996. The dynamics of treefall gaps in alluvial flat coast redwood (*Sequoia sempervirens* [D. Don] Endl.) forests. Pp. 94–95 in J. LeBlanc, ed., *Proceedings of the conference on coast redwood forest ecology and management*. University of California Cooperative Extension Forestry Publication. Arcata, CA: Humboldt State University.

Swanson, F. J., J. A. Jones, and G. E. Grant. 1997. The physical environment as a basis for managing ecosystems. Pp. 229–238 in K. A. Kohm and J. F. Franklin, eds., *Creating a forestry for the twenty-first century*. Washington DC: Island Press.

Tanai, T. 1967. Tertiary floral changes of Japan. Pp. 317–333 in *Jubilee publication commemorating Professor Sasa's sixtieth birthday* (reprinted 1972 in *Tertiary Floras of Japan*, vol. 2, 1972, Association of Paleobotanical Research in Japan).

Tappeiner, J. C., P. M. McDonald, D. F. Roy. 1990. *Lithocarpus densiflorus* (Hook and Arn.) Rehd. Pp. 414–425 in R. M. Barnes and B. H. Honkala eds., *Silvics of North America*. Vol. 2, *Hardwoods*. Agriculture Handbook 654. Washington, DC: USDA Forest Service.

Tappeiner, J. C., D. Lavender, J. Walstad, R. O. Curtis, and D. S. DeBell. 1997. Silvicultural systems and regeneration methods: Current practices and new alternatives. Pp. 151–164 in K. A. Kohm and J. F. Franklin, eds., *Creating a forestry for the twenty-first century*. Washington, DC: Island Press.

Taylor, A. A. 1912. *California Redwood Park*. Sacramento, CA: California Redwood Park Commission.

Thomas, J. W., E. D. Forsman, J. B. Lint, E. C. Meslow, B. R. Noon, and J. Verner. 1990. *A conservation strategy for the northern spotted owl*. USDI Bureau of Land Management, U.S. Fish and Wildlife Service, and National Park Service. Portland, OR: USDA Forest Service.

Thompson, I. D. 1986. Diet choice, hunting behaviour, activity patterns and ecological energetics of marten in natural and logged areas. Ph.D. diss., Queen's University, Kingston, Ontario.

Thompson, I. D., and P. W. Colgan. 1987. Numerical responses of martens to a food shortage in northcentral Ontario. *Journal of Wildlife Management* 51:824–835.

Thompson, I. D., and A. S. Harestad. 1994. Effects of logging on American martens with models for habitat management. Pp. 355–367 in S. W. Buskirk, A. S. Harestad, M. G. Raphael, and R. A. Powell, eds., *Martens, sables and fishers: Biology and conservation*. Ithaca, NY: Cornell Press.

Tidwell, W. D. 1998. *Common fossil plants of western North America*. 2d ed. Washington, DC: Smithsonian Institution Press.

Townsend, C. R. 1989. The patch dynamics concept of stream community ecology. *Journal of the North American Benthological Society* 8:36–50.

Trappe, J. P. 1998. List of *Glomus* species associated with coast redwood. Unpublished report. Corvallis, OR: Department of Botany, Oregon State University.

Triska, F. J., V. C. Kennedy, R. J. Avanzino, G. W. Zellweger, and K. E. Bencala. 1989a. Retention and transport of nutrients in a third-order stream in northwestern California: Hyporheic processes. *Ecology* 70:1893–1905.

———. 1989b. Retention and transport of nutrients in a third-order stream: Channel processes. *Ecology* 70:1877–1892.

Trombulak, S. C., and C. A. Frissell. 2000. Review of the ecological effects of roads on terrestrial and aquatic communities. *Conservation Biology* 14: in press.

Trotter, P. C. 1997. Sea-run cutthroat trout: Life history profile. Pp. 7–15 in J. D. Hall, P. A. Bisson, and R. E. Gresswell, eds., *Sea-run cutthroat trout: Biology, management, and future conservation*. Corvallis, OR: Oregon Chapter, American Fisheries Society.

Tuchmann, E., K. P. Connaughton, L. E. Freedman, and C. B. Moriwaki. 1996. *The Northwest Forest Plan: A report to the President and Congress.* Washington, DC: USDA Office of Forestry and Economic Assistance.

Tufuor, K. 1973. Comparative growth performance of seedlings and vegetative propagules of *Pinus radiata* and *Sequoia sempervirens.* Ph.D. diss., University of California, Berkeley.

Turner, M. G., and R. H. Gardner, eds. 1991. *Quantitative methods in landscape ecology.* New York, NY: Springer-Verlag.

Twining, H., and A. Hensley. 1947. The status of pine martens in California. *California Fish and Game* 33:133–137.

Twitty, V. C. 1964. *Taricha rivularis* (Twitty), red-bellied newt. *Catalogue of American Amphibians and Reptiles* 9.1–9.2.

Tyree, M. T., and J. S. Sperry. 1989. Vulnerability of xylem to cavitation and embolism. *Annual Review of Plant Physiology and Plant Molecular Biology* 40:19–38.

UNESCO. 1974. *Task force on criteria and guidelines for the choice and establishment of biosphere reserves.* Man and the Biosphere Report No. 22. Paris, France.

U.S. Departments of Agriculture and Interior. 1993. Forest ecosystem management: An ecological, economic and social assessment. Report for the Forest Ecosystem Management Assessment Team, USDA Forest Service, U.S. Fish and Wildlife Service, National Marine Fisheries Service, National Park Service, Bureau of Land Management, and Environmental Protection Agency. Portland, OR.

U.S. Fish and Wildlife Service. 1992. Endangered and threatened wildlife and plants; determination of threatened status for the Washington, Oregon, and California population of the marbled murrelet. U.S. Fish and Wildlife Service Federal Register 57:45328–45337.

———. 1996. Endangered and threatened wildlife and plants; final designation of critical habitat for the marbled murrelet. U.S. Fish and Wildlife Service Federal Register 61:26256–26320.

———. 1997. Recovery plan for the marbled murrelet *(Brachyramphus marmoratus)* in Washington, Oregon and California. Portland, OR: U.S. Fish and Wildlife Service.

van Geen, L. A., S. N. Luoma, C. C. Fuller, C. C. R. Anima, H. E. Clifton, and S. Trumbore. 1992. Evidence from Cd/Ca ratios in foraminifera for greater upwelling off California 4,000 years ago. *Nature* 358:54–56.

Vannote, R. L., G. W. Minshall, G. W. Cummins, J. R. Sedell, and C. E. Cushing. 1980. The river continuum concept. *Canadian Journal of Fisheries and Aquatic Sciences* 37:130–137.

Vasey, R. B. 1970. Relative physiological capacity of coast redwood and Douglas-fir to tolerate saturated soil. Ph.D. diss., University of California, Berkeley.

Vaudois, N., and C. Prive. 1971. Revision des bois fossiles de Cupressaceae. *Palaeontographica Abteilung B,* 134:61–86.

Veirs, S. D. 1975. Redwood vegetation dynamics. *Bulletin of the Ecological Society of America* 56:34–35.

———. 1980a. The role of fire in northern coast redwood forest dynamics. Pp. 190–209 in *Proceedings, conference on scientific research in the national parks.* Vol. 10, *Fire Ecology.* San Francisco, CA. Washington, DC: National Park Service.

———. 1980b. The influence of fire in coast redwood forests. Pp. 93–95 in *Proceedings, fire history workshop.* Tucson, AZ. USDA Forest Service, General Technical Report RM-81. Fort Collins, CO: Rocky Mountain Forest and Range Experimental Station.

———. 1982. Coast redwood forest: Stand dynamics, successional status, and the role of fire. Pp. 119–141 in J. E. Means, ed., *Forest succession and stand development research in the northwest.* Corvallis, OR: Forest Research Laboratory, Oregon State University.

———. 1996. Ecology of the coast redwood. Pp. 9–12 in J. LeBlanc, ed., *Proceedings of the conference on coast redwood forest ecology and management.* University of California Cooperative Extension Forestry Publication. Arcata, CA: Humboldt State University.

Vlug, H., and J. H. Borden. 1973. Soil Acari and Collembola populations affected by logging and slash burning in a coastal British Columbia coniferous forest. *Environmental Entomology* 2:1016–1023.

Voeghtlin, D. J. 1982. *Invertebrates of the H. J. Andrews Experimental Forest, Western Cascade Mountains, Oregon: A survey of arthropods associated with the canopy of old-growth* Pseudotsuga menziesii. Special Publication 4. Corvallis, OR: Forest Research Laboratory, Oregon State University.

Waide, J. B. 1995. Ecosystem stability: Revision of the resistance-resilience model. Pp. 372–396 in B. C. Patten and S. E. Jorgensen, eds., *Complex ecology: The part-whole relation in ecosystems.* Englewood Cliffs, NJ: Prentice Hall.

Walters, C. J. 1986. *Adaptive management of renewable resources.* New York: McGraw-Hill.

Ward, J. V. 1994. The structure and dynamics of lotic ecosystems. Pp. 195–218 in R. Margalef, ed., *Limnology now: A paradigm of planetary problems.* Amsterdam: Elsevier.

Waring, R. H., and J. F. Franklin. 1979. Evergreen coniferous forest of the Pacific Northwest. *Science* 204:1380–1386.

Waring R. H., and J. Major. 1964. Some vegetation of the California coastal redwood region in relation to gradients of moisture, nutrients, light, and temperature. *Ecological Monographs* 34:167–215.

Watson, R. A., and H. E. Wright Jr. 1980. The end of the Pleistocene: A general critique of chronostratigraphic classification. *Boreas* 9:153–163.

Weaver, W. E., D. K. Hagans, and J. H. Popenoe. 1995. Magnitude and causes of gully erosion in the lower Redwood Creek drainage basin. Pp. I1–I21 in K. M. Nolan, H. M. Kelsey, and D. C. Marron, eds., *Geomorphic processes and aquatic habitat in Redwood Creek Basin, northwestern California.* U.S. Geological Survey Professional Paper 1454. Washington, DC: U.S. Government Printing Office.

Webb, S. D., and A. D. Barnosky. 1989. Faunal dynamics of Pleistocene mammals. *Annual Reviews of Earth and Planetary Sciences* 17:413–438.

Welsh, H. H., Jr. 1988. An ecogeographic analysis of the herpetofauna of the Sierra San Pedro Martir Region, Baja California, with a contribution to the biogeography of the Baja California herpetofauna. *Proceedings of the California Academy of Science* 46(1):1–72.

———. 1990. Relictual amphibians and old-growth forests. *Conservation Biology* 4:309–319.

———. 1993. A hierarchical analysis of the niche relationships of four amphibians from forested habitats of northwestern California. Ph.D. diss., University of California, Berkeley.

Welsh, H. H. Jr., and A. J. Lind. 1991. The structure of the herpetofaunal assemblage in the Douglas-fir/hardwood forests of northwestern California and southwestern Oregon. Pp. 394–413 in L. F. Ruggiero, K. B. Aubry, A. B. Carey, and M. H. Huff, tech. coordinators, *Wildlife and vegetation of unmanaged Douglas-fir forests.* USDA Forest Service General Technical Report PNW-GTR-285. Portland, OR: USDA Forest Service.

———. 1992. Population ecology of two relictual salamanders from the Klamath Mountains of northwestern California. Pp. 419–437 in D. R. McCullough and R. H. Barrett, eds., *Wildlife 2001: Populations.* London: Elsevier.

———. 1995. Habitat correlates of the Del Norte salamander *Plethodon elongatus* (Caudata: Plethodontidae) in northwestern California. *Journal of Herpetology* 29:198–210.

———. 1996. Habitat correlates of the southern torrent salamander *(Rhyacotriton variegatus)* (Caudata: Rhyacotritonidae) in northwestern California. *Journal of Herpetology* 30:385–398.

———. In Review. The stream amphibian assemblage of the mixed conifer-hardwood forests of northwestern California and southwestern Oregon: Relationships with forest and stream environments.

Welsh, H. H., Jr., and L. M. Ollivier. 1998. Stream amphibians as indicators of ecosystem stress: A case study from California's redwoods. *Ecological Applications* 8:1118–1132.

Welsh, H. H., Jr., and R. A. Wilson. 1995. *Aneides ferreus* (clouded salamander) reproduction. *Herpetological Review* 26:196–197.

Welsh, H. H., Jr., L. M. Ollivier, and D. G. Hankin. 1997. A habitat-based design for sampling and monitoring stream amphibians with an illustration from Redwood National Park. *Northwestern Naturalist* 78:1–116.

Welsh, H. H., Jr., A. J. Lind, and G. A. Hodgson. In prep. Landscape patterns of the herpetofaunal assemblage of a northwestern California watershed: Relationships with forest and stream processes.

Wemple, B. C., and J. A. Jones. 1996. Channel network extension by logging roads in two basins, western Cascades, Oregon. *Water Resources Bulletin* 32:1195–1207.

West, G. J. 1986. *Pollen analysis of sediments from Elkhorn Slough, Monterey County, California: A late Holocene record of sedimentation rates and the local and upland vegetation.* Abstracts. American Quaternary association. Ninth biennial meeting, University of Illinois, Champaign-Urbana, IL.

———. 1988. Exploratory pollen analysis of sediments from Elkhorn Slough. Pp. 23–40 in S. A. Dietz, W. Hildebrandt, and T. Jones, eds., *Archaeological investigations at Elkhorn Slough: CA-MNT-229, a middle period site on the central California coast.* Papers in northern California anthropology. Berkeley, CA: Northern California Anthropological Group.

———. 1990. Holocene fossil pollen records of Douglas-fir in northwestern California: Reconstruction of past climate. Pp. 119–122 in J. L. Betancourt and A. M. MacKay, eds., *Proceedings of the sixth annual Pacific Climate (PACLIM) workshop.* March 5–8, 1989. Interagency Ecological Studies Program Technical Report 23. Sacramento, CA: California Department of Water Resources.

———. 1993. The late Pleistocene-Holocene pollen record and prehistory of

California's North Coast ranges. Pp. 219–236 in G. White, P. Mikkelsen, M. E. Hildebrandt, and M. E. Basgall, eds., *There grows a green tree: Papers in honor of David A. Fredrickson.* Publication No. 11. Davis, CA: Center for Archaeological Research at Davis.

———. 1998. A late glacial age pollen biozone for central California. Pp. 1–13 in R. C. Wilson and V. L. Tharp, eds., *Proceedings of the thirteenth annual Pacific Climate (PACLIM) workshop.* April 7–9, 1997. Interagency Ecological Program, Technical Report 57. Sacramento, CA: California Department of Water Resources.

———. N.d. Late Holocene successional change in a coast redwood (*Sequoia sempervirens* [D. Don] Endl.) forest, Sonoma County, California. Ms.

Westman, W. E., and R. H. Whittaker. 1975. The pygmy forest region of northern California: Studies on biomass and primary production. *Ecological Monographs* 63:453–520.

Whitney Survey. 1865. *Geological survey.* Vol. 1, *Geology of the coast ranges in California.* Philadelphia, PA: Paxton Press.

Whittaker, R. H., ed., 1978. *Classification of plant communities.* The Hague: Junk.

Wilcove, D. S., and D. D. Murphy. 1991. The spotted owl controversy and conservation biology. *Conservation Biology* 5:261–262.

Wilcove, D. S., C. H. McLellan, and A. P. Dobson. 1986. Habitat fragmentation in the temperate zone. Pp. 237–256 in M. E. Soulé, ed., *Conservation biology: The science of scarcity and diversity.* Sunderland, MA: Sinauer.

Wilson, E. O. 1987. The little things that run the world (The importance and conservation of invertebrates). *Conservation Biology* 1:344–345.

———. 1992. *The diversity of life.* Cambridge, MA: Belknap Press of Harvard University Press.

Wilson, M. F., and K. C. Halupka. 1995. Anadromous fish as keystone species in vertebrate communities. *Conservation Biology* 9:489–497.

Winchester, N. N. 1993. Coastal Sitka spruce canopies: Conservation of biodiversity. *Bioline* 11:9–14.

———. 1996. Canopy arthropods of coastal Sitka spruce trees on Vancouver Island, British Columbia, Canada. Pp. 151–168 in N. E. Stork, J. A. Adis, and R. K. Didham, eds., *Canopy arthropods.* London, UK: Chapman and Hall.

———. 1997. The arboreal superhighway: Arthropod and landscape dynamics. *Canadian Entomologist* 129:595–599.

Winchester, N. N., and R. A. Ring. 1996a. Centinelan extinctions: Extirpation of northern temperate old-growth rainforest arthropod communities. *Selybana* 17:50–57.

———. 1996b. Northern temperate coastal Sitka spruce forests with special emphasis on canopies: Studying arthropods in an unexplored frontier. *Northwest Science* (Special Issue) 70:94–103.

Wirth, Conrad L. 1980. *Parks, politics, and the people.* Norman, OK: University of Oklahoma Press.

Wolfe, J. A., and R. W. Brown. 1964. Age and correlation. Pp. 22–25 in D. L. Peck, A. B. Griggs, H. G. Schlicker, F. G. Wells, and H. M. Dole, eds., *Geology of the central and northern parts of the western Cascade Range in Oregon.* U.S. Geological Survey Professional Paper 449. Washington, DC: U.S. Government Printing Office.

Wolfe, J. A., D. M. Hopkins, and E. B. Leopold. 1966. *Tertiary stratigraphy and paleob-*

otany of the Cook Inlet region, Alaska. U.S. Geological Survey Professional Paper 398-A:A1–A29. Washington, DC: U.S. Government Printing Office.

Wolfe, J. A., H. E. Schorn, C. E. Forest, and P. Molnar. 1997. Paleobotanical evidence for high altitudes in Nevada during the Miocene. *Science* 276:1672–1675.

Wootton, J. T., M. S. Parker, and M. E. Power. 1996. Effects of disturbance on river food webs. *Science* 273: 1558–1561

Wright, H. E., Jr. ed. 1983. *Late-Quaternary environments of the United States.* Vol. 2, *The Holocene.* Minneapolis, MI: University of Minnesota Press.

Wroble, J., and D. Waters. 1989. Summary of tailed frog *(Ascaphus truei)* and Olympic salamander *(Rhyacotriton olympicus variegatus)* stream surveys for the Pacific Lumber Company, October 1987 to September 1988. Report submitted to Pacific Lumber Co., January 9, 1989.

Yahner, R. H. 1988. Changes in wildlife communities near edges. *Conservation Biology* 2:333–339.

Yanov, K. P. 1980. Biogeography and distribution of three parapatric salamander species in coastal and borderland California. Pp. 531–550 in D. M. Power, ed., *The California islands: Proceedings of a multidisciplinary symposium.* Santa Barbara, CA: Santa Barbara Museum of Natural History.

Yao, X., T. N. Taylor, and E. L. Taylor. 1997. A taxodiaceous seed cone from the Triassic of Antarctica. *American Journal of Botany* 84:343–354.

Yoshiyama, R. M., F. W. Fisher, and P. B. Moyle. 1998. Historical abundance and decline of Chinook salmon in the Central Valley Region of California. *North American Journal of Fisheries Management* 18(3):487–521

Yount, J. D., and G. J. Niemi. 1990. Recovery of lotic communities and ecosystems from disturbance: A narrative review of case studies. *Environmental Management* 14:547–570.

Yung, Y. L., T. Lee, C-H. Wang, and Y-T. Shieh. 1996. Dust: A diagnostic of the hydrologic cycle during the last glacial maximum. *Science* 271:962–963.

Zeiner, D. C., W. F. Laudenslayer Jr., and K. E. Mayer. 1988. *California's wildlife.* Vol. 1, *Amphibians and reptiles.* Sacramento, CA: California Department of Fish and Game.

Zeiner, D. C., W. F. Laudenslayer Jr., K. E. Mayer, and M. White. 1990a. *California's wildlife.* Vol. 2, *Birds.* Sacramento, CA: California Department of Fish and Game.

———. 1990b. *California's wildlife.* Vol. 3, *Mammals.* Sacramento, CA: California Department of Fish and Game.

Zielinski, W. J., and S. T. Gellman. 1999. Bat use of remnant old-growth redwood stands. *Conservation Biology* 13:160–167.

Zielinski, W. J., and R. T. Golightly. 1996. The status of marten in redwoods: Is the Humboldt marten extinct? Pp. 115–119 in John LeBlanc, ed., *Conference on coast redwood forest ecology and management, June 18–20, 1996.* Arcata, CA: Humboldt State University.

Zielinski, W. J., and T. E. Kucera. 1995. *American martens, fisher, lynx, and wolverine: Survey methods for their detection.* General Technical Report PSW-GTR-157. USDA Forest Service, Pacific Southwest Research Station, Albany, CA. 163 p.

Zielinski, W. J., R. L. Truex, C. V. Ogan, and K. Busse. 1997. Detection surveys for fishers and American martens in California, 1989–1994: Summary and interpretations.

Pp. 372–392 in G. Proulx, H. N. Bryant, and P. M. Woodard, eds., *Martes: taxonomy, ecology, techniques, and management*. Edmonton, Alberta, Canada: Provincial Museum of Alberta.

Ziemer, R. 1998a. Flooding and stormflows. Pp. 15–24 in R. Ziemer, ed., *Proceedings of the conference on coastal watersheds: The Caspar Creek story*. USDA FS PS-GTR 165.

Ziemer, R., ed. 1998b. *Proceedings of the conference on coastal watersheds: The Caspar Creek story*. USDA FS PS-GTR 165.

Ziemer, R. R., and T. E. Lisle. 1998. Hydrology. Pp. 43–68 in R. J. Naiman and R. E. Bilby, eds., *River ecology and management: Lessons from the Pacific Coastal Ecoregion*. New York, NY: Springer Verlag.

Zimmerman, U., A. Hasse, D. Langbein, and F. Meinzer. 1993. Mechanisms of long-distance water transport in plants: A re-examination of some paradigms in the light of new evidence. *Philosophical Transactions of the Royal Society of London* B 342:19–31.

Zimmermann, U., F. C. Meinzer, R. Benkert, J. J. Zhu, H. Schneider, G. Goldstein, E. Kuchenbrod, and E. Haase. 1994. Opinion: Xylem water transport is the available evidence consistent with the cohesion theory? *Plant, Cell and Environment* 17:1169–1181.

Zinke, P. J. 1988. The redwood forest and associated north coast forests. Pp. 679–699 in M. G. Barbour and J. Major, eds., *Terrestrial vegetation of California*. Exp. edn. Sacramento, CA: California Plant Society Press.

Zinke, P. J., A. G. Stangenberger, and J. L. Bertenshaw. 1996. Pattern and process in forests of *Sequoia sempervirens* (D. Don) Endl. Pp. 26–41 in *Proceedings of the conference on coast redwood forest ecology and management*. University of California Cooperative Extension Forestry Publication. Arcata, CA: Humboldt State University.

Zweifel, R. G. 1955. Ecology, distribution, and systematics of frogs of the *Rana boylii* group. *University of California Publications in Zoology* 54:207–292.

About the Contributors

DEAN P. ANGELIDES is vice-president of VESTRA Resources, a geographic information systems (GIS) consulting firm in Redding, California. He has more than twenty years' experience in applying GIS technologies to natural resources management issues in California.

CARLOS CARROLL is a research biologist with the Conservation Biology Institute in Corvallis, Oregon. He received his M.S. in wildlife science from Oregon State University in 1997 for research focusing on the distribution of the Pacific fisher in the Klamath/Siskiyou region. His current doctoral research focuses on developing spatial habitat models for multispecies conservation planning in the Rocky Mountains and Sierra Nevada.

ALLEN COOPERRIDER is a conservation biologist with more than thirty years' experience dealing with natural resource issues, both as a federal government biologist and as a private consultant. He recently retired from the U.S. Fish and Wildlife Service and lives with his wife among the redwoods west of Ukiah in rural Mendocino County.

TODD E. DAWSON, formerly at Cornell University, is a professor of plant physiological ecology in the Department of Integrative Biology at the University of California, Berkeley. He has researched the relationships between plants and water in many different environments. Most recently, he is exploring how fog influences the physiology of redwood trees and the ecology of redwood forests.

PETER DEL TREDICI holds a B.A. from the University of California, Berkeley (1968), and a Ph.D. in biology from Boston University (1991). He is currently Director of Living Collections at the Arnold Arboretum of Harvard University. His primary area of research is secondary sprouting in trees (including *Ginkgo biloba* and *Sequoia sempervirens*) following catastrophic disturbance or in old age.

JOSEPH H. ENGBECK JR. is the author of five books and many other publications about the history and natural history of California and the American West. He has been actively involved in the environmental movement since 1960 and is currently a councilor of the Save-the-Redwoods League and vice-president of Save San Francisco Bay Association. He was a trustee of the California Historical Society for six years and helped organize People for Open Space in the 1960s before joining the California Department of Parks and Recreation, where he was a writer, editor, and publications program manager.

FRED EUPHRAT is a forest scientist who conducts management and research in northern California. Through his business—Forest, Soil and Water—Fred manages forestland both for clients and family and conducts projects in watershed management, social forestry, and resource education. His work is generally in Sonoma and Mendocino Counties but has ranged as far as the Upper Mississippi and Nepal. Fred currently serves as a director on the Sonoma County Agricultural Preservation and Open Space Authority, and as a director on both the Ernie Carpenter Fund for the Environment and the Friends of the Russian River.

CHRISTOPHER A. FRISSELL is a research associate professor at the Flathead Lake Biological Station, University of Montana. His research interests include the effects of human land use and natural events on stream and river ecosystems, and the conservation ecology of aquatic biota. His dissertation research focused on the influence of forests and logging practices on salmon habitat in the Oregon portion of the Klamath Mountains region.

PAMELA A. FROST earned a B.S. in environmental science from the University of Maine, Machias, in 1984 and an M.S. in natural resource information systems from Ohio State University, Columbus. She is vice-president for Earth Design Consultants, Inc., and serves as secretary/treasurer for the Conservation Biology Institute in Corvallis, Oregon. Her areas of expertise are geographic information systems and conservation planning.

JANE GRAY, professor of biology at the University of Oregon, received a B.A. in geology from Harvard University (Radcliffe College) and a Ph.D. in paleontology from the University of California, Berkeley. Following graduation from Radcliffe, she spent a year as an NSF predoctoral in the pollen laboratory of Johs. Iversen in Copenhagen, then pursued doctoral research on pollen analysis of Tertiary floras of the Pacific Northwest with Ralph W. Chaney at Berkeley. Her current interests are nonmarine paleoecology and evolution, with emphasis on the evolution of early terrestrial and freshwater ecosystems.

GERALD E. HEILMAN JR. is a conservation biologist with the Conservation Biology Institute in Corvallis, Oregon. He received his B.S. in terrestrial ecology at Western Washington University and his M.S. in natural resources: land, water, and air at the University of Connecticut. His current research interests include design, analysis, and implementation of nature reserves and the use of geographic information systems as a management and decision analysis support tool.

ANTHONY LaBANCA is a botanist in the Resource Management and Science Division of Redwood National and State Parks. His primary responsibilities include the study of redwood forest disturbance and fire history, fire effects monitoring, and assisting the park staff with other natural resource, fire management, and research projects. He also continues the study of coastal dune systems, on which he received an M.A. in plant ecology from Humboldt State University, California

DAVID L. LARGENT is a professor of botany at Humboldt State University, California, where he began teaching in 1968. He teaches general botany, biology of the microfungi, biology of the Ascomycetes and Basidiomycetes, biology of the fleshy fungi, and forest pathology. His current research interests include biosystematics of the Entolomataceae, systematics and ecology of fungi, especially fleshy fungi, fungi associated with old-growth redwoods, fungi of the Redwood National Park, fungi of northern California, and odors and tastes of fungi, with special emphasis on their function in fruiting bodies.

WILLIAM J. LIBBY has worked professionally with coast redwoods since 1962, having been in awe of them since his first encounter in 1951. He specializes in the genetic architecture of redwoods and other western conifers, and in the selection and cloning of redwoods both for amenity and timber purposes. He is a councilor of Save-the-Redwoods League.

BRUCE G. MARCOT is a research wildlife ecologist with the Ecosystems Processes Research Program, USDA Forest Service, Portland, Oregon. He participates in technology application projects dealing with forest planning, specifically on wildlife population viability, biodiversity conservation, and ecologically sustainable forest planning. He has served as scientific leader for the Terrestrial Ecology Assessment Staff of the Interior Columbia Basin Ecosystem Management Project, and on the Forest Ecosystem Management Assessment Team, the Interagency Scientific Committee on the Northern Spotted Owl, the Scientific Panel on Late-Successional Forest Ecosystems, and other projects dealing with biodiversity, old-growth forest ecosystems, and population viability. He has extensive international experience, including serving as chief ecologist on a sustainable land-use planning project spanning Far East Russia and China, and serving on a biodiversity planning team in India.

S. KIM NELSON is a Research Wildlife Biologist and Senior Research Assistant with the Oregon Cooperative Wildlife Research Unit and Department of Fisheries and Wildlife at Oregon State University. She has been studying the distribution, nesting behavior, nest-site characteristics, and vocalizations of the threatened marbled murrelet, and modeling its stand and landscape habitat associations, in western Oregon since 1989. She also has participated in or been an advisor to research on the marbled murrelet in the redwood forests of northern and central California.

REED F. NOSS is an international consultant on biodiversity issues, former editor of *Conservation Biology*, science editor for *Wild Earth* magazine, and president of the Society for Conservation Biology. He has authored more than 150 publications. Now engaged largely in conservation planning at regional and broader scales, he is, at heart, a field naturalist.

CRAIG M. OLSON is a consulting forester with over twenty years experience in forest research and forest resource management. Dr. Olson is a biometrician and quantitative ecologist whose work has included numerous studies in restoration, mitigation, and silviculture in California, Oregon, Washington, and Alaska, from desert riparian systems to the boreal forest.

DAVID OLSON, a senior scientist with the World Wildlife Fund, focuses on identifying conservation priorities for the organization's global programs. He has been involved in invertebrate conservation issues in the Pacific Northwest for the past decade.

JAMES H. POPENOE holds a B.A. in chemistry from Occidental College and an M.S. in biology from San Diego State University. He was a mapper with the California Soil-Vegetation Survey. Since 1979, he has been a plant ecologist and soil scientist in Redwood National Park.

ROLAND RAYMOND, after receiving a B.S. degree from Humboldt State University, California, has served as the Yurok tribe's Forest Program Manager. Being a Yurok tribal member, he has a vested interest in the land—past, present, and future. He is married to a member of the Hupa tribe, and together they have three children.

TERRY D. ROELOFS is a professor of fisheries at Humboldt State University in Arcata, California, where he has taught since 1970. His primary academic interests are stream ecology, watershed processes and management, and salmonid biology.

JOHN O. SAWYER is a professor of botany at Humboldt State University, California. He has been studying the flora and vegetation of northern California for more than thirty years.

TERESA SHOLARS is a tenured faculty member at the Mendocino Coast Campus of College of the Redwoods. Having taught courses in biology, ecology, forestry, and mushroom and plant identification for twenty-four years, she is currently a research associate at University of California, Berkeley, where she continues her studies in systematics of perennial lupines. She is a member of the California Native Plant Society's Rare Plant Scientific Advisory Committee.

STEPHEN C. SILLETT is an assistant professor of botany in the Department of Biological Sciences at Humboldt State University, California. He began studying old-growth forest canopies using rope techniques in 1988 and is the first person to study redwood forests from a canopy perspective.

MARK SORENSEN is founder and president of Geographic Planning Collaborative, Inc. He has been involved in the planning, design, and implementation of geographic information system technology for urban and regional planning and resource management for more than seventeen years.

JAMES R. STRITTHOLT is currently executive director of the Conservation Biology Institute in Corvallis, Oregon, which is dedicated to collaborative conservation biology research and education. He has worked with large mammals, has conducted field research on forests and vertebrates, and has more than six years' teaching experience. Over the past six years, he has worked on a variety of conservation planning projects throughout North America, including Canada, Central America, and the United States. His areas of expertise include nature reserve design, conservation planning, landscape ecology, geographic information systems, and remote sensing.

DALE A. THORNBURGH teaches sustainable forest management in the Forestry Department at Humboldt State University, Arcata, California. He is the chief forester for the Pacific Forest Trust, Mendocino, California, developing Conservation Easements for small-forest landowners who practice sustainable forestry.

ROBERT VAN PELT lives in Seattle and is currently a postdoctoral researcher at the University of Washington, where he teaches several field classes on Pacific Northwest old-growth and forestry issues in the Pacific Northwest. He received his M.S. in 1991 and his Ph.D. in 1995 from the University of Washington. His main research interests are old-growth ecology, canopy structure and its control of the understory environment, spatial patterns in old-growth forests, and tree plant geography.

STEPHEN D. VEIRS Jr. worked for the National Park Service in research and resource management at Redwood National Park from 1970 to 1986. His research was focused on the role of fire in the redwood forest, succession, and stand dynamics. He retired from the Biological Resources Division of the U.S. Geological Survey in 1997.

HARTWELL H. WELSH JR. is a research wildlife ecologist with the USDA Forest Service, Pacific Southwest Experiment Station, Redwood Sciences Laboratory, in Arcata, California. His research focuses on amphibian and reptile ecology and the relationships between their population viability and landscape processes.

G. JAMES WEST is Regional Archeologist for the Bureau of Reclamation, Sacramento, and a research associate at the Department of Anthropology, University of California, Davis. His specialization is Quaternary palynology, with special research interest in the study of human interaction with the environment and the reconstruction of past vegetation and climate. His current research is directed toward the late Pleistocene and Holocene vegetation and climate history of northern and central California.

WILLIAM ZIELINSKI is a research wildlife ecologist with the Pacific Southwest Research Station of the U.S. Forest Service. His research interests include the conservation and ecology of mammalian carnivores and the design and implementation of ecosystem monitoring programs. He received his Ph.D. in zoology from North Carolina State University and an M.S. in wildland resource science from the University of California, Berkeley.

INDEX